# Insect Ecology

# Insect Ecology

Edited by **Christopher Fleming**

**SYRAWOOD**
PUBLISHING HOUSE

New York

Published by Syrawood Publishing House,
750 Third Avenue, 9th Floor,
New York, NY 10017, USA
www.syrawoodpublishinghouse.com

**Insect Ecology**
Edited by Christopher Fleming

© 2016 Syrawood Publishing House

International Standard Book Number: 978-1-68286-093-9(Hardback)

# Contents

# Preface

This book has been an outcome of determined endeavour from a group of educationists in the field. The primary objective was to involve a broad spectrum of professionals from diverse cultural background involved in the field for developing new researches. The book not only targets students but also scholars pursuing higher research for further enhancement of the theoretical and practical applications of the subject.

Insects form an extremely significant part of our ecosystem. They help in maintaining the ecological balance. This book on insect ecology specifically focuses on studying the behavior and interactions of insects with their surroundings. It discusses topics like evaluation of insects, colony performance, ant communities, competitive interactions, termite studies, etc. This book is a collective contribution of a renowned group of international experts. It is ideal for graduate and postgraduate students pursuing entomology and associated disciplines.

It was an honour to edit such a profound book and also a challenging task to compile and examine all the relevant data for accuracy and originality. I wish to acknowledge the efforts of the contributors for submitting such brilliant and diverse chapters in the field and for endlessly working for the completion of the book. Last, but not the least; I thank my family for being a constant source of support in all my research endeavours.

**Editor**

# Social Wasps on *Eugenia uniflora* Linnaeus (Myrtaceae) Plants in an Urban Area

GK Souza[1], TG Pikart[1], GC Jacques[1], AA Castro[1], MM De Souza[2], JE Serrão[1], JC Zanuncio[1]

1 - Universidade Federal de Viçosa, Viçosa - MG, Brazil

2 - Instituto Federal de Educação, Ciência e Tecnologia Sul de Minas Gerais, Pouso Redondo-MG, Brazil

**Keywords**
Biological control, diversity, Polistinae, urban forestry, Vespidae

**Corresponding author**
Tiago Georg Pikart
Departamento de Entomologia
Universidade Federal de Viçosa
Viçosa, Minas Gerais, Brazil
36570-000
E-Mail: tiago.florestal@gmail.com

**Abstract**

Social wasps (Hymenoptera: Vespidae) of the subfamily Polistinae can be effectively incorporated into IPM systems for urban forestry. This study, conducted in Viçosa, Minas Gerais State, Brazil, in May 2011, identified species of this group foraging on *Eugenia uniflora* Linnaeus (Myrtaceae) plants. The study area was monitored once a week and data collected included daily activity pattern, diversity, dominance and overlap of temporal niches by social wasps. Data analysis revealed that *E. uniflora* plants were visited by 217 individuals of 16 species of the subfamily Polistinae. Foraging behavior of social wasps bore no relationship with sampling time, but overlap of temporal niche was high. Wasps were not observed damaging healthy fruits, but were probably searching for Lonchaeidae and Tephritidae larvae. This study highlights the need for conservation of predator diversity in order to provide a partial alternative to the environmentally degrading chemical pesticides currently used in urban forestry for pest control.

## Introduction

Social wasps (Hymenoptera: Vespidae) belong to the subfamilies Polistinae, Stenogastrinae and Vespinae (Schmitz & Moritz, 1998). Among these families, only Polistinae can be found in the Neotropical region, with approximately 319 species occurring in Brazil (Carpenter & Marques, 2001). Social wasps are an important part of food webs, preying on insects of the orders Coleoptera, Diptera, Hemiptera, Hymenoptera and Lepidoptera (Prezoto et al., 2005) in terrestrial ecosystems (Richter, 2000; Torezan-Silingardi, 2011).

Studies of social wasps have focused on their habitat preferences (Da Cruz et al., 2006; De Souza et al., 2010b), nest density (Diniz & Kitayama, 1994), seasonal number of colonies, nesting habits (Strassmann et al., 1997; Diniz & Kitayama, 1998; Santos et al., 2009a), insecticide selectivity (Galvan et al., 2002; Bacci et al., 2009), and floral visitation (Da Silva-Pereira & Santos, 2006). However, diversity of these insects in urban areas has not been well studied.

Floristic diversity and vegetation structure can deter-mine composition of social wasp communities (Santos et al., 2007) by affecting foraging activity during their search for water, vegetable fiber, carbohydrates, and proteins (Richter, 2000; Elisei et al., 2010). Social wasps, like other generalist organisms, forage predominantly on the most abundant resource, without preference and/or selective behavior (Santos et al., 2007), but they can focus their resource collection activities on a small group of plants (Aguiar & Santos, 2007). Their generalist condition allows them to have low dependence on specific or constant food resources (Real, 1981).

Brazilian Myrtaceae includes trees and shrubs with potential for fruit production and use in urban forestry (Donadio & Moro, 2004). *Eugenia uniflora* Linnaeus (Myrtaceae) is a semi-deciduous tree native to the south and southeast regions of Brazil used in urban forestry (Alves et al., 2008). Its popularity is mainly due to the pleasant and refreshing taste of its fruits, which resemble cherries. However, the occurrence of insect pests such as *Eugeniamyia dispar* Maia, Mendonça & Romanowski (Diptera: Cecidomyiidae) and *Anastrepha fraterculus* (Wiedemann) (Diptera: Tephritidae) (Bierhals et al., 2012) has been discouraging the use of *E. uniflora* for landscaping and urban forestry in Brazil.

Social wasps are important natural enemies of insect pests (Prezoto et al., 2005; De Souza et al., 2010a; Picanço et al., 2010, 2011) and their diversity is similar or even higher in urban areas than in natural environments (Jacques et al., 2012). In this study we identified the social wasp species foraging on *E. uniflora* (Myrtaceae) plants in order to evaluate the potential of this group to provide biological control in an integrated pest management program.

**Material and Methods**

Diversity of social wasps (Vespidae: Polistinae) on *E. uniflora* plants was evaluated at the Universidade Federal de Viçosa (UFV) in an urban area of Viçosa, Minas Gerais State, Brazil, in May 2011. Two people monitored four heavily fruiting *E. uniflora* plants used in the urban forestry of the university once a week between 9:00 h and 15:00 h over a period of four weeks, and living specimens of social wasps foraging on these plants were collected with an insect net and preserved in a vial with 92.8% ethanol. Collected specimens were taken to the Laboratory of Biological Control of Insects at UFV, mounted and identified. Collection data were used to analyze daily activity pattern, diversity, dominance and temporal niche overlap of social wasp species.

The diversity of social wasps was calculated with the Shannon index (Shannon, 1948) using the formula $H' = \Sigma_{pk} x \ln_{pk}$. The evenness of visits by each wasp species to *E. uniflora* plants was calculated with the formula $J' = H'/H'_{max}$ (Pielou, 1969). The dominance index was calculated with Berger Parker (May, 1975) with the formula $d = N_{max}/N_T$.

The temporal niche overlap per pair of wasp species was determined by using the Schoener index (Schoener, 1986) with the formula $NO_{ih} = 1 - 1/2\Sigma_k |p_{ik} - p_{hk}| d$. Species with very small numbers of individuals (n <10) were excluded from the analysis. A Kolmogorov-Smirnov test for two samples was used to evaluate interspecific differences between activity patterns per pair of social wasp species (Siegel, 1956).

**Results and Discussion**

Plants of *E. uniflora* were visited by 217 individuals of 16 species of social wasps of the subfamily Polistinae (Table 1). Species diversity was similar to that of the Atlantic Forest (Santos et al., 2007), but it was lower than that of Cerrado (De Souza & Prezoto, 2006; Elpino-Campos et al., 2007) and Amazon Forest (Silveira et al., 2008; Silva & Silveira, 2009) areas. The similarity between the fauna of these insects and the species commonly found in the urban area sampled and the Atlantic Forest region is important. Viçosa is included in this biome, which shows that social wasps have a high capacity to adapt to urban environments. Besides occurring in different ecosystems, differences in collection methods and sample

sizes may have contributed to differences in diversity between this study and those from other regions. On the other hand, surveys considering only one plant species showed similar or lower diversity than that of the present study (De Souza et al., 2010a; Santos & Presley, 2010; De Souza et al., 2011), suggesting that social wasp diversity may be determined more by variation in tolerance levels between species than by habitat complexity (Bomfim & Antonialli Junior, 2012).

**Table 1.** Number of species and individuals of social wasps (Hymenoptera: Vespidae: Polistinae) collected on *Eugenia uniflora* (Myrtaceae) fruits in Viçosa, Minas Gerais State, Brazil

| | Number of individuals | | | | |
|---|---|---|---|---|---|
| | 04 May | 18 May | 25 May | 30 May | Total |
| *Agelaia multipicta* | 4 | 14 | 20 | 10 | 48 |
| *Agelaia vicina* | 2 | 0 | 0 | 0 | 2 |
| *Brachygastra lecheguana* | 0 | 0 | 1 | 6 | 7 |
| *Mischocyttarus atramentarius* | 1 | 0 | 0 | 0 | 1 |
| *Mischocyttarus cassununga* | 7 | 6 | 8 | 8 | 29 |
| *Mischocyttarus drewseni* | 0 | 0 | 1 | 0 | 1 |
| *Polistes actaeon* | 0 | 1 | 3 | 0 | 4 |
| *Polistes simillimus* | 1 | 1 | 6 | 3 | 11 |
| *Polistes versicolor* | 9 | 5 | 4 | 3 | 21 |
| *Polybia bifasciata* | 1 | 0 | 1 | 0 | 2 |
| *Polybia fastidiosuscula* | 7 | 9 | 8 | 7 | 31 |
| *Polybia ignobilis* | 1 | 0 | 0 | 0 | 1 |
| *Polybia jurinei* | 1 | 0 | 0 | 0 | 1 |
| *Polybia platycephala* | 5 | 5 | 23 | 18 | 51 |
| *Polybia sericea* | 0 | 2 | 2 | 1 | 5 |
| *Protopolybia exigua* | 0 | 1 | 1 | 0 | 2 |
| Total | 39 | 44 | 78 | 56 | 217 |

Species of the dipteran families Lonchaeidae and Tephritidae, and *Trigona spinipes* (Fabricius) (Hymenoptera: Apidae) foraged on *E. uniflora* fruits. *Apis mellifera* Linnaeus (Hymenoptera: Apidae) and *T. spinipes* visited flowers of this plant, but social wasps were not observed in these structures. These last insects can collect nectar to feed their larvae and adults (Da Silva et al., 2011), but this survey was conducted during the dry season, when nectar production could have been insufficient to attract them. Social wasps have no specialized structures for transporting pollen (corbicula) as do *A. mellifera* and *T. spinipes*, which were visiting *E. uniflora* flowers to collect pollen.

All social wasp species exploited mainly green fruits

**Table 2.** Dominance (Berger-Parker) of social wasp species (Hymenoptera: Vespidae: Polistinae) collected on *Eugenia uniflora* (Myrtaceae) fruits as function of time of the day and total period. Viçosa, Minas Gerais State, Brazil.

|  | 09:45 | 10:45 | 11:45 | 12:45 | 13:45 | 14:45 | Total |
|---|---|---|---|---|---|---|---|
| *Agelaia multipicta* | 0.194 | 0.180 | 0.177 | 0.243 | 0.333 | 0.172 | 0.221 |
| *Agelaia vicina* | - | - | 0.059 | - | - | - | 0.009 |
| *Brachygastra lecheguana* | - | - | 0.029 | 0.054 | 0.048 | 0.069 | 0.032 |
| *Mischocyttarus atramentarius* | - | - | - | - | 0.024 | - | 0.005 |
| *Mischocyttarus cassununga* | 0.056 | 0.051 | 0.235 | 0.162 | 0.191 | 0.103 | 0.134 |
| *Mischocyttarus drewseni* | - | - | - | - | 0.024 | - | 0.005 |
| *Polistes actaeon* | 0.028 | 0.026 | - | - | 0.024 | 0.035 | 0.018 |
| *Polistes simillinus* | 0.056 | 0.077 | 0.029 | 0.027 | 0.071 | 0.035 | 0.051 |
| *Polistes versicolor* | 0.139 | 0.103 | 0.088 | 0.081 | 0.071 | 0.103 | 0.097 |
| *Polybia bifasciata* | - | 0.026 | 0.029 | - | - | - | 0.009 |
| *Polybia fastidiosuscula* | 0.222 | 0.128 | 0.177 | 0.108 | 0.071 | 0.172 | 0.143 |
| *Polybia ignobilis* | 0.028 | - | - | - | - | - | 0.005 |
| *Polybia jurinei* | - | - | - | 0.027 | - | - | 0.005 |
| *Polybia platycephala* | 0.222 | 0.333 | 0.177 | 0.270 | 0.119 | 0.310 | 0.235 |
| *Polybia sericea* | 0.056 | 0.077 | - | - | - | - | 0.023 |
| *Protopolybia exigua* | - | - | - | 0.027 | 0.024 | - | 0.009 |
| Number of individuals | 36 | 39 | 34 | 37 | 42 | 29 | 217 |
| Shannon-Winner index *(H')* | 0.843 | 0.828 | 0.847 | 0.820 | 0.869 | 0.806 | 0.911 |
| Pielou Evenness index (J') | 0.884 | 0.868 | 0.888 | 0.859 | 0.835 | 0.892 | 0.757 |

with damaged areas and pre-existing holes caused by *T. spinipes*, and by oviposition behavior of fruit flies, as observed in *Anacardium occidentale* Linnaeus (Anacardiaceae) plantations (Santos & Presley, 2010), and on *Myrciaria* sp. (Myrtaceae) plants (De Souza et al., 2010a). *Brachygastra lecheguana* (Latreille), *Polistes simillimus* Zikán, *Polistes versicolor* (Olivier), *Polybia ignobilis* (Haliday), *Polybia platycephala* Richards, and *Polybia sericea* (Olivier) prey on larval forms of coleopteran, dipteran, and lepidopteran pests of agricultural and forest crops (Prezoto et al., 2005; Prezoto et al., 2006; Bichara et al., 2009; Elisei et al., 2010). Foraging behavior of social wasp species may be related to the presence of larvae of fruit flies in *E. uniflora*.

Polybia platycephala (n = 51, d= 0.235) and *Agelaia multipicta* (Haliday) (n= 48, d= 0.221) were the social wasp species with the higher values of frequency and for the dominance index, followed by *Polybia fastidiosuscula* Saussure (n= 31, d= 0.143), *Mischocyttarus cassununga* (Von Ihering) (n= 29, d= 0.134), *P. versicolor* (n= 21, d= 0.097), *P. simillimus* (n= 11, d= 0.051), *B. lecheguana* (n= 7, d= 0.032), *P. sericea* (n= 5, d= 0.023), and *Polistes actaeon* Haliday (n= 4, d= 0.018) (Tables 1 and 2). *Polybia bifasciata* Saussure, *Protopolybia exigua* Saussure, and *Agelaia vicina* (Saussure) (n= 2, d= 0.009), and *Mischocyttarus atramentarius* Zikán, *Mischocyttarus drewseni* (Saussure), *P. ignobilis* and *Polybia jurinei* (Saussure) (n= 1, d= 0.005) were less frequent (Tables 1 and 2) and of accidental occurrence. The active collection

of social wasps is an efficient process and can include most species in a survey (De Souza & Prezoto, 2006; De Souza et al., 2011; Jacques et al., 2012), but some of them, such as *A. vicina* and *P. bifasciata,* may only be collected in baited traps (De Souza & Prezoto, 2006; Jacques et al., 2012), which may explain their low frequencies in this study.

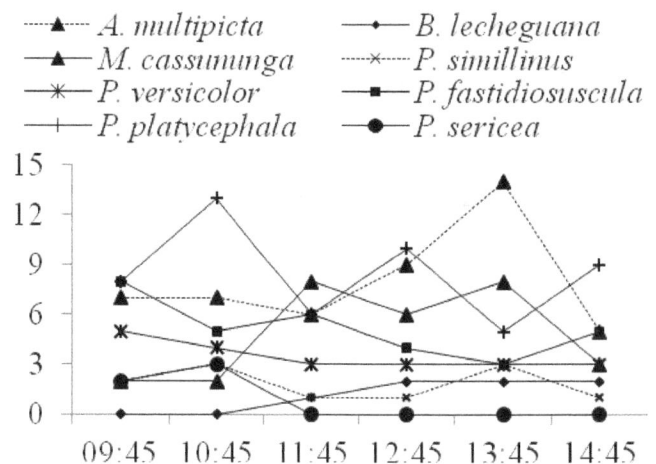

Fig 1. Temporal foraging activity of social wasps (Hymenoptera: Vespidae: Polistinae) on *Eugenia uniflora* (Myrtaceae) fruits. Viçosa, Minas Gerais State, Brazil.

The foraging behavior of social wasps showed no relationship with sampling time (Fig 1), differing from that in *A. occidentale* plantations, where species of this group foraged more frequently between 09:00 h and 12:00 h, and from

14:00 h to 16:00 h (Santos & Presley, 2010), or the higher foraging activity of the social wasp *Parachartergus fraternus* (Gribodo) on sunny days, between 12:00 h and 14:00 h (Santos et al., 2009b). Climatic factors, such as light, temperature and wind speed, directly affect the foraging behavior of social wasps (Kasper et al., 2008; Santos et al., 2009b; De Castro et al., 2011), so changes in these factors during the collection periods may have affected the results. Diversity and evenness values for social wasp species were also similar during all collecting periods (Table 2).

The temporal niche overlap between pairs of species of social wasps was high, 0.65-0.91 (Schoener index) (Table 3), with the highest value for *P. fastidiosuscula* and *P. versicolor* ($NO_{ih}$ index= 0.9109). Only five of the 15 comparisons

between pairs of species differed statistically, that is, with low value of temporal niche overlap (Table 3). These results are similar to those of social wasps sharing the same food resource (Santos & Presley, 2010), again suggesting a tendency of co-existence between species of this group.

Social wasps are important predators of insect herbivores and preserving their diversity can contribute significantly to reduced application of chemical pesticides in urban forestry. Social wasps can be easily found in anthropized environments (Zanette et al., 2005; Alvarenga et al., 2010; Jacques et al., 2012), but small changes like adding floral resources to gardens or even the creation of public squares will enhance their presence and efficiency.

**Table 3.** Interspecific pairwise comparison of activity patterns of social wasps (species with more than 10 individuals) collected on *Eugenia uniflora* (Myrtaceae) fruits in Viçosa, Minas Gerais State, Brazil. Values above the diagonal correspond to statistical significance by the Kolmogorov-Smirnov test for two samples (significant values in bold, P< 0.05). Values below the diagonal correspond to temporal niche overlap between pairs of species of social wasps (Schoener index).

|  | *A. m.* | *M. c.* | *P. s.* | *P. v.* | *P. f.* | *P. p.* |
|---|---|---|---|---|---|---|
| *Agelaia multipicta* | - | 0.8096 | **0.0122** | **0.0361** | 0.2090 | 0.6974 |
| *Mischocyttarus cassununga* | 0.8297 | - | 0.2090 | 0.2090 | 0.9996 | 0.2090 |
| *Polistes simillinus* | 0.8371 | 0.6834 | - | 0.2090 | **0.0361** | **0.0361** |
| *Polistes versicolor* | 0.8065 | 0.6700 | 0.7879 | - | 0.2090 | **0.0361** |
| *Polybia fastidiosuscula* | 0.7466 | 0.6607 | 0.7126 | 0.9109 | - | 0.2090 |
| *Polybia platycephala* | 0.7990 | 0.6531 | 0.7825 | 0.8487 | 0.8229 | - |

## Acknowledgements

The authors thank Dr. Tobias O. de Oliveira (Museu Paraense Emílio Goeldi, Belém, Pará) for the identification of social wasps species. To "Conselho Nac. de Desenv. Científico e Tecnológico (CNPq)", "Coord. de Aperfeiçoamento de Pessoal de Nível Superior (CAPES)" and "Fundação de Amparo à Pesquisa do Estado de Minas Gerais (FAPEMIG)" for financial support. Global Edico corrected and edited the English of this manuscript. We thank two anonymous reviewers for valuable comments that improved this article.

## References

Aguiar, C.M.L. & Santos, G.M.M. (2007). Compartilhamento de recursos florais por vespas sociais (Hymenoptera: Vespidae) e abelhas (Hymenoptera: Apoidea) em uma área de Caatinga. Neotrop. Entomol., 36: 836-842. doi: 10.1590/S1519-566X2007000600003

Alvarenga, R.B., Castro, M.M., Santos-Prezoto, H.H. & Prezoto, F. (2010). Nesting of social wasps (Hymenoptera, Vespidae) in urban gardens in southeastern Brazil. Sociobiology, 55: 445-452.

Alves, E.S., Tresmondi, F. & Longui, E.L. (2008). Leaf anatomy of *Eugenia uniflora* L. (Myrtaceae) in urban and rural

environments, São Paulo State, Brazil. Acta Bot. Bras., 22: 241-248.

Alves-Silva, E., Barônio, G.J., Torezan-Silingardi, H.M. & Del-Claro, K. (2012). Foraging behavior of *Brachygastra lecheguana* (Hymenoptera: Vespidae) on *Banisteriopsis malifolia* (Malpighiaceae): Extrafloral nectar consumption and herbivore predation in a tending ant system. Entomol. Sci., 16: 162- 169. doi: 10.1111/ens.12004

Bacci, L., Picanço, M.C., Barros, E.C., Rosado, J.F., Silva, G.A., Silva, V.F. & Silva, N.R. (2009). Physiological selectivity of insecticides to wasps (Hymenoptera: Vespidae) preying on the diamondback moth. Sociobiology, 53: 151-167.

Bichara, C.C., Santos, G.M.D., Resende, J.J., Da Cruz, J.D., Gobbi, N. & Machado, V.L.L. (2009). Foraging behavior of the swarm-founding wasp, *Polybia* (*Trichothorax*) *sericea* (Hymenoptera, Vespidae): prey capture and load capacity. Sociobiology, 53: 61-69.

Bierhals, A.N., Nava, D.E., Costa, V.A., Maia, V.C. & Diez-Rodriguez, G.I. (2012). *Eugeniamyia dispar* in surinam cherry: associated parasitoids, population dynamics and distribution of plant galls. Rev. Bras. Frut., 34: 109-115.

Bomfim, M.G.C.P. & Antonialli Junior, W.F. (2012). Community structure of social wasps (Hymenoptera: Vespidae) in riparian forest in Batayporã, Mato Grosso do Sul, Brazil. Sociobiology, 59: 755-765.

Carpenter, J.M. & Marques, O.M. (2001). Contribuição ao estudo dos vespídeos do Brasil (Insecta, Hymenoptera, Vespidae) [CD - ROM]. Série publicações digitais, 2. Cruz das Almas, Bahia, Brasil: Universidade Federal da Bahia, Escola de Agronomia, Departamento de Fitotecnia/Mestrado em Ciências Agrárias.

Da Cruz, J.D., Giannotti, E., Santos, G.M.M., Bichara-Filho, C.C. & Rocha, A.A. (2006). Nest site selection and flying capacity of neotropical wasp *Angiopolybia pallens* (Hymenoptera: Vespidae) in the Atlantic Rain Forest, Bahia State, Brazil. Sociobiology, 47: 739-749.

Da Rocha, A.A., Giannotti, E. & Bichara, C.C. (2009). Resources taken to the nest by *Protopolybia exigua* (Hymenoptera, Vespidae) in different phases of the colony cycle, in a region of the medio Sao Francisco River, Bahia, Brazil. Sociobiology, 54: 439-456.

Da Silva, E.R., Togni, O.C., Locher, G.A. & Giannotti, E. (2011). Distribution of resources collected among individuals from colonies of *Mischocyttarus drewseni* (Hymenoptera, Vespidae). Sociobiology, 58: 135-147.

Da Silva-Pereira, V. & Santos, G.M.M. (2006). Diversity in bee (Hymenoptera: Apoidea) and social wasp (Hymenoptera: Vespidae, Polistinae) community in "Campos Rupestres", Bahia, Brasil. Neotrop. Entomol., 35, 165-174. doi: 10.1590/S1519-566X2006000200003

De Castro, M.M., Guimarães, D.L. & Prezoto, F. (2011). Influence of environmental factors on the foraging activity of *Mischocyttarus cassununga* (Hymenoptera, Vespidae). Sociobiology, 58: 133-141.

De Souza, M.M. & Prezoto, F. (2006). Diversity of social wasps (Hymenoptera, Vespidae) in Semideciduous Forest and Cerrado (Savanna) regions in Brazil. Sociobiology, 47: 135-147.

De Souza, A.R., Venâncio, D.F.A. & Prezoto, F. (2010a). Social wasps (Hymenoptera: Vespidae: Polistinae) damaging fruits of *Myrciaria* sp. (Myrtaceae). Sociobiology, 55: 297-299.

De Souza, M.M., Louzada, J., Serrão, J.E. & Zanuncio, J.C. (2010b). Social wasps (Hymenoptera: Vespidae) as indicators of conservation degree of riparian forests in Southeast Brazil. Sociobiology, 56: 387-396.

De Souza, A.R., Venâncio, D.F.A., Zanuncio, J.C. & Prezoto, F. (2011). Sampling methods for assessing social wasp species diversity in a eucalyptus plantation. J. Econ. Entomol., 104: 1120-1123.

Diniz, I.R. & Kitayama, K. (1994). Colony densities and preferences for nest habitat of some social wasps in Mato Grosso State, Brazil (Hymenoptera, Vespidae). J. Hymenopt. Res., 3: 133-143.

Diniz, I.R. & Kitayama, K. (1998). Seasonality of vespid species (Hymenoptera, Vespidae) in Central Brazilian Cerrado. Rev. Biol. Trop., 46: 15-22.

Donadio, L.C. & Moro, F.V. (2004). Potential of Brazilian *Eugenia* (Myrtaceae) - as ornamental and as a fruit crop. Acta Hort., 632: 65-68.

Elisei, T., Vaz e Nunes, J., Ribeiro-Junior, C., Fernandes-Junior, A.J. & Prezoto, P. (2010). Uso da vespa social *Polistes versicolor* no controle de desfolhadores de eucalipto. Pesq. Agropec. Bras., 45: 958-964. doi: 10.1590/S0100-204X2010000900004

Elpino-Campos, A., Del-Claro, K. & Prezoto, F. (2007). Diversity of social wasps (Hymenoptera: Vespidae) in *Cerrado* fragments of Uberlândia, Minas Gerais State, Brazil. Neotrop. Entomol., 36: 685-692. doi: 10.1590/S1519-566X2007000500008

Galvan, T.L., Picanço, M.C., Bacci, L., Pereira, E.J.G. & Crespo, A.L.B. (2002). Selectivity of eight insecticides to predators of citrus caterpillars. Pesq. Agropec. Bras., 37: 117-122.

Jacques, G.C., Castro, A.A., Souza, G.K., Silva-Filho, R., De Souza, M.M. & Zanuncio, J.C. (2012). Diversity of social wasps on the *campus* of the "Universidade Federal de Viçosa" in Viçosa, Minas Gerais State, Brazil. Sociobiology, 59: 1053-1062.

Kasper, M.L., Reeson, A.F., Mackay, D.A. & Austin, A.D. (2008). Environmental factors influencing daily foraging activity of *Vespula germanica* (Hymenoptera, Vespidae) in Mediterranean Austrália. Insectes Soc., 55: 288-295.

Loomans, A.J.M., Murai, T. & Greene, I.D. (1997). Interactions with hymenopterous parasitoids and parasitic nematodes. In T. Lewis (Ed.), Thrips as Crop Pests (pp. 355-397). Wallingford, UK: CAB International.

May, R.M. (1975). Patterns of species abundance and diversity. In M.L. Cody & J.M. Diamond (Eds.), Ecology and Evolution of communities (pp. 81-120). Harvard University, Massachusetts: Belknap Press.

Picanço, M.C., De Oliveira, I.R., Rosado, J.F., Da Silva, F.M., Gontijo, P.D. & Da Silva, R.S. (2010). Natural biological control of *Ascia monuste* by the social wasp *Polybia ignobilis* (Hymenoptera: Vespidae). Sociobiology, 56: 67-76.

Picanço, M.C., Bacci, L., Queiroz, R.B., Silva, G.A., Miranda, M.M.M., Leite, G.L.D. & Suinaga, F.A. (2011). Social wasp predators of *Tuta absoluta*. Sociobiology, 58: 621-633.

Pickett, K.M. & Carpenter, J.M. (2010). Simultaneous analysis and the origin of eusociality in the Vespidaea (Insecta: Hymenoptera). Arthropod Syst. Phyl., 68: 3-33.

Pielou, E.C. (1969). An introduction to mathematical ecology. New York: Willey-Interscience.

Prezoto, F., Lima, M.A.P. & Machado, V.L.L. (2005). Survey of preys captured and used by *Polybia platycephala* (Richards) (Hymenoptera: Vespidae: Epiponini). Neotrop. Entomol., 34: 849-851. doi: 10.1590/S1519-566X2005000500019

Prezoto, F., Santos-Prezoto, H.H., Machado, V.L.L. & Zanuncio, J.C. (2006). Prey captured and used in *Polistes versicolor* (Olivier) (Hymenoptera: Vespidae) nourishment. Neotrop. Entomol., 35: 707-709. doi: 10.1590/S1519-566X2006000500021

Real, L.A. (1981). Uncertainty and pollinator-plant interactions: the foraging behavior of bees and wasps on artificial flowers. Ecology, 62: 20-26.

Richter, M.R. (2000). Social wasp (Hymenoptera: Vespidae) foraging behavior. Annu. Rev. Entomol., 45: 121-150.

Santos, G.M.M., Silva, S.O.C., Bichara-Filho, C.C. & Gobbi, N. (1998). Influencia del tamaño del cuerpo en el forrajeo de avispas sociales (Hymenoptera, Polistinae) visitantes de *Syagrus coronata* (Martius) (Arecacea). Rev. Gayana Zool., 62: 167-170.

Santos, G.M.M., Aguiar, C.M.L. & Gobbi, N. (2006). Characterization of the social wasp guild (Hymenoptera, Vespidae) visiting flowers in the Caatinga (Itatim, Bahia, Brazil). Sociobiology, 47: 1-12.

Santos, G.M.M., Cruz, J.D., Bichara-Filho, C.S.C., Marques, O.M. & Aguiar, C.M.L. (2007). Utilização de frutos de cactos (Cactaceae) como recurso alimentar por vespas sociais (Hymenoptera, Vespidae, Polistinae) em uma área de caatinga (Ipirá, Bahia, Brasil). Rev. Bras. Zool., 24: 1052-1056. doi: 10.1590/S0101-81752007000400023

Santos, G.M.M., Bispo, P.C. & Aguiar, C.M.L. (2009a). Fluctuations in richness and abundance of social wasps during the dry and wet seasons in three phyto-physiognomies at the tropical dry forest of Brazil. Environ. Entomol., 38: 1613-1617. doi: 10.1603/022.038.0613

Santos, G.M.M. & Presley, S.J. (2010). Niche overlap and temporal activity patterns of social wasps (Hymenoptera: Vespidae) in a Brazilian cashew orchard. Sociobiology, 56: 121-131.

Santos, G.P., Zanuncio, J.C., Pires, E.M., Prezoto, F., Pereira, J.M.M. & Serrão, J.E. (2009b). Foraging of *Parachartergus fraternus* (Hymenoptera: Vespidae: Epiponini) on cloudy and sunny days. Sociobiology, 53: 431-441.

Schoener, T.W. (1986). Resource partitioning. In J. Kikkawa & D.J. Anderson (Eds.), Community Ecology - Pattern and Process (pp. 91-126). Melbourne: Blackwell Scientific Publications.

Schmitz, J. & Moritz, R.F.A. (1998). Molecular phylogeny of Vespidae (Hymenoptera) and the evolution of sociality in wasps. Mol. Phylogenet. Evol., 9, 183-191.

Shannon, C.E. (1948). The mathematical theory of communication. In C.E. Shannon & W. Weaver (Eds.), The mathematical theory of communication (pp. 3-91). Urbana: University Illinois Press.

Siegel, S. (1956). Nonparametric statistics for behavioral sciences. New York: McGraw-Hill Book Company.

Silva S.S. & Silveira, O.T. (2009). Vespas sociais (Hymenoptera, Vespidae, Polistinae) de floresta pluvial Amazônica de terra firme em Caxiuanã, Melgaço, Pará. Iheringia, Sér. Zool., 99: 317-323. doi: 10.1590/S0073-47212009000300015

Silveira, O.T., Costa Neto, S.V. & Silveira, O.F.M. (2008). Social wasps of two wetland ecosystems in Brazilian Amazonia (Hymenoptera, Vespidae, Polistinae). Acta Amaz., 38, 333-344. doi: 10.1590/S0044-59672008000200018

Strassmann, J.E., Solis, C.R., Hughes, C.R., Goodnight, K.F. & Queller, D.C. (1997). Colony life history and demography of a swarm-founding social wasp. Behav. Ecol. Sociobiol., 40, 71-77.

Torezan-Silingardi, H.M. (2011). Predatory behavior of *Pachodynerus brevithorax* (Hymenoptera: Vespidae, Eumeninae) on endophytic herbivore beetles in the Brazilian Tropical Savanna. Sociobiology, 57: 181-189.

Zanette, L.R.S., Martins, R.P. & Ribeiro, S.P. (2005). Effects of urbanization on Neotropical wasp and bee assemblages in a Brazilian metropolis. Landscape Urban Plan., 71: 105-121. doi: 10.1016/j.landurbplan.2004.02.003

# Observation of *Trigona recursa* Smith (Hymenoptera: Apidae) Feeding on *Crotalaria micans* Link (Fabaceae: Faboideae) in a Brazilian Savanna Fragment

TMR Santos[1], JT Shapiro[1], PS Shibuya[1], C Aoki[2]

1 - Universidade Federal de Mato Grosso do Sul, Campo Grande – MS, Brazil

2 - Universidade Federal de Mato Grosso do Sul, Aquidauana – MS, Brazil

**Keywords**

Entomotoxicity, Feeding behavior, Geographic distribution, Monocrotaline, Stingless bee

**Corresponding author**

Julie Teresa Shapiro
Laboratório de Zoologia
Universidade Federal de Mato Grosso do Sul, Cidade Universitária s/n,
Campo Grande, MS, Brazil
79090-900
julie.teresa.shapiro@post.harvard.edu

**Abstract**

In this paper we present observations of individuals of the bee species *Trigona recursa* feeding on the fruits of *Crotalaria micans*. This plant, which contains pyrrolizidine alkaloids, is known to be toxic to humans, mammals and poultry. Over the course of three days, we observed a large number of bees feeding on many individual *Crotalaria micans* plants in an urban fragment of Brazilian Savanna. The bees preferred greener fruits, which are the softest and most toxic. Consumption of the plant had no immediately apparent fatal effect on the bees, since we did not find any dead individuals near the observation site. Some insect species are known to use pyrrolizidine and alkaloids for defense by incorporating them into their body or using them as precursors to pheromones. *Trigona recursa* and other bee species have not been previously recorded consuming *Crotalaria micans* and it is unclear what their motivation may be. We present these observations as a novel finding of the feeding behavior of *Trigona recursa*.

## Introduction

Plants of the genus *Crotalaria* (L.) contain pyrrolizidine alkaloids, which are known to be toxic to humans and animals, particularly livestock and poultry (Rose et al., 1957; Alfonso et al., 1993). Monocrotaline, the primary toxin of this genus, has been shown to damage hepatocytes, astrocytes, and glial cells, interfere with cell growth, cytoskeleton protein expression, and ATP production, damage DNA and cause apoptosis (Silva-Neto et al., 2010;Pitanga et al., 2011).

*Trigona jurine* 1807 is a Neotropical stingless bee genus that occurs from Mexico to northern Argentina, Paraguay and Uruguay (Camargo & Pedro, 2012). The greatest diversity of the genus is found in the Amazon and the central region of Brazil and a total of 19 species can be found within Brazilian territory (Rebêloet al., 2003).Relatively little is known of the feeding behavior of *T. recursa*. Individuals mark their paths and food sources for their nestmates to follow, using pheromones produced in the labial glands (Jarau et al., 2003).

They tend to exploit floral resources, although they have also been observed using non-floral sources and gathering sweat (Lorenzon & Matrangolo, 2005). Other species of the genus are known to be generalists feeding on pollen from a variety of plants (Oliveira et al., 2009). Three Neotropical Trigona species are necrophagous (Noll et al.,1996).

Previous studies of *Crotalaria retusa* have found that few insects visit these plants, which contain toxins throughout all their parts, (Kissmann & Groth, 1999) with two carpenter bee species, *Xylocopa grisescens* and *X. frontalis* making up for 90% of visits (Jacobi et al., 2005). *Trigona spinipes* has been observed visiting occasionally in order to obtain nectar by perforating the base of the flowers (Jacobi et al., 2005).

## Materials and Methods

During three consecutive days in February 2012, individuals of the bee species *T. recursa* were observed eating the fruits of *C. micans* in an urban Cerrado fragment on the campus of the Universidade Federal de MatoGrosso do Sul (Federal University of Mato Grosso do Sul), Campo Grande, Mato Grosso do Sul, Brazil (20º 30' S, 54º 36' W).

The first observation was carried out in the afternoon before sunset at approximately 18:00. On the following days, the observations were repeated twice per day, the first at 12:00 and the second at 18:00. Each observation period lasted 30 minutes.

At the end of the second day of observation, seven individual bees, as well as a sample from the plant were collected. These specimens were processed and given to specialists for identification. The bees were deposited in the Entomological Collection Pe. Jesus S. Moure (DZUP) of the Department of Zoology at the Universidade Federal do Paraná (Federal University of Paraná).

## Results

We observed the presence of a large number of individuals of *T. recursa* (n>100) consuming the fruit of several dozen individual *C. micans* plants during all observations. Apparently, the number of *T. recursa* was lower during the observations at 12:00 than at 18:00. However, because of their large numbers and frequent, fast movements around and between the plants, it was not possible to quantify the exact number of individuals and therefore it is impossible to be certain of the quantity of individuals consuming the fruits during the different hours of observation.

The fruits of *C. micans* are dry. The bees scraped the velvety-textured external layer of the fruits with their tongues, apparently marking them (Fig. 1).

The bees demonstrated a preference for the younger

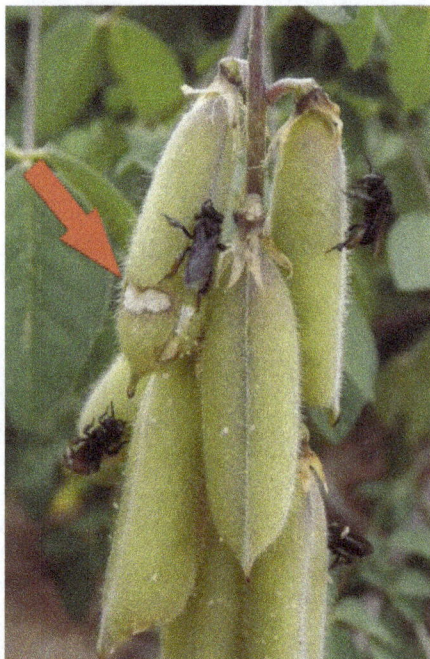

*Fig. 1.Bees feeding on the fruit ofCrotalariamicans. The arrow indicates the scraped portion of the fruit.*

and softer fruits. Older and drier fruits were discarded after partial consumption. Each plant had over one hundred fruits and the ratio between consumed and not consumed fruits was approximately 1:1, although a few individual plants had almost 100% of their fruits consumed. All plants had at least some of their fruits eaten.

We did not observe any indication of fatal intoxication in the bees. Over the three consecutive days of observation, no dead bees were found in the area of the plants. Possible intoxication later could not be analyzed.

## Discussion

Our observations of *T. recursa* feeding on plants of *C. micans* are notable because it is not a known food source for this bee speciesand toxins are found throughout the entire plant, including the fruit (Kissmann&Groth, 1999). Although *T. spinipesis* known to occasionally collect nectar from *Crotalaria* species (Jacobi et al., 2005), the individuals of *T. recursa* observed did not appear to collect nectar. Furthermore, the scraping method that we observed them using has not been previously recorded. It is possible that the bees scraped the fruits in order to differentiate the consumed fruits from those that had not yet been consumed. They may have also been marking them to recruit nest-mates to this food source (Jarau et al., 2003).

Studies have shown that secondary metabolites of plants can be toxic to bee species. For example, nicotine at high levels can reduce bees' fitness, although at lower, naturally occurring levels, there are no apparently detrimental effects and the substance can even be beneficial (Kohler et al., 2012). Other toxins produced in flowers have also been shown to reduce the life span of bees in laboratory settings, depending on the dose (Santoro et al., 2004; Rother et al.,2009; Rocha-Neto et al.,2011). Rother et al. (2009) tested the effects of ricinine (a toxic compound of *Ricinus communis* - Euphorbiaceae) on bees of the species *Apis mellifera* and *Scaptotrigona postica* and observed that *A. mellifera* had a high mortality within 72 hours for two of three concentrations tested (0.05 and 0.1%), while *S. postica* presented high mortality rate only after 14 days. Santoro et al. (2004) demonstrated that *A. mellifera* was also susceptible to tannins of *Stryphnodendron* spp. with mortality rates increasing on day 3 at all concentrations tested (1.25, 2.5 and 3.75%). These two studies showed that *A. mellifera* seems to be sensitive and have a rapid response to the toxins of native plant species, while the same did not occur with *S. postica* native bee species.

We note that for the above studies, bees were exposed to toxic treatments with extracts in laboratory experiments, whilehere we report that *T. recursa* spontaneously consumed *C. micans*. However, in more similar circumstances, Del Lama and Peruquetti (2006) observed that the consumption of the toxic plant *Caesalpinia peltophoroides* in a natural setting caused a large number of mortalities in bees, with 273 individuals of 20 different species dying after visiting the plant. Toxicity seemed to vary by time and individual tree. Although the long term effects could not be observed, most bees dropped to the ground in narcosis and died immediately

Species of the orders Lepidoptera, Coleoptera, Diptera, and Orthoptera are known to sequester pyrrolizidine alkaloids, including those found in *Crotalaria*, for purposes of defense against predators and as precursors to pheromones (Boppré, 1990). Notably, studies have shown that the moth species Utetheisa ornatrix obtains the toxins from plants during its larval stage, retaining the compounds as an adult and then passing it on to their eggs (Eisner & Eisner, 1991). Laboratory and field experiments have confirmed that the consumption of pyrolizidine alkaloids protects individuals from predation (Dussourd et al., 1988; Boppré, 1990; Eisner & Eisner, 1991).

In this case, there was no immediately apparent negative effect on *T. recursa* from consuming *C. micans*. Although we cannot be sure of the longer-term effects on the bees' fitness, it is possible that *T. recursa* benefits from the protection from monocrotaline. The toxins could also have a longer-term negative effect or intoxicate the bees after several days, but we were unable to confirm or rule out any of these possibilities.

We also note that this is the first record of *T. recursa* in the state of Mato Grosso do Sul. The species has been previously recorded in the neighboring states of São Paulo, Goiás, and Mato Grosso (Camargo & Pedro, 2012).

## Acknowledgements

We would like to thank Gabriel Melo for assisting in insect identification, Thales Henrique Dias Leandro for plant identification, and the Fulbright Program for the support of Julie Shapiro. We thank two anonymous reviewers for valuable comments that improved this article.

## References

Alfonso, H.A., Sanchez, L.M., Figeurdo, M.A. & Gomez, B.C. (1993). Intoxication due to *Crotalaria retusa* and *C. spectabilis* in chickens and geese.Vet. Hum. Toxicol., 35(6): 539.

Boppré, M. (1990).Lepidoptera and pyrrolizidine alkaloids. Exemplification of complexity in chemical ecology.J. Chem. Ecol., 16(1): 165–185.

Camargo, J.M.F.& Pedro, S.R.M. (2012). Meliponini Lepeletier, 1836. In Moure, J. S., Urban, D. &Melo, G. A. R. (Orgs). Catalogue of bees (Hymenoptera, Apoidea) in the neotropical region - online version. Available at http://www.moure.cria.org.br/catalogue. Accessed 1 March 2013.

Del Lama, M.A. & Peruquetti, R.C. (2006). Mortalidade de abelhas visitantes de flores de *Caesalpinia peltophoroi-des* Benth (Leguminosae) no estado de São Paulo, Brasil. Rev. Bras. Entomol., 50(4): 547–549. doi: 10.1590/S0085-56262006000400017

Eisner, T. & Eisner, M. (1991). Unpalatability of the pyrrolizidine alkaloid-containing moth, Utetheisaornatrix, and its larva, to wolf spiders. Psyche, 98: 111–118. doi: 10.1155/1991/95350

Jacobi, C.M., Ramalho, M. & Silva, M. (2005). Pollination Biology of the Exotic Rattleweed *Crotalaria retusa* L. (Fabaceae) in NE Brazil. Biotropica, 37(3):357–363. doi: 10-111/j.1744-7429.2005.00047.x

Jarau, S., Hrncir, M., Schmidt, V.M., Zucchi, R. & Barth, F.G.(2003).Effectiveness of recruitment behavior in stingless bees (Apidae, Meliponini).Insectes Soc., 50(4): 365–374. doi: 10.1007/s00040-003-0684-2

Kissmann, K.G.& Groth, D. (1999). Plantas infestantes e nocivas, Tomo II (2°ed.). São Paulo: Basf.

Köhler, A., Pirk, W.W.C. & Nicolson, W.S. (2012). Honeybees and nectar nicotine: Deterrence and reduced survival versus potential health benefits. J. Insect Physiol., 58:286–292. doi: 10.1016/j.bbr.2011.03.031

Lorenzon, M.C.A.I. & Matrangolo, C.A.R. (2005). Foraging on some nonfloral resources by stingless bees (Hymenoptera, Meliponini) in a caatinga region. J. Biol., 65(2):291–298. doi: 10.1590/S1519-69842005000200013

Neto, R.T.J., Leite, D.T., Maracajá, P.B., Pereira-Filho, R.R. & Silva, O.S.D.(2011). Toxicidade de flores de Jatropha gossypiifolia L. a abelha africanizada em condições controladas. Rev. Verde, 6 (2): 64–68.

Noll, F.B., Zucchi, R., Jorge, J.A. & Mateus, S. (1996). Food collection and maturation in the necrophagous stingless bee, *Trigona hypogea* (Hymenoptera: Meliponinae). J. Kans. Entomol. Soc., 69(4): 287–293.

Oliveira, F.P.M., Absy, M.L. & Miranda, I.S. (2009). Recurso polínico coletado por abelhas sem ferrão (Apidae, Meliponinae) em um fragment de floresta na região de Manaus- Amazonas. Acta Amazon., 39(3): 505–518.

Pitanga, S.P.B., Silva, A.D.V., Souza, S.C., Junqueira, H.A., Fragomeni, N.O.B., Nascimento, P.R., Silva, R.A., Costa, D.F.M., El-Bacha, R.S. & Costa, S.L. (2011). Assessment of neurotoxicity of monocrotaline, an alkaloid extracted from *Crotalaria retusa* in astrocyte/neuron co-culture system. NeuroToxicology, 32: 776–784. doi: 10.1016/j.neuro.2011.07.002

Rebêlo, M.M.J., Rêgo, M.M.C. & Albuquerque, C.M.P. (2003).Abelhas (Hymenoptera, Apoidea) da região setentrional do Estado do Maranhão, Brasil. In G.A.R. Melo& I. Alves-dos-Santos (Eds.),Apoidea Neotropica: Homenagem aos 90 Anos de Jesus Santiago Moure (pp. 265–278).

Criciúma:  Editora UNESC.

Rose, A.L., Gardner, C.A., McConnell, J.D. & Bull, L.B. (1957). Field and experimental investigation of "walkabout" disease of horses (Kimberley horse disease) in northern Australia: Orota-Laria Poisoning in horses. Part II. Aust. Vet. J., 33(3): 49–62. doi: 10.1111/j.1751-0813.1957.tb08270.x

Rother, C.D., Souza, F.T., Malaspina, O., Bueno, C.O., Silva, F.G.F.M., Vieira, C.P. & Fernandes, J.B.(2009). Suscetibilidade de operárias e larvas de abelhas sociais em relação à ricinina. Iheringia, 9(1): 61–65. doi: 10.1590/S0073-47212009000100009

Santoro, K.R.,Vieira, M.E.Q., Queiroz M.L., Queiroz M.C. & Barbosa S.B.P. (2004). Efeito do tanino de *Stryphnodendron* spp.sobre a longevidade de abelhas *Apis mellifera* L. (abelhas africanizadas). Arch. Zootec., 53: 281–291.

Silva-Neto, J.P., Barreto, R.A., Pitanga, B.P.S., Souza, C.S., Silva, V.D., Silva, A.R., Velozo, E.S., Cunha, S.D., Batatinha, M.J.M, Tardy, M., Ribeiro, C.S.O., Costa, M.F.D., El-Bachá, R.S. & Costa, S.L. (2010). Genotoxicity and morphological changes induced by the alkaloid monocrotaline, extracted from *Crotalaria retusa*, in a model of glial cells. Toxicon, 55 (1): 105–117. doi: 10.1016/j.toxicon.2009.07.007.

# 3

# Insecticidal effect of volatile compounds from fresh plant materials of *Tephrosia vogelii* against *Solenopsis invicta* workers

WS Li [1], Y Zhou [1], H Li [1], K Wang [1], DM Cheng [1,3], ZX Zhang [1,2*]

1. *Key Laboratory of Natural Pesticide and Chemical Biology, Ministry of Education, South China Agricultural University, Guangzhou, China, 510642*
2. *State Key Laboratory for Conservation and Utilization of Subtropical Agro-Bioresources, Guangzhou, China, 510642*
3. Department of Plant Protection, Zhongkai University of Agriculture and Engineering, Guangzhou, China, 510225

**Keywords**
volatile compounds, Red imported fire ant

**Corresponding Author**
Zhixiang Zhang
Key Laboratory of Natural Pesticide and
Chemical Biology, Ministry of Education
South China Agricultural University
Guangzhou, China, 510642
E-mail: zdsys@scau.edu.cn

**Abstract**
The effect of volatile compounds from the mashed fresh bean pods (B) as well as the branches and leaves (L) of *Tephrosia vogelii* on the behavior of *Solenopsis invicta* workers was investigated by fumigation toxicity bioassay. Gas chromatography–mass spectrometry analysis was used to identify and quantify the volatile compounds. α-pinene, thujene, caryophyllene, and d-limonene were identified as major components of the volatile compounds, which were found toxic to workers when applied by fumigation. Responses varied according to worker size, exposure time, and plant material. An increase in exposure time from 1h to 12h led to increases in mortality from 18.33% to 100.00% (B) and 13.33% to 100.00% (L) in minor workers as well as increases from 1.67% to 95.00% (B) and 15.00% to 98.33% (L) in major workers. The volatile compounds were also found to exert a behavioral effect against *S. invicta* in an A4 paper test. Walking and grasping abilities decreased at exposure times ranging from 40 min to 280 min. These findings suggest that the volatile compounds of *T. vogelii* can be used to control *S. invicta*.

## Introduction

The red imported fire ant (RIFA), *Solenopsis invicta*, is one of over 280 species in the widespread genus *Solenopsis*. Although RIFA is native to South America, it has become a pest in the southern United States, Australia, Thailand, Taiwan, the Philippines, Hong Kong, and the southern Chinese Provinces of Guangdong, Guangxi and Fujian. There are also reports of ant hills in Macau. RIFA are known to have a strong, painful and persistent irritating sting that often lead a pustule on the skin (Laura & Rudolf, 2001). The ant stings humans, pets, farm animals and wildlife, as well as damaging farm, electrical equipments and irrigation systems. Moreover, besides destroying crops and fruits directly or indirectly, they negatively affect the local biodiversity and cause approximately US$5 billion losses in urban and agricultural areas yearly in the USA (Cheng et al., 2008). Many botanical insecticidal compounds possessed good toxicity against RIFA, and it is potential for applying botanical insecticides to control RIFA.

*Tephrosia vogelii* is native to West Africa, but is found in India, Asia and other tropical regions (Dalziel, 1937; Lambert et al., 1993). It is widespread in tropical Africa from Sierra Leone and Ethiopia southwards to Angola, the Flora Zambesiaca area and the Comoro Islands, also from Assam to Indonesia. It was introduced into China by 1986 and planted over a large area in Guangdong province. It is a shrubby plant used as a fallow plant to improve soil fertility and to reduce erosion, particularly in higher areas. The leaf macerate is purgative and emetic (Walker, 1961; Burkill, 1995). Powders of *T. vogelii* are effectively used in the Congo against the stored ground nut pest *Caryedon serratus* (Delobel, 1987). It is also applied directly to treat head lice, fleas, scabies and other ectoparasites (Klaassen, 1996; Nwude, 1997). The water extract of the dried leaf possessed molluscicidal activity against *Bulinus globosus* (Chiotha, 1986). Water extract of oven dried stem, leaf and seed showed weak molluscicidal activity against *Biomphalaria pfeifferi* (Kloos, 1987). Acetone extract of the leaf showed feeding deterrent activity against the insect *Pieris rapae* (Shin, 1989). This study is the first to report on the toxicity of the volatile compounds released from the mashed fresh bean pods (B) as well as the branches and leaves (L) of *T. vogelii* against RIFA workers. In addition, this study aims to investigate the effects of the volatile compounds from the fresh plant materials of *T. vogelii* on the RIFA workers and analyze the volatile compounds by gas chromatography–mass spectrometry (GC–MS).

## Materials and Methods

### Plant materials

*T. vogelii* (Family: Legume, Papilionacea; Genus: *Tephrosia*) was planted in the sample region of insecticidal plants at South China Agricultural University in April 2010. The B and L were cut off in April 2013 and immediately sent to the laboratory.

### Insects

*S. invicta* colonies were obtained from a suburb in Guangzhou and maintained in the laboratory for bioassays in plastic containers under the following conditions: $25 \pm 2$ °C, $65\%\pm5\%$ relative humidity and constant darkness. Colonies were fed with a mixture of live insects (*Tenebrio molitor*) and 10% honey. A test tube (25mm × 200 mm) partially filled with water and plugged with cotton was used as the water source. Experiments were performed under similar environmental conditions.

### Fumigation toxicity bioassay

**Indoor test of the toxicity of mashed fresh bean pods (B) as well as branches and leaves (L) of *T. vogelii* against RIFA workers**

B (500 g) and L (500 g) *T. vogelii* were mashed with a high-speed organization stamp mill for 5 min. Mashed plant materials (5 g) were weighed and placed at the bottom of the beaker (1000 mL). Red imported fire ants were classified into minor workers (body length = 2.8 mm to 3.0 mm, head width = 0.6 mm to 0.7 mm) and major workers (body length = 4.3 mm to 4.5 mm, head width = 1.0 mm to 1.1 mm). Twenty workers settled on the bottom of the beaker (100 mL), which had an inside vertical wall coated with Fluon emulsion that was allowed to dry for 24 h to prevent the ants from escaping. The 100 mL beaker was allowed to settle on the bottom of the 1000 mL beaker without covering the mashed plant materials. The 1000 mL beaker was sealed with a plastic wrap. The

worker was maintained at $25\pm2$ °C and $65\%\pm5\%$ relative humidity. Experiments 1 (fumigant toxicity bioassay), 2 (walking ability test), and 3 (grasping ability test) were conducted. All treatments were replicated thrice. The contrast was the absence of mashed plant materials in the 1000mL beaker.

**Outdoor test of the toxicity of live bean pods (B) as well as branches and leaves (L) of *T. vogelii* against RIFA workers**

Twenty workers settled on the bottom of the beaker (100mL) sealed with gauze. One live branch of *T. vogelii* (not cut off from the shrub) with about 10 bean pods or 100 leaves was slipped into a transparent plastic bag (40cm × 29 cm × 10 cm) with two open ends. The beaker (100 mL) was tied upside down in the bag, and the other open end was sealed. Experiments 1 (fumigant toxicity bioassay), 2 (walking ability test), and 3 (grasping ability test) were conducted. The average temperature was set between 25°C and 30°C, average humidity was 70% to 90%, and average wind velocity was <0.5m/s during the test.

### Physiological index observation of RIFA workers

The cumulative mortalities of fumigant toxicity were determined 1, 2, 4, 6, 8, 10, and 12h after testing. Behavioral observation on the walking ability of the workers was determined 40, 80, 120, 160, 200, 240, and 280min after testing. The workers were placed on an A4 paper. If the workers could walk continuously for 10 cm without falling, these workers were regarded to possess walking ability. We used the following equation: walking rate = (number of workers possessing walking ability/number of workers per replicate) × 100.

Behavioral observation on the grasping ability of the workers was determined 40, 80, 120, 160, 200, 240, and 280 min after testing. The workers were placed on an A4 paper, which was softly rotated in 180° after 10s. The workers that did not fall from the A4 paper were regarded to possess grasping ability. We used the following equation: grasping rate = (number of workers possessing grasping ability/number of workers per replicate) × 100.

### Chemical analysis of the volatile compounds

Isolation, identification, and quantification of the component of the volatile compounds from the mashed fresh plant materials of *T. vogelii* were analyzed using GC–MS. The indoor test method was conducted following the procedure in 2.3; however, the 20 workers were replaced with a 2g adsorbents. The adsorbents were silicone (200 mesh to 300 mesh). The adsorbent was inserted into the injector (length = 20cm, diameter = 0.5cm) 6h after the treatment. The volatile compounds were absorbed by the silicone, which was eluted with 5ml petroleum ether or acetone. The analytical conditions of the GC–MS are shown in Table 1. Most compounds were identified based on GC–MS retention times, Kovats indices, and mass spectra.

*Statistical analysis*

Data were transformed into arcsine square root values for a three-way analysis of variance (ANOVA) to determine the significance of the effects of plant material, exposure time, and worker size on the mortalities, walking rate, and grasping rate of the minor and major workers as well as various interactions. The differences in the data were also assessed using Duncan's multiple range test, with P < 0.05 considered statistically significant. The figures were generated using the Microsoft Office Excel 2007 program.

Table1. The analytical conditions of GC-MS

| Gas chromatography | HP 6890N-5957 series plus (Agilent) | | |
|---|---|---|---|
| Carrier Gas | Helium | | |
| Injector Temp | 280℃ | | |
| Detector Temp | 230℃ | | |
| Capillary column | HP-5MS (Agilent) or DB-5 (J&W) | | |
| Head Pressure | 100 kPa | | |
| Oven Program | | | |
| Initial Temp | 60℃ | | |
| Initial Time | 1 min | | |
| | Rate (℃/min) | Final Temp. (℃) | Final Time (min) |
| | 25 | 210 | 5 |
| | 10 | 280 | 10 |
| Mass spectrometer | HP 5972 (Agilent) | | |
| EI voltage | 70 Ev | | |
| Mass spectrum database | NBSLI-BRARY | | |

**Results**

*Indoor test*

When B and L of *T. vogelii* were bioassayed, the mortalities, walking rate, and grasping rate of minor and major workers varied significantly according to plant material, exposure time, and worker size (ANOVA, P < 0.05). The results reveal that the three main effects and the partial interactions were significant (Table 2). Marked variations in the mortality of fumigant toxicity as well as the walking and grasping abilities were observed at different plant material, exposure time, and worker size (Table 3). The mortality of RIFA was significant at different plant material (F = 10.272, P =0.0022), exposure time(F = 235.652, P < 0.0001), and worker size(F = 54.450, P = 0.0001). However, there was no difference in the percentage of mortality between the interaction exposure time×worker size (F = 1.341, P =0.2547) and the interaction plant material×exposure time×worker size (F = 1.230, P = 0.3052). In addition, significant differences were found at walking rate and grasping rate of RIFA: developing from different exposure time (F = 312.108, P < 0.0001; F = 286.326, < 0.0001) and worker size (F = 49.371, P = 0.0001; F = 9.496, P =0.0032).

*Fumigant toxicity*

The indoor test of the mashed B and L of *T. vogelii* showed effective fumigant activity against the RIFA workers; by contrast, the outdoor test showed no such activity (Fig. 1). The volatile compounds released from the live B and L of *T. vogelii* caused no mortality in the fumigant toxicity assay, even after 12h exposure,there were no dead individuals. However, in all cases of the indoor test, a marked variation in insect mortalities was observed with the increase in exposure time. A significant difference in the mortalities between the minor and the major workers was indicated, with the volatile compound of the B and L of *T. vogelii* being significantly more toxic to the RIFA minors than the majors. After a 1h exposure, the volatile compounds of B and L resulted in 13.33% (B) and 18.33% (L) mortality rates in minor workers as well as

Table 2. ANOVA for the main factors of the indoor test affecting the behaviors of RIFA

| Factors | Mortality | | | Walking ability | | | Grasping ability | | |
|---|---|---|---|---|---|---|---|---|---|
| | DF | F values | P values | DF | F values | P values | DF | F values | P values |
| A | 1 | 10.272 | 0.0022 | 1 | 0.610 | 0.4383 | 1 | 4.845 | 0.0319 |
| B | 6 | 235.652 | 0.0001 | 6 | 312.108 | 0.0001 | 6 | 286.326 | 0.0001 |
| C | 1 | 54.450 | 0.0001 | 1 | 49.371 | 0.0001 | 1 | 9.496 | 0.0032 |
| A×B | 6 | 3.000 | 0.0130 | 6 | 0.876 | 0.5183 | 6 | 1.390 | 0.2347 |
| A×C | 1 | 5.339 | 0.0246 | 1 | 0.343 | 0.5605 | 1 | 0.628 | 0.4315 |
| B×C | 6 | 1.341 | 0.2547 | 6 | 2.660 | 0.0243 | 6 | 0.796 | 0.5772 |
| A×B×C | 6 | 1.230 | 0.3052 | 6 | 0.521 | 0.7902 | 6 | 1.008 | 0.4296 |

A: Plant material, B: Exposure time, C: Worker size (P=0.05)

1.67% (B) and 15.00% (L) mortality rates in major workers. However, the mortality rates were 100% (B) and 100.00% (L) for the minors as well as 95.00% (B) and 98.33% (L) for the majors at 12h of treatment.

*Walking and grasping abilities*

The volatile compound from the mashed B and L exhibited good fumigant activity observed at different exposure times against the RIFA workers. However, the volatile compounds released from the live B and L of *T. vogelii* caused no reduction in the walking and grasping abilities of the RIFA workers (Figs. 2 and 3). The minor and major workers showed good walking and grasping abilities with the live plant materials in outdoor test. And the walking and grasping abilities of the RIFA workers decreased with longer exposure in indoor test (Table 3). At exposure times ranging from 40 min to 280 min, the walking abilities decreased from 98.33% to 20.00% (B) and 96.67% to 21.67% (L) for the major workers and from 96.67% to 13.33% (B) and 95.00% to 16.67% (L) for the minor workers; the grasping abilities was reduced from 96.67% to 15.00% (B) and 95.00% to 13.33% (L) for the major workers and from 91.67% to 15.00% (B) and 96.67% to 15.00% (L) for the minor workers (Table 3).

*Chemical components of the volatile compounds*

The data from the GC–MS analysis demonstrated that the volatile compound from the mashed B and L of *T. vogelii* contains 17 and 14 major constituents, respectively (Table 4). The major components comprising 88.46% (B) and 75.37% (L) of the total volatile compound were identified as α-pinene, thujene, caryophyllene, and d-limonene. Other major components of the volatile compound included eucalyptol, α-cubebene, β-pinene, sabinene, and α-guaiene. α-Pinene is the major component, accounting for 63.44%(B) and 65.82%(L).

**Discussion**

*T. vogelii* is a shrubby plant used as a fallow plant in southern China. Our data show that the volatile compounds released from live B and L show no toxicity against RIFA workers and can not reduce the walking and grasping abilities of the RIFA workers despite abundant insecticidal contents. However, the volatile compounds released from the mashed B and L exhibited high toxicity against the RIFA workers and evidently reduced the walking and grasping abilities of the RIFA workers. This reduction indicated that the volatile

Table 3. Influences of mashed B and L of *T. vogelii* on mortality, walking and grasping abilities of workers (Mean ± SE, %)

| Material | Worker size | Mortality | | Walking ability | | Grasping ability | |
|---|---|---|---|---|---|---|---|
| | | Time (h) | Percentage | Time (min) | Percentage | Time (min) | Percentage |
| B | Major | | 1.67±1.67b | | 98.33 ±1.67a | | 96.67±1.67ab |
| | Minor | 1 | 13.33±1.67a | 40 | 96.67±1.67a | 40 | 91.67±1.67 b |
| L | Major | | 15.00±2.89a | | 96.67±1.67a | | 95.00±2.89ab |
| | Minor | | 18.33±1.67a | | 95.00±2.89a | | 96.67±3.33ab |
| B | Major | | 6.67±1.67b | | 91.67±1.67b | | 90.00±2.89ab |
| | Minor | 2 | 30.00±5.77a | 80 | 86.67±1.67bc | 80 | 83.33±4.41b |
| L | Major | | 30.00±7.64a | | 88.33±3.33bc | | 85.00±5.77b |
| | Minor | | 35.00±2.89a | | 81.67 ±3.33c | | 81.67±3.33b |
| B | Major | | 16.67±4.41b | | 80.00±2.89bc | | 76.67±4.41b |
| | Minor | 4 | 40.00±5.77a | 120 | 73.33±1.67bc | 120 | 71.67±1.67b |
| L | Major | | 35.00±7.64a | | 81.67 ±4.41b | | 80.00±5.00b |
| | Minor | | 45.00±2.89a | | 71.67 ±3.33c | | 73.33±4.41b |
| B | Major | | 36.67±4.41b | | 65.00 ±5.77b | | 60.00±2.89b |
| | Minor | 6 | 61.67±4.41a | 160 | 58.33 ±3.33b | 160 | 58.33±3.33b |
| L | Major | | 51.67±4.41a | | 70.00 ±5.77b | | 70.00±5.77b |
| | Minor | | 60.00±5.77a | | 61.67 ±4.41b | | 61.67±4.41b |
| B | Major | | 66.67±4.41a | | 60.00 ±2.89b | | 43.33±4.41bc |
| | Minor | 8 | 76.67±6.67a | 200 | 40.00 ±2.89c | 200 | 41.67±4.41c |
| L | Major | | 63.33±4.41a | | 60.00±5.77b | | 53.33±3.33b |
| | Minor | | 76.67±3.33a | | 45.00±2.89c | | 48.33±1.67bc |
| B | Major | | 85.00±5.77ab | | 41.67 ±1.67b | | 23.33±1.67c |
| | Minor | 10 | 95.00±2.89a | 240 | 23.33±4.41d | 240 | 21.67±4.41c |
| L | Major | | 76.67±1.67b | | 40.00±2.89b | | 36.67±1.67b |
| | Minor | | 91.67±4.41a | | 31.67 ±1.67c | | 20.00±2.89c |
| B | Major | | 95.00±2.89ab | | 20.00±2.89b | | 15.00±2.89b |
| | Minor | 12 | 100.00±0.00a | 280 | 13.33±1.67c | 280 | 15.00±2.89b |
| L | Major | | 98.33±1.67ab | | 21.67±1.67b | | 13.33±3.33b |
| | Minor | | 100.00±0.00a | | 16.67±1.67bc | | 15.00±2.89b |

Values sharing same letters means they are not significantly different from each other (P>0.05, Duncan's Multiple Range Test).

**Table 4.** The volatile compounds from mashed B or L (5 g) of *T. vogelii* adsorbed on silicon(2g) of indoor test.

| The volatile compounds from 5 g mashed fresh bean pods(B) | | | The volatile compounds 5 g mashed fresh branches and leaves(L) | | |
| --- | --- | --- | --- | --- | --- |
| Volatile compound | Retention time (min) | Relative content (%) | Volatile compound | Retention time (min) | Relative content (%) |
| thujene | 3.379 | 0.88 | 1R-α-pinene | 3.459 | 65.82 |
| 1R-α-pinene | 3.454 | 63.44 | β-terpinene | 3.746 | 0.23 |
| β-phellandrene | 3.741 | 2.21 | 2-propyl pyridine | 3.827 | 0.69 |
| 4-methylene-1-(1-methylethyl)-cyclohexene | 3.784 | 2.56 | d-limonene | 4.145 | 0.74 |
| β-pinene | 3.822 | 5.66 | eucalyptol | 4.178 | 1.28 |
| d-limonene | 4.146 | 0.96 | decamethylcyclopentasiloxane | 4.842 | 3.99 |
| tridecane | 5.799 | 0.17 | 2,3-dihydro-5-methyl-1H-indene | 4.945 | 0.50 |
| cyclohexene | 6.123 | 0.39 | azulene | 5.285 | 0.41 |
| α-cubebene | 6.371 | 1.59 | copaene | 6.366 | 0.21 |
| aristolene | 6.436 | 1.07 | α.-guaiene | 6.431 | 0.11 |
| (-)-isocaryophyllene | 6.560 | 0.44 | caryophyllene | 6.636 | 0.69 |
| caryophyllene | 6.641 | 5.77 | calarene | 6.674 | 0.16 |
| 1H-cyclopenta[1,3]cyclopropal[1,2] benzene,octahydro-7-methyl-3-methylene-4-(1-methylethyl)[3aS(3a.α.3b.β.4.β.7.α.7aS*]] | 6.679 | 0.64 | germacrene | 6.955 | 0.39 |
| (-)-g-cadinene | 6.820 | 0.79 | hexadecane | 7.376 | 0.15 |
| β-sesquiphellandrene | 6.955 | 1.30 | | | |
| germacrene | 7.036 | 0.31 | | | |
| 10s,11s-himachala-3(12),4-diene | 7.063 | 0.28 | | | |

compounds could produce and release insecticidal compounds that could be defensive compounds, with functions varying from those in the B and L. The volatile compounds can be used to ward off pests when branches and leaves were harmed.The fumigant activity as well as walking and grasping abilities were influenced by the worker size, exposure time, and plant material.

To the best of our knowledge, no previous studies have been reported regarding the effects of volatile compounds from fresh plant materials of *T. vogelii* on the fumigant activity as well as the walking and grasping abilities of RIFA workers. However, earlier studies indicated that the insecticidal activity of the acetone extract of the leaf showed feeding deterrent activity against the insect *Pieris rapae* (Shin, 1989).

Plant-derived essential oils may contain hundreds of

different constituents but only few components predominate (Rajendran & Sriranjini, 2008). The results of the GC–MS analysis showed thujene, α-pinene, caryophyllene, eucalyptol, and d-limonene were found in the compounds from the mashed B and L. Eucalyptol, one of the monoterpenoids of the volatile compounds, is widely known for its high insecticidal property. α-pinene(64.63%)as its major components, is also found in other plants exhibiting biological activity against various insect species (Ojimelukwe & Alder, 1999; Choi et al., 2006).

The L of *T. vogelii* could have abundant insecticidal compounds and aboveground biomass (dry matter) produced by *T. vogelii* during the six-month growth period of 5.3 t/ ha with an L:stem ratio = 1:2 (Venant, 1999). This finding indicates that *T. vogelii* can potentially control RIFA. The dry matter of *T. vogelii* is often prepared in the production of rotenone-based insecticides. However, this study suggests that the mashed B and L of *T. vogelii* are more suitable than dry matter for producing insecticides.

The volatile compounds from the mashed B and L of *T. vogelii* possess high toxicity against the RIFA workers and can reduce the walking and grasping abilities of these workers. Compared with dry matter, the mashed B and L of *T. vogelii* exhibit greater potential for producing insecticides.

## Acknowledgments

This study was supported by the Science and Technology Planning Project of Guangdong Province, China (No.2012B020309004). The authors acknowledge the technical support provided by Mr. Hai Hong (South China Agricultural University,China) for GC–MS.

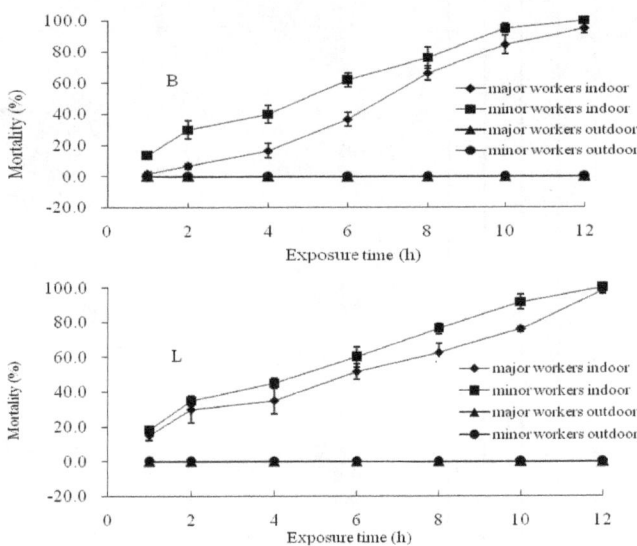

Fig. 1 Mortalities of workers after treated with fresh B or L of *T. vogelii* (Indoor test5g mashed fresh plant material, Outdoor test one live branch with about 10 bean pods or 100 leaves)

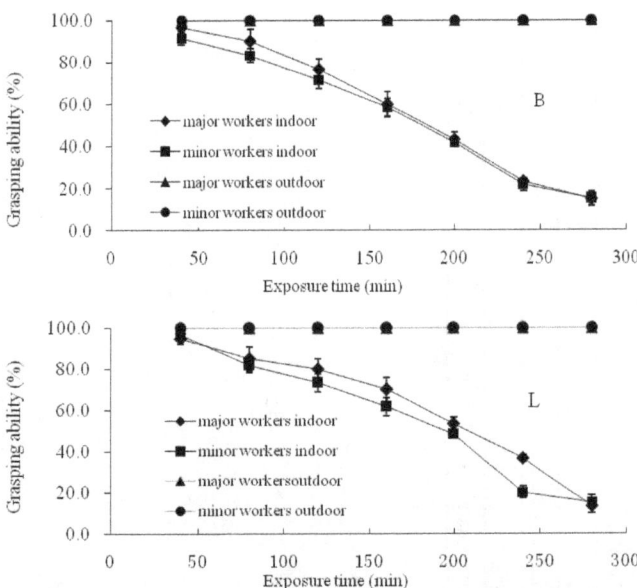

Fig. 2 - The grasping abilities of workers after treated with fresh B or L of *T. vogelii* (Indoor test: 5g mashed fresh plant materials, Outdoor test: one live branch with about 10 bean pods or 100 leaves)

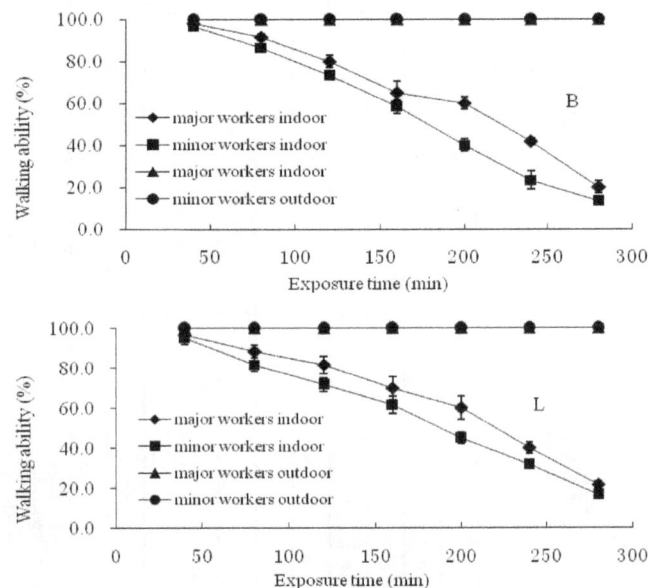

Fig. 3 - The walking abilities of workers after treated with fresh B or L of *T. vogelii* (Indoor test: 5g mashed fresh plant materials, Outdoor test: one live branch with about 10 bean pods or 100 leaves)

## References

Burkill, H.M. Ed.(1995). Useful plants of west tropical africa. Kew: Royal Botanical Gardens. http://www.nhbs.com/useful_plants_of_west_tropical_africa_sefno_39683.html

Cheng, S. S., J. Y. Lin, C. Y. Lin, Y. R. Hsui, M. C. Lu, W. J. Wu. & S. T. Chang. (2008). Terminating red imported fire ants using *Cinnamomum osmophloeum* leaf essential oil. Bioresource Technology, 99: 889-893. doi: 10.1016/j.biortech.2007.01.039

Chiotha, S. S., & J. D. Msonthi. (1986). Screening of indigenous plants for possible use in controlling bilharzias transmitting snails in Malawi. Fitoterapia, 57: 193-7.

Choi WS, Park BS, Lee YH, Jang DY, Yoon HY, Lee SE (2006). Fumigant toxicities of essential oils and monoterpenes against *Lycoriella mali* adults. Crop Protection, 2: 398-401. doi: 10.1016/j.cropro.2005.05.009

Dalziel, J. M. (1937). Useful plants of west tropical africa. London Crown Agents. http://www.nhbs.com/useful_plants_of_west_tropical_africa_sefno_39683.html

Delobel, A., & P. Malonga.(1987). Insecticidal properties of six plant materials against *Caryedon serratus*. Journal of Stored Products Research, 23: 173-6. doi: 10.1016/0022-474X(87)90048-8

Klaassen, C. D. (1996). Nonmetallic environmental toxicants: Air pollutants, solvents, vapors and pesticides. Pharmacological Basis of Therapeutics 9th edn. New York: McGraw-Hill. 1651-1652.

Kloos, H., F. W. Thiongo, J. H. Ouma & A. E. Butterworth. (1987). Preliminary evaluation of some wild and cultivated plants for snail control in Machakos district, Kenya. Journal of Tropical Medicine and Hygiene, 90: 197 -204.

Lambert, N., M. F. Trouslot, C. Nef-Campa. & H. Chrestin. (1993). Production of rotenoids by and photomixotrophic cell cultures of *Tephrosia vogelii*. Phytochemistry, 34: 15-20. doi: 10.1016/S0031-9422(00)90838-0

Nwude, N. (1997). Ethnoveterinary pharmacology and ethnoveterinary practices in Nigeria: an overview. Tropical Veterinary, 15: 17-23.

Ojimelukwe P.C., & C. Alder. (1999). Potential of zimtaldehyde,4-allyl-anisol, linalool, terpineol and other phytochemicals for the control of the confused flour beetle (*Tribolium confusum* J.d. V.) (Col: Tenebrionidae). Journal of Pesticide Science, 72: 81-86.

Rajendran, S., & V. Sriranjini. (2008). Plant products as fumigants for stored product insect control. Journal of Stored Products Research, 44:126-135. doi:10.1016/j.jspr.2007.08.003

Shin, F. C. (1989). Studies on plants as a source of insect growth regulators for crop protection. Journal of Applied Entomology, 107: 85-92. doi: 10.1111/j.1439-0418.1989.tb00247.x

Laura, C. & Rudolf, H. S.(2001). Red imported fire ant, *Solenopsis invicta* Buren. UF/IFAS Featured Creatures. http://entnemdept.ufl.edu/creatures/urban/ants/red_imported_fire_ant.htm

Walker, A. R., & R. Sillans. (1961). Plants used in Gabon. Paris: Encyclopedie Biologique.

Venant R., K. K. Nancy, K. K Charles, Gachene. & P. Cheryl. (1999). Biomass production and nutrient accumulation by *Tephrosia vogelii* (Hemsley) A. Gray and *Tithonia diversifolia* Hook F. fallows during the six-month growth period at Maseno, Western Kenya. Biotechnology, Agronomy, Society and Environment, 3: 237-246.

# 4

# Effects of Hematoporphyrin Monomethyl Ether on Worker Behavior of Red Imported Fire Ant *Solenopsis invicta*

ZX ZHANG [1,2], Y ZHOU [1], DM CHENG [1,3,*]

1 - Key Laboratory of Natural Pesticide and Chemical Biology, Ministry of Education, South China Agricultural University, Guangzhou, China

2 - State Key Laboratory for Conservation and Utilization of Subtropical Agro-Bioresources, Guangzhou, China

3 - Department of Plant Protection, Zhongkai University of Agriculture and Engineering, Guangzhou, China

**Keywords**

*Solenopsis invicta*, Red imported fire ant, Hematoporphyrin Monomethyl Ether, Behavior.

**Corresponding author**

Dongmei Cheng
Key Laboratory of Natural Pesticide and Chemical Biology, Ministry of Education
South China Agricultural University
Guangzhou, China, 510642
E-Mail: zdsys@scau.edu.cn

**Abstract**

The effect of hematoporphyrin monomethyl ether (HMME) activated under visible light on worker behavior of *Solenopsis invcita* was studied with the potter spray tower method. The results showed that greater than 10 mg/L HMME activated under visible light could reduce the walking, grasping, aggregation, and water and food recognition abilities of red imported fire worker ant significantly, but 100 mg/L HMME in darkness could not affect their abilities or behaviors significantly. Therefore, HMME may be a potential novel insecticide that can be used as a substitute for toxic insecticides for controlling red imported fire ants.

## Introduction

The red imported fire ant, a pest newly introduced to mainland China in 2005 (Zhang et al., 2007) and widely distributed in South China, threatens households, agriculture and wildlife. During foraging, worker ants leave a chemical pheromone trail to guide additional worker ants to the food source. These additional worker ants retrieve the food and return to the colony, also marking the pheromone trail laid down by the first group of ants (Xu et al., 2007). The abilities to walk, grasp, aggregate and recognize food and water are important for forager ants or other worker ants that leave the nest. If they lose the ability to move and recognize, they will have difficulty living in the complicated external circumstances, which will directly cause a decrease in the food in their nest. This occurrence will eventually affect their population, and may even induce nestmates to become aggressive and bite each other.

Photosensitizer such as α-Terthienyl (α-T) can affect the walking behavior of worker ants and even kill them directly (Yan et al., 2012; Liu et al., 2011). However, α-T activated under UV light is not suitable for controlling worker ants because the worker ants leaving the nest usually wander under visible light.

Hematoporphyrin monomethyl ether (HMME), a novel photosensitizer activated under visible light, is the second generation of porphyrin-related photosensitizer; it consists of two monomer porphyrins, namely, 3-(1-methyloxyethyl)-8-(1-hydroxyethyl) deuteropor-phyrin IX and 8-(1-methyloxyethyl)-3-(1-hydroxyethyl) deuteropor-phyrin IX, that are mutually locational isomers (Chen et al., 2000). HMME possesses some excellent properties, such as strong photodynamic effect and fast removal from organs or cells. Moreover, HMME has been reported for the treatment of some cancers (Anderson et al., 2002; Ascencio et al., 2007; Moghissi et al., 2000; Yoshihiro et al., 1993). HMME may be useful for controlling worker ants.

To our knowledge, no research has been conducted on

the effect of HMME on red imported fire worker ants. This study aimed to investigate the effects of HMME activated under visible light on the behavior of red imported fire ants.

## Materials and methods

*Solenopsis invicta* colonies were collected from the suburb of Guangzhou and maintained in the laboratory for bioassays. The collected ants were fed with a mixture of 10% honey and live insects (*Tenebrio molitor*). A test tube (25 mm×200 mm), which was partially filled with water and plugged with cotton, was used as a water source. The ants were maintained in the laboratory at 25±2 °C. Large worker ants used in the test were about 6 mm in length, whereas small worker ants were about 3 mm in length.

HMME was provided by the Institute of Red and Green Photosensitizer (Shanghai, China). The stock solution was prepared in ethanol at a concentration of 10 mg/mL and kept in the dark at -20 °C.

A 75 W bromine tungsten lamp source (provided by Zolix Instrument Co. Ltd., Beijing, China) provided spectral radiation from 350 nm to over 2500 nm, which falls within the entire visible range of wavelengths.

The stock solution was reconstituted at 5, 10, 25, 50 and 100 mg/L in acetone–water mixtures (3/7, v/v). Freshly prepared HMME solutions were kept from light at all times, except during actual measurement.

HMME solution was applied to the worker ants with a potter spray tower (Burkard Manufacturing Co. Ltd., UK) using methods similar to those described by Harris *et al* (1962). Worker ants were placed in a glass Petri dish (120 mm in diameter) whose vertical wall was coated with a Fluon emulsion. The ants were then placed in the spray tower and sprayed with 2 ml HMME solution. The treated worker ants were transferred into a clean 500 ml beaker, whose vertical wall was coated with a Fluon emulsion (the same as below), and then placed in darkness immediately and incubated at 25 °C for 30 h. They were then exposed to visible light emitted by the bromine tungsten lamp (20 cm in height above the 500 ml beaker) for 10 min. The treated worker ants were placed in darkness immediately and incubated at 25 °C for 1h, and then their behaviors were observed. Each treatment was replicated three times, and each replicate included 30 to 40 worker ants.

In the following methods used to observe the behavior of worker ants regarding water and food recognition, the worker ants were placed in darkness without food and water for 10 h before treatment.

Controls were similar to the above steps, except that 10 min of light treatment was replaced with 10 min of dark treatment.

### Behavior observation on walking ability of worker ants

Worker ants were placed on an A4 paper. They were regarded as possessing walking ability if they could walk continuously for 10 cm and did not fall down.

The formula was as follows: walking rate = number of worker ants possessing walking ability/number of worker ants per replicate × 100.

### Behavior observation on worker ants' aggregation

Worker ants were placed in a 500 ml beaker, and 20 min later, worker ant aggregation was observed. Aggregation was considered present if over five worker ants gathered into an aggregate mass.

The formula was as follows: aggregation rate = number of worker ants in aggregate mass/number of worker ants per replicate × 100.

### Behavior observation on grasping ability of worker ants

Worker ants were placed on an A4 paper, and 10 s later, the A4 paper was turned over 180 degrees gently. They were regarded as possessing grasping ability if they would not fall down from the A4 paper.

The formula was as follows: grasping rate = number of worker ants possessing grasping ability/number of worker ants per replicate × 100.

### Behavior observation on water recognition of worker ants

A water-soaked cotton ball (1 g) and 20 worker ants (the distance between the cotton ball and the ants was 25 cm) were placed on the midcourt line of a porcelain tray (20 cm × 30 cm × 5 cm) whose vertical wall was coated with a Fluon emulsion (the same as below). Then, the water drinking behavior of worker ant was observed within 30 min. Worker ants were regarded as having water recognition ability if they continuously touched the cotton ball with their mouth for greater than 10 s. A worker ant was removed from the porcelain tray if it had drunk water.

The formula was as follows: drinking water rate = number of worker ants drank water/number of worker ants per replicate × 100.

### Behavior observation on food recognition of worker ants

The treatment method was the same as that used for the behavior observation on the water recognition of worker ants, but the cotton ball was replaced with a dead larva of *Tenebrio molitor*. Worker ants were regarded as possessing food recognition ability if they continuously touched the dead larvae with their mouth for greater than 10 s.

The formula was as follows: food recognition rate = number of worker ants could recognize the dead larvae/number of worker ants per replicate × 100.

*Statistical Analysis*

Data were reported as means ± SE based on three independent experiments. The percentage values were transformed into arc sin of square root of the percentages prior to the analysis, and three-factor ANOVA with worker size, light treatment, and concentration as the main effects were conducted. Moreover, the differences between the data were assessed by Duncan's multiple range test (SAS 1989) with $P < 0.05$ regarded as statistically significant. The figures were produced using Microsoft Office Excel 2007.

**Results**

Three-factor ANOVA with worker size, light treatment, and concentration as the main effects, as well as two and three-way interactions between these effects, was conducted. Results show that all three main effects and partial interactions were significant (Table 1.). Significant differences in walking, aggregation, grasping, and water and food recognition abilities were observed from the different treatment concentrations, light treatment, and size of the worker ants (Table 2.).

Small and big worker ants possessed good walking, aggregation, and grasping abilities after treated with HMME in darkness. After treated with 5, 10, 25 and 50 mg/L HMME, the walking rates of small and big worker ants were greater than 96.0%. The walking rates were 86.67% and 98.89% at 100 mg/L, respectively (Fig. 1A, Table 2.). All the aggregation rates of small and big worker ants were greater than 92.0% (Fig. 1B, Table 2.), and all their grasping rates were greater than 82.0% at the treatment concentration in darkness (Fig. 1C, Table 2.).

HMME activated under visible light could significantly affect the walking, aggregation, and grasping abilities of red imported worker fire ants. After treated with 10, 25, 50 and 100 mg/L HMME activated under visible light, the walking rates of small worker ants were 82.22%, 53.33%, 14.44% and 0.0%, respectively, and those of big worker ants were 92.22%, 81.11%, 34.44% and 0.0%, respectively, which were significantly different from the values obtained from the treatments in darkness ($P < 0.05$) (Fig. 1A, Table 2.). The aggregation rates of small worker ants were 70.83%, 50.83%, 26.67% and 2.50%, respectively, and those of big worker ants were 92.50%, 80.00%, 52.50% and 8.33%, respectively, which were significantly different from the values obtained from the treatments in darkness ($P < 0.05$) (Fig. 1B, Table 2.). The climbing rates of small worker ants were 45.56%, 21.11%, 2.22% and 0.0%, and those of big worker ants were 68.89%, 35.56%, 6.67%, and 0.0%, respectively, which were significantly different from the values obtained from the treatments in darkness ($P < 0.05$) (Fig. 1C, Table 2.).

Small and big worker ants possessed good water and food recognition abilities after treated with HMME in dark-

ness. After treated with HMME at test concentration in darkness, the drinking water rates of small and big worker ants were greater than 73.00%, and their food recognition rates were greater than 90.00% (Fig. 2, Table 2.).

The visible light-activated HMME could affect the water and food recognition abilities of worker ants significantly. After treated with 5, 10, 25, 50 and 100 mg/L HMME activated under visible light, the drinking water rates of small worker ants were 51.11%, 43.33%, 34.44%, 13.33% and 5.56%, respectively, and those of big worker ants were 84.44%, 78.89%, 61.11%, 32.22% and 14.44%, respectively, which were significantly different from the values obtained from the treatments in darkness ($P < 0.05$) (Fig. 2A, Table 2.). The food recognition rates of small worker ants were 80.00%, 56.67%, 44.44%, 20.00% and 12.22%, respectively, and those of big worker ants were 93.33%, 86.67%, 58.89%, 27.78% and 10.00%, respectively, which were significantly different from the values obtained from the treatments in darkness ($P < 0.05$) (Fig. 2B, Table 2.).

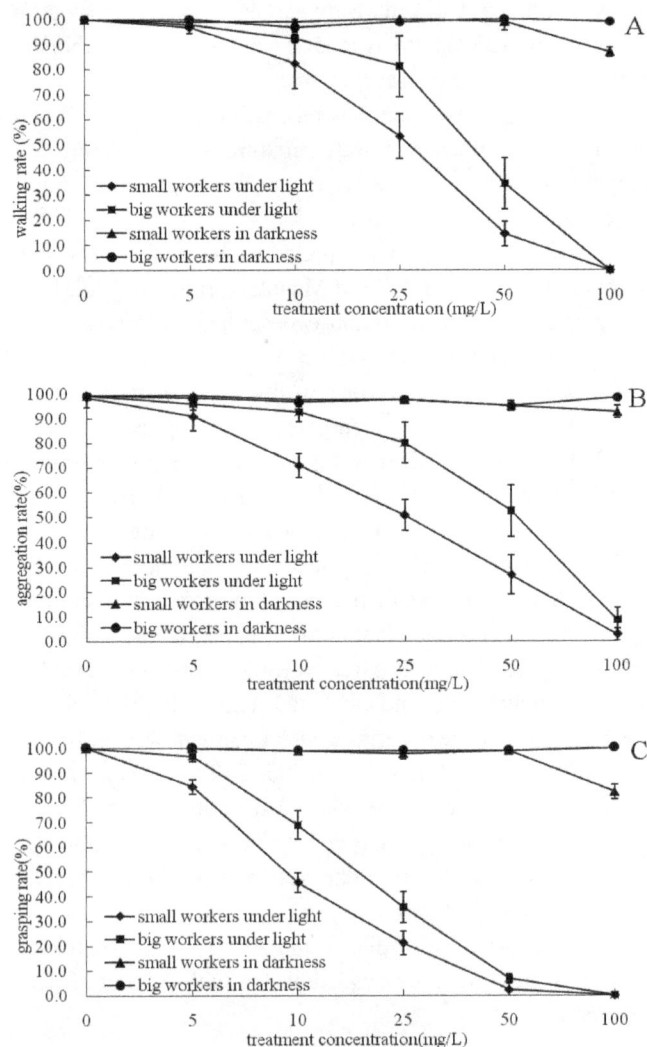

Fig. 1 The effect of visible light-activated HMME on walking ability (A), aggregation (B), and grasping ability (C) of worker ants.

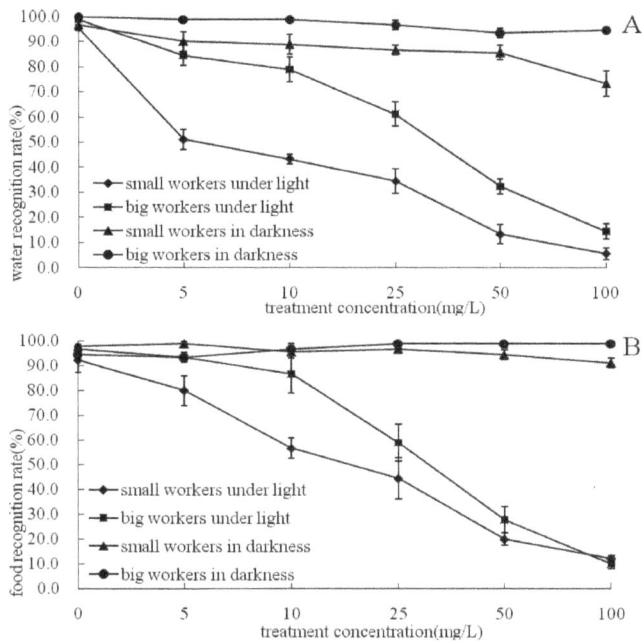

Fig. 2 The effect of visible light-activated HMME on water (A) and food (B) recognition of worker ants.

## Discussion

This study showed that HMME concentration greater than 10 mg/L activated under visible light could reduce the walking, grasping, aggregation, and water and food recognition abilities of red imported worker fire ants. However, 100 mg/L HMME in darkness does not affect these abilities or behaviors of worker ants. This finding suggests that HMME could be transmitted successfully in an ant nest, which is the key factor for effectively controlling red imported fire ants.

HMME activated under visible light is a novel photosensitizer that has been effectively used for solid tumors, which means that it is relatively safe for humans. HMME is also safe for the environment because of its photodegradation characteristics. As a formulation for controlling red imported fire ants, HMME can be used as bait or as powder applied in the morning or late afternoon. Thousands of worker ants living in one nest will then carry it into the nest, resulting in effectively decreasing the labor and photodegradation of HMME, and thus, producing good controlling effect.

In conclusion, this study showed that HMME is a potential novel alternative for moderately and highly toxic insecticides to control red imported fire ant.

## Reference

Anderson, G. S., Miyagi, K., Sampson, R. W. & Sieber, F. (2002). Anti-tumor effect of Merocyanine 540-mediated photochemotherapy combined with Edelfosine: potential implications for the ex vivo purging of hematopoietic stem cell grafts from breast cancer patients. J. Photochem. Photobiol. B, 68: 101-108. doi: 10.1016/S1011-1344(02)00377-9.

Ascencio, M., Delemer, M., Farine, M. O., Jouve, E., Collinet, P. & Mordon, S. (2007). Evaluation of ALA-PDT of ovarian cancer in the Fisher 344 rat tumor model. Photodiagn. Photodyn., 4: 254-260. doi: 10.1016/j.pdpdt.2007.07.003.

Chen, W. H., Yu, J. X., Yao, J. Z., Shen, W. D., Liu, J. F., & Xu, D. Y. (2000). Pharmacokinetic studies on hematoporphyrin monomethyl ether: a new promising drug for photodynamic therapy of tumors. Chin. J. Laser Med. Surg., 9: 105-108.

Harris, C. R., Manson, G. F., Mazurek, J. H. (1962). Development of insecticidal resistance by soil insects in Canada. J. Econ. Entomol., 55: 777-780.

Liu, N., Cheng, D. M., Xu, H. H. & Zhang Z. X. (2011). Behavioral and Lethal Effects of α-terthienyl on the Red Imported Fire Ant (RIFA). Chin. Agri. Sci., 44: 4815-4822. http://211.155.251.135:81/Jwk_zgnykx/EN/Y2011/V44/I23/4815.

Moghissi, K., Dixon, K., Thorpe, J. A., Stringer, M. & Moore, P. J. (2000). The role of photodynamic therapy (PDT) in inoperable oesophageal cancer. Eur. J. Cardiothorac. Surg., 17: 95-100.

Xu Y. J., Lu Y. Y., Zeng L. & Liang G. W. (2007). Foraging behavior and recruitment of red imported fire ant Solenopsis invicta Buren in typical habitats of South China. Acta Ecol. Sin. , 27: 855-860. doi: 10.1016/S1872-2032(07)60022-5.

Yan W. W., Zhang H., Dong Y. L., Xu H. H. & Zhang Z. X. (2012). Effect of α-terthienyl on the foraging behavior and feeler responses of Solenopsis invicta Buren. Chin. J. Pest Sci., 14, 277-282. http://www.nyxxb.com.cn/nyaoxxb/ch/reader/create_pdf.aspx?file_no=20120306&flag=1&journal_id=nyaoxxb&year_id=2012.

Yoshihiro, H., Harubumi, K., Chimori, K. & Tetsuya, O. (1993). Photodynamic therapy (PDT) in early stage lung cancer. Lung Cancer, 9: 287-293. doi: 10.1016/0169-5002-(93)90683-O.

Zhang, R., Li, Y., Liu, N. & Porter, S. D. (2007). An overview of the Red Imported Fire Ant (Hymenoptera: Formicidae) in Mainland China. Fla. Entomol., 90: 723-731. http://journals.fcla.edu/flaent/article/view/75725/73383.

Table 1. ANOVA for main factors that affect the behaviors of red imported fire ants.

| Factors | Walking ability | | Aggregation | | Grasping ability | | Water recognition ability | | Food recognition ability | |
|---|---|---|---|---|---|---|---|---|---|---|
| | F values | P values | F values | P values | F values | P values | F values | P values | F values | P values |
| A | 235.870 | 0.0001 | 82.473 | 0.0001 | 204.860 | 0.0001 | 96.551 | 0.0001 | 88.786 | 0.0001 |
| B | 990.869 | 0.0001 | 384.242 | 0.0001 | 1505.817 | 0.0001 | 607.761 | 0.0001 | 668.217 | 0.0001 |
| C | 22.714 | 0.0001 | 20.820 | 0.0001 | 30.094 | 0.0001 | 129.990 | 0.0001 | 18.087 | 0.0001 |
| A×B | 159.554 | 0.0001 | 51.779 | 0.0001 | 134.497 | 0.0001 | 31.030 | 0.0001 | 51.749 | 0.0001 |
| A×C | 2.450 | 0.0468 | 0.879 | 0.5023 | 2.144 | 0.0761 | 1.764 | 0.1383 | 3.606 | 0.0075 |
| B×C | 4.233 | 0.0451 | 13.727 | 0.0005 | 2.496 | 0.1207 | 0.981 | 0.3268 | 9.766 | 0.0030 |
| A×B×C | 7.790 | 0.0001 | 1.418 | 0.2350 | 7.290 | 0.0001 | 1.010 | 0.4222 | 1.301 | 0.2793 |

A: Concentration, B: Light treatment, C: Worker size (P=0.05)

Table 2. Influences of different treatments on behaviors of red imported fire ants.

| Treatments | | | Percentage (Means ± SE, %) | | | | |
|---|---|---|---|---|---|---|---|
| Concentration (mg/L) | Light | Worker size | Walking ability | Aggregation | Grasping ability | Water recognition | Food recognition |
| 0 | Light | Small | 100.00±0.96a | 98.33±3.91abc | 100.00±0.00a | 95.56±1.11bcd | 92.22±5.09bcde |
| | | Big | 100.00±0.00a | 99.17±1.38a | 100.00±0.00a | 98.89±1.11ab | 96.67±1.57abc |
| | Darkness | Small | 100.00±0.96a | 99.17±0.72a | 100.00±0.00a | 96.67±1.92abc | 97.78±0.96ab |
| | | Big | 100.00±0.00a | 99.17±1.38a | 100.00±0.00a | 100.00±0.00a | 97.78±1.11ab |
| 5 | Light | Small | 96.67±2.48b | 90.83±5.83cd | 84.44±2.94b | 51.11±4.01ij | 80.00±5.98f |
| | | Big | 97.78±1.57ab | 95.83±2.47abc | 96.67±1.92a | 84.44±4.01efg | 93.33±1.84bcde |
| | Darkness | Small | 98.89±0.96ab | 99.17±0.72a | 100.00±0.00a | 90.00±3.85cdef | 95.56±0.96abcd |
| | | Big | 100.00±0.00a | 98.33±1.38ab | 100.00±0.00a | 98.89±1.11ab | 98.89±1.84a |
| 10 | Light | Small | 82.22±10.11d | 70.83±4.77e | 45.56±4.01d | 43.33±1.92jk | 56.67±4.16gh |
| | | Big | 92.22±1.84c | 92.50±3.80bc | 68.89±5.77c | 78.89±4.84fg | 86.67±7.86ef |
| | Darkness | Small | 98.89±0.96ab | 97.50±1.18abc | 98.89±1.11a | 88.89±4.01def | 93.33±2.48bcde |
| | | Big | 96.67±1.84b | 96.67±1.86abc | 98.89±1.11a | 98.89±1.11ab | 95.56±2.22abcd |
| 25 | Light | Small | 53.33±8.80e | 50.83±6.28f | 21.11±4.84e | 34.44±4.84k | 44.44±8.37h |
| | | Big | 81.11±12.11d | 80.00±8.28de | 35.56±6.19d | 61.11±4.84hi | 58.89±7.43g |
| | Darkness | Small | 100.00±0.00a | 97.50±1.18abc | 97.78±2.22a | 86.67±1.92def | 97.78±1.11ab |
| | | Big | 98.89±0.96ab | 97.50±1.18abc | 98.89±1.11a | 96.67±1.92abc | 96.67±1.84abc |
| 50 | Light | Small | 14.44±4.81g | 26.67±7.91g | 2.22±4.84g | 13.33±3.85lm | 20.00±2.48ij |
| | | Big | 34.44±10.23f | 52.50±10.37f | 6.67±1.92f | 32.22±2.94k | 27.78±5.30i |
| | Darkness | Small | 98.89±3.64ab | 95.00±1.18abc | 98.89±1.11a | 85.56±2.94efg | 96.67±1.84abc |
| | | Big | 100.00±0.00a | 95.00±1.86abc | 98.89±1.11a | 93.33±1.92cde | 94.44±1.57bcde |
| 100 | Light | Small | 0.00±0.00h | 2.50±2.50h | 0.00±0.00g | 5.56±2.22m | 12.22±1.11j |
| | | Big | 0.00±0.00h | 8.33±5.07h | 0.00±0.00g | 14.44±2.94l | 10.00±1.92j |
| | Darkness | Small | 86.67±1.92cd | 92.50±2.50bc | 82.22±2.94b | 73.33±5.09gh | 90.00±1.92def |
| | | Big | 98.33±1.11ab | 98.33±0.83ab | 100.00±0.00a | 94.44±1.11cde | 91.11±1.11cde |

[a] Sharing same letters means not significantly different from each other (P>0.05, Duncan's Multiple Range Test).

# 5

# Ant assemblages (Hymenoptera: Formicidae) in three different stages of forest regeneration in a fragment of Atlantic Forest in Sergipe, Brazil

ECF Gomes, GT Ribeiro, TMS Souza, L Sousa-Souto

*Universidade Federal de Sergipe (UFS), São Cristóvão, Sergipe, Brazil*

**Keywords**

bioindicators, degraded areas, environmental monitoring, species composition, Forest recovery.

**Corresponding author**

Genésio Tâmara Ribeiro
Universidade Federal de Sergipe
Cidade Univ. Prof. José Aloisio de Campos
Av. Mal. Rondon, s/n, Jardim Rosa Elze
São Cristovão-SE, Brazil
49100-000
E-Mail: genesiotr@hotmail.com

**Abstract**

In this study we compared the epigeic ant assemblages in forest fragments with three different status of plant recovery (an area reforested in 2007, another reforested in 2005 and another one of secondary forest, with over 35 years of plant regeneration), located in the municipality of Laranjeiras, Sergipe, Brazil. The ants were sampled in February (dry season) and June (rainy season) of 2012. We tested the following hypotheses: (1) the species richness of ants increases with time after the process of forest restoration; and (2) there are significant changes in species composition of ants among the three stages of forest regeneration. Twenty-five pitfall traps were installed in each area. A total of 82 morphospecies of ants were sampled, distributed in 31 genera and seven subfamilies. The richness of ants was similar among the three sites (F = 1.71, p = 0.19). The composition of ant species, however, was different in the area of late regeneration (35 years) compared to other areas of early reforestation (p <0.05). Thus, epigeic ants were partially sensitive to changes in the habitat studied in response to reforestation, presenting changes in species composition but no differences in ant species richness among areas. We conclude that seven years after reforestation are not enough to restore the same ant diversity in disturbed environments.

## Introduction

Remaining forests are important for the maintenance of favorable environmental conditions for the establishment and persistence of native fauna (Gibson et al., 2011; Ulyshen, 2011). Several studies have shown that part of the worldwide decline in biodiversity, threatening the functioning of ecosystems, is related to anthropogenic modification of the landscape (Dirzo & Raven, 2003; Colombo & Joly, 2010; Tabarelli et al., 2010), including the Atlantic Forest, one of the main hotspots in the world (Myers et al., 2000).

Deforestation of Atlantic Forest is considered a constant threat to biological diversity (Melo et al., 2009; Oliveira et al., 2004), including ant assemblages (Leal et al., 2012) and the monitoring of areas in process of plant recovery can be an important tool in the diagnosis of these threats (Conceição et al., 2006; Delabie et al., 2006).

Due to its abundance in most terrestrial ecosystems ants are considered ecologically dominant and play complex ecological roles such as ecosystem engineers, predators, herbivores and seed dispersal agents (Hölldobler & Wilson,

1990; Folgarait, 1998). In tropical ecosystems the importance of ants is more evident because they can represent up to 60% of all arthropod biomass, and approximately 90% of their abundance (Hölldobler & Wilson, 1990; Floren & Linsenmair, 1997).

Species richness and structure of ant assemblages can be used as response variables in environmental monitoring, as these insects are sensitive to anthropogenic activities, including agricultural practices (Hernández-Ruiz & Castaño-Meneses, 2006) and reforestation (Pais & Varanda de 2010; Schmidt et al., 2013). Therefore, the study of these insects is useful to assess the success of forest restoration practices (Sobrinho et al., 2003; Silva & Silvestre, 2004; Holway & Suarez, 2006; Wetterer, 2012).

Although recovery of degraded areas is commonly used to reduce the negative environmental impacts on forest remnants (Metzger, 2009; Calmon et al., 2011) and, in spite of several studies on the role of the replanting of native species in accelerating the recovery of degraded environments, there are still many questions about the time required for the recovering of ant fauna along a gradient of forest regeneration of fragments dominated previously by an agricultural matrix (Neves et al., 2010; Teodoro et al., 2010; Leal et al., 2012).

Processes that influence the structure and species diversity of epigeic ants in agroecosystems are still poorly known (Neves et al., 2010; Teodoro et al., 2010), despite the increasing conversion of forest fragments in less diverse and structurally simple habitats (Primack & Corlett, 2005; Barona et al., 2010).

In this study, we investigated the response of epigaeic ant assemblages in forest fragments with three different status of plant recovery, aiming to test the following hypotheses: (1) Species richness of ants increases with time after the process of forest restoration (following an increase in habitat complexity) and (2) the composition of ant species undergoes changes along a gradient of regeneration of reforested area, with reduction of generalist species.

## Material and Methods

The study was conducted in three sites: two sites were previously plantations of sugar-cane that were reforested, one with 32ha in 2005 (RF1) and another with 30.7ha and reforested with native species in 2007 (RF2). The third area is a secondary Atlantic Forest fragment with 55ha (FF) used as "Area of permanent preservation" (APP). All sites are located at Fazenda Boa Sorte, a large sugar-cane company, located in the municipality of Laranjeiras (10° 48' 44"S, 37° 10' 16" W), state of Sergipe, Brazil.

The studied region is dominated by agricultural land with altitude ranging from 30 to 68 m a.s.l. The mean annual temperature is 25.5 °C and annual average rainfall of 1,200 mm. The rainy season usually lasts from May to October. The original vegetation was dominated by Atlantic forest and all remnants are embedded in a 20 year-old, homogeneous matrix of sugar-cane fields (Cuenca & Mandarino, 2007). The soil type is Spodozol, mainly sandy clay, deep, with low fertility and high porosity (draining rainfall). The area reforested in 2005 (RF1) with 32 ha, is at an intermediate stage of development, with seven years of planting and composed of trees with canopy of approximately 4-6 meters (10°49'15,8"S; 37°09'41"W). The other area, reforested in 2007 (RF2), with 30.7 ha is in early stage of development, with five years of planting and composed by sparse patches of woody vegetation, shrubs, herbs and grasses with a single layer of treetops with up to 4 m tall (10°49'01,6"S; 37°09'40,7"W).

Fourteen species of trees native to the Atlantic Forest were used in reforestation: *Tapirira guianensis, Caesalpinia echinata, Genipa amerciana, Spondias lutea, Schinus terebinthifolius, Erythrina velutina, Enterolobium contorsiliquum, Cleome tapia, Caesalpinia leiostachya, Inga marginata, Cassia grandis, Lonchocarpus sericeus, Anadenanthera macrocarpa* and *Hymenaea courbaril*

The fragment (FF) of Atlantic Forest is an area of secondary forest, protected from logging for over 35 years and consists of trees with 7-20 meters in height that forms a closed canopy (10°49'17"S; 37°11'13"W).

Epigeic ants were sampled in 15 transects of 50 m, being five transects per area. We established a minimum distance of 150 m between each transect. Ant sampling was conducted using pitfall traps on the ground surface. In each transect, five pitfalls were installed at a distance of 10 m, totaling 25 pitfalls/site. Pitfalls consisted of 1,000 cm³ plastic pots containing approximately 120 cm³ water with detergent and were kept for 48 h in the field (Schmidt & Solar, 2010).

Sampling was conducted in two periods, one during the dry season (February 2012) and another during the rainy season (June 2012). All ants collected were sorted to species level when possible or morphospecies, using identification keys from Bolton (1994) and Fernandez (2003) and later the identification was confirmed through comparison with specimens from the collection of the Laboratório de Ecologia de Comunidades (Ant collection), of the Universidade Federal de Viçosa and Laboratório de Mirmecologia of the CEPEC/CEPLAC, Ilhéus, Bahia, Brazil. Voucher specimens of all species are deposited in Laboratório de Pragas Florestais of the Universidade Federal de Sergipe.

To verify the effect of habitat type (RF1, RF2, and FF) and sampling period (wet or dry season) (response variables) on the species richness of ants (explanatory variable) the linear mixed effect (LME) was used, followed by residuals analysis to verify the adequacy of the error distribution and the fit of the model. Fixed factors were sampling sites (RF1, RF2, and FF), while samples (nested within sites) were treated as random factors. A minimum adequate model (MAM) was obtained by extracting non-significant terms (P < 0.05) from the full model arranged by all variables and their interaction (Crawley, 2007), using the software R (R Development Core Team, 2009).

A non-metric multidimensional scaling (NMDS) analysis was carried out to verify differences in the composition of ant fauna among the forest regeneration types (Neves et al., 2010). The ordination was conducted using the Jaccard index. Additionally, similarity analysis (ANOSIM; Clarke, 1993) were conducted to compare the difference between two or more groups of sampling units among sites. Differences between R-values were used to determine similarity patterns among ant assemblages in the three sites. The analysis were conducted using the software PAST (Hammer et al., 2001).

## Results

We collected 82 ant morphospecies, distributed in 31 genera (Table 1). The subfamilies Myrmicinae and Formicinae presented 66% of all ant species sampled, with 42 and 12 morphospecies, respectively. The genera *Pheidole* and *Camponotus* presented the higher richness with 11 (13.5%) and 10 (12%) morphospecies, respectively. Twelve species were restricted to RF1 site, six species were found exclusively in RF2 site while 34 species were restricted to FF site (Table 1).

**Table 1.** Relative frequency of epigeic ant species collected in pit-falls, during the wet and dry season of 2012 in three sites of different forest regeneration stages: a fragment of secondary forest (FF) one area of reforestation with five years (RF2) and other with seven years of reforestation (RF1).

| Ant Subfamilies | FF Dry | FF Wet | RF1 Dry | RF1 Wet | RF2 Dry | RF2 Wet |
|---|---|---|---|---|---|---|
| **DOLICHODERINAE** | | | | | | |
| *Dolichoderus lutosus* | - | 0.2 | - | - | - | - |
| *Dolichoderus diversus* | - | 0.2 | - | - | - | - |
| *Dolichoderus attelaboides* | - | 0.2 | - | - | - | - |
| *Dorymyrmex biconis* | - | - | - | - | 0.2 | - |
| *Azteca* sp. 1 | 0.8 | 0.4 | - | - | - | - |
| *Azteca* sp. 2 | 0.2 | - | - | - | - | - |
| *Azteca* sp. 3 | - | 0.2 | - | - | - | - |
| **ECITONINAE** | | | | | | |
| *Labidus praedator* | - | 0.4 | - | - | - | - |
| *Labidus coecus* | - | - | 0.4 | - | 0.4 | - |
| *Nomamyrmex esenbeckii* | - | - | - | 0.2 | - | - |
| *Neivamyrmex diana* | - | - | - | 0.2 | - | - |
| **FORMICINAE** | | | | | | |
| *Brachymyrmex* pr. *patagonicus* | - | - | - | 0.2 | - | - |
| *Camponotus trapezoideus* | 0.2 | - | - | - | - | - |
| *Camponotus renggeri* | 1 | 0.4 | - | - | - | - |
| *Camponotus bispinosus* | 0.2 | - | - | - | - | - |
| *Camponotus novogranadensis* | 0.6 | 1 | - | - | - | - |
| *Camponotus fastigatus* | 1 | - | - | - | - | - |
| *Camponotus arboreus* | 0.2 | - | 0.2 | - | - | - |
| *Camponotus cingulatus* | 0.8 | - | - | - | - | - |
| *Camponotus vittatus* | - | 0.4 | 0.8 | 1 | 0.6 | 0.8 |
| *Camponotus* (*Myrmaphaenus*) sp.9 | - | - | 1 | 0.8 | 0.8 | - |
| *Camponotus rufipes* | - | - | 0.4 | - | 0.2 | 1 |
| *Nylanderia* pr. *fulva* | - | - | 0.4 | - | 0.2 | - |
| **MYRMICINAE** | | | | | | |
| *Piramica* pr. *perpava* | 0.4 | 0.4 | - | - | - | - |
| *Piramica* sp. 2 | - | 0.2 | - | - | - | - |
| *Cephalotes atratus* | 0.6 | - | - | - | - | - |
| *Cephalotes umbraculatus* | 0.2 | - | - | - | - | - |
| *Cephalotes minutus* | - | - | - | 0.2 | - | - |
| *Cephalotes maculatus* | 0.2 | - | - | - | - | - |
| *Cephalotes pusillus* | - | - | - | 0.2 | - | - |
| *Cephalotes depressus* | - | - | - | - | - | 0.4 |
| *Acromyrmex balsani* | - | - | 0.6 | - | 0.4 | 0.2 |
| *Acromyrmex rugosos rugosos* | - | 0.6 | 0.4 | 0.2 | 0.2 | - |
| *Atta sexdens rubropilosa* | - | - | 0.2 | - | 0.2 | - |
| *Cyphomyrmex minutus* | - | 0.2 | - | - | - | - |
| *Cyphomyrmex transversus* | - | - | 0.6 | 0.2 | 0.8 | 0.4 |
| *Mycetosoritis* sp. 1 * | - | - | - | - | - | 0.2 |
| *Mycocepurus obsoletus* | - | 0.4 | - | - | - | - |
| *Sericomyrmex* sp. 1 | 0.2 | 0.2 | - | - | - | - |
| *Sericomyrmex* sp. 2 | - | 0.4 | - | - | - | - |
| *Trachymyrmex* sp. 1 | 0.2 | - | - | - | - | - |
| *Monomorium floricula* | - | - | 0.2 | - | 0.8 | - |
| *Solenopsis tridens* | - | - | 0.2 | 0.2 | - | - |
| *Solenopsis* sp. 2 | 0.2 | 0.6 | - | - | - | - |
| *Solenopsis* sp. 3 | 0.6 | - | - | - | - | - |
| *Solenopsis saevissima* | 0.2 | - | 0.6 | 0.6 | 0.8 | 0.4 |
| *Solenopsis globularia* | - | - | 0.2 | 1 | - | 0.8 |
| *Hylomyrma balzani* | 0.2 | 0.2 | - | - | - | - |
| *Pheidole radoszkowskii* | 0.4 | 0.4 | 1 | 1 | 0.6 | 1 |
| *Pheidole fimbriata* | - | - | 0.2 | 0.4 | - | - |
| *Pheidole* (gr. *Diligens*) sp. 3 | 0.4 | 0.8 | - | - | - | - |
| *Pheidole* sp. 4 | 0.6 | 1 | - | - | - | - |
| *Pheidole* (gr. *Flavens*) sp. 5 | 0.2 | 0.4 | - | 0.4 | - | - |
| *Pheidole* (gr. *Tristis*) sp. 6 | 0.8 | 0.4 | 0.2 | - | - | - |
| *Pheidole* (gr. *Fallax*) sp. 7 | - | - | 1 | 0.4 | 0.8 | 1 |
| *Pheidole* (gr. *Diligens*) sp. 8 | - | - | - | 0.2 | 1 | 0.2 |
| *Pheidole* sp. 9 | - | - | - | - | - | 0.4 |
| *Pheidole* (gr. *Fallax*) sp. 10 | - | - | 0.4 | 0.6 | 0.2 | - |
| *Pheidole* (gr. *Fallax*) sp. 11 | 0.6 | 0.8 | 0.8 | 0.4 | 0.8 | 0.4 |
| *Crematogaster abstinens* | - | - | 1 | 0.4 | 0.4 | 1 |
| *Crematogaster* sp. 2 | - | - | 0.2 | - | - | - |
| *Crematogaster* pr. *Distans* | - | - | - | - | 0.6 | - |
| *Crematogaster* sp. 4 | 0.4 | - | - | - | - | - |
| *Crematogaster* sp. 5 | 0.2 | - | - | - | - | - |
| *Cardiocondyla emeryi* | - | - | 0.2 | - | 0.2 | - |
| **PONERINAE** | | | | | | |
| *Odontomachus haematodos* | 1 | 1 | 1 | 0.6 | 0.8 | 1 |
| *Leptogenys unistimulosa* | 1 | 0.6 | 0.6 | 0.8 | 1 | 0.4 |
| *Hypoponera* sp. 1 | 0.2 | - | - | - | - | - |
| *Pachycondyla venerae* | 0.4 | 0.4 | - | - | - | - |
| *Pachycondyla harpax* | 0.4 | - | - | - | 1 | - |
| **ECTATOMMINAE** | | | | | | |
| *Gnamptogenys acuminata* | 0.2 | - | - | - | - | - |
| *Gnamptogenys sulcata* | - | - | 0.4 | - | 0.4 | - |
| *Ectatoma bruneunn* | - | - | 0.4 | - | 0.6 | 0.6 |
| *Ectatoma tuberculatum* | 0.2 | - | 0.6 | 0.2 | - | - |
| *Ectatoma edentatum* | 0.4 | 0.8 | 0.2 | 0.6 | - | - |
| **PSEUDOMYRMECINAE** | | | | | | |
| *Pseudomyrmex tenuis* | 0.8 | 0.6 | - | - | - | - |
| *Pseudomyrmex termitarius* | - | - | 0.2 | - | 0.2 | - |
| *Pseudomyrmex* sp. (gr. *Pallidus*) | - | - | - | - | 0.2 | - |
| *Pseudomyrmex* sp. 4 | - | - | - | 0.4 | - | - |
| *Pseudomyrmex gracilis* | - | - | 0.2 | - | - | - |
| *Pseudomyrmex* sp. 6 | - | - | 0.2 | - | - | - |
| *Pseudomyrmex* sp. 7 | - | - | 0.2 | - | - | - |

FF34 ve/ RF1.12 vd/ RF2.6 p/ Comuns30; * New genus sp.1, R. Feitosa (personal communication, 19 September 2012)

There was no significant difference in species richness of ants among the three sites of forest regeneration ($F_{2,22}$ = 2.26, p = 0.12). However, the species richness of ants was lower in the wet season in the RF1 and RF2 sites, compared to FF area ($F_{1,22}$ = 11.19, p = 0.002) (Fig.1).

The NMDS analysis indicates the formation of two distinct groups (stress = 0.15) with one group represented by FF and another group formed by RF1 And RF2 (Fig.2). Besides, the analysis of similarity (ANOSIM) indicated significant difference in the structure of ant assemblages between FF *versus* RF1 (p = 0.003) as well as FF *versus* RF2 (p = 0.001) (Table 2).

The SIMPER analysis indicated that the morphospecies that contributed most to the differentiation among sites were *Pheidole* (group *Fallax*) sp.7, *Camponotus* (*Myrmaphaenus*) sp 9, *Crematogaster abstinens*, *Camponotus vittatus*, *Solenopsis saevissima*, *Pheidole* sp. 4, *Cyphomyrmex transversus*, *Solenopsis globularia*, *Ectatoma edentatum*, *Camponotus renggeri* and *Pseudomyrmex tenuis*. These morphospecies together contributed to 31.5% of cumulative dissimilarity among stages of plant recovery (Table 3).

**Table 2.** Analysis of similarity (ANOSIM) among three sites of different forest regeneration stages: a fragment of secondary forest (FF) one area of reforestation with five years (RF2) and other with seven years of reforestation (RF1).

|       | FMN       | RF1       | RF2       |
|-------|-----------|-----------|-----------|
| FMN   | -         | 0.0003**  | 0.0001**  |
| RF1   | 0.0003**  | -         | 0.0732    |
| RF2   | 0.0001**  | 0.0732    | -         |

** significant difference, p< 0,01.

## Discussion

In our study the species richness did not differ with time of restoration and, on the one hand, it suggests that five years are enough for the recovery of ant species richness. This time can be considered short compared to that from other studies conducted by Vasconcelos (1999) and Roth et al. (1994) (10 and 25 years, respectively). On the other hand, however, the differences found in species composition among FF and RF1 or RF2 indicate that other parameters, rather than species richness, are important to make decisions about the use of ants as bioindicators. An increase in species richness in FF might be correlated with a more complex environment, leading to an increase in availability of resources (Matos et al., 1994, Oliveira et al., 1995).

The restoration of RF1 and RF2 sites with native species of trees might have created favorable conditions for colonization of ants and have led to comparable values of species richness in FF area. The colonization of ants, however, seems to have been carried out by ant species from adjacent agroecosystems and not from nearby forested areas, since the composition of species between FF and the other two areas differ greatly. Difference in ant species composition between

more complex areas and regenerating ones have been found by other authors (Vasconcelos, 1999; Schmidt et al., 2013). In general, generalist species have higher colonization rate of disturbed fragments than do specialist ants (Schoereder et al., 2004).

Although there were no differences in species richness among sites, there were differences between season of sampling, with higher values in the dry season of RF1 and RF2 (seasonality was not tested as a hypothesis in this study, and thus we used this variable only as a source of variation in the statistical model).

**Table 3.** SIMPER Analysis among three sites of different forest regeneration stages: a fragment of secondary forest (FF) one area of reforestation with five years (RF2) and other with seven years of reforestation (RF1).

| Species | Cumulative percent of dissimilarity (%) |
|---------|-----------------------------------------|
| *Pheidole* (group *Fallax*) sp.7 | 3.407 |
| *Camponotus (Myrmaphaenus)* sp.9 | 6.709 |
| *Crematogaster abstinens* | 9.905 |
| *Camponotus vittatus* | 12.89 |
| *Solenopsis saevissima* | 15.77 |
| *Pheidole* sp. 4 | 18.58 |
| *Cyphomyrmex transversus* | 21.25 |
| *Solenopsis globularia* | 23.92 |
| *Ectatoma edentatum* | 26.59 |
| *Camponotus renggeri* | 29.05 |
| *Pseudomyrmex tenuis* | 31.51 |

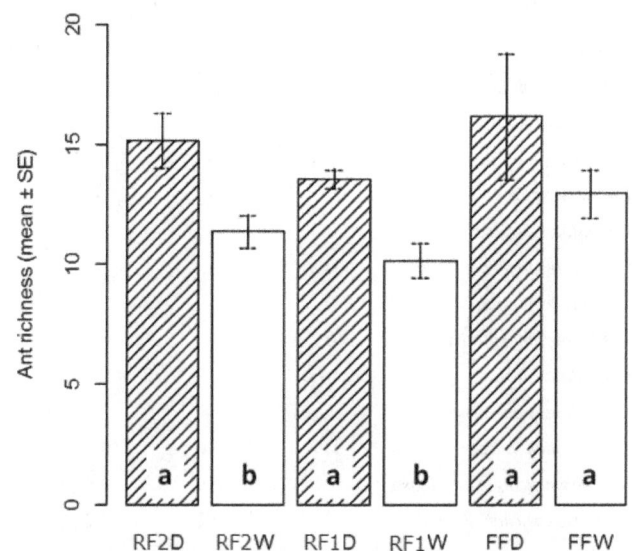

**Figure 1.** Species richness of ants in the dry season (white bars) and rainy season (hatched bars) (mean ± SE) sampled in three areas with different stages of forest recovery. RF1 = Fragment with 7 years of reforestation; RF2 = Fragment of 5 years of reforestation and FF = forest fragment with 35 years of plant recovery (D and W indicate sampling in dry and wet seasons, respectively). Different letters on bars indicate significant difference within the same site (p <0.05).

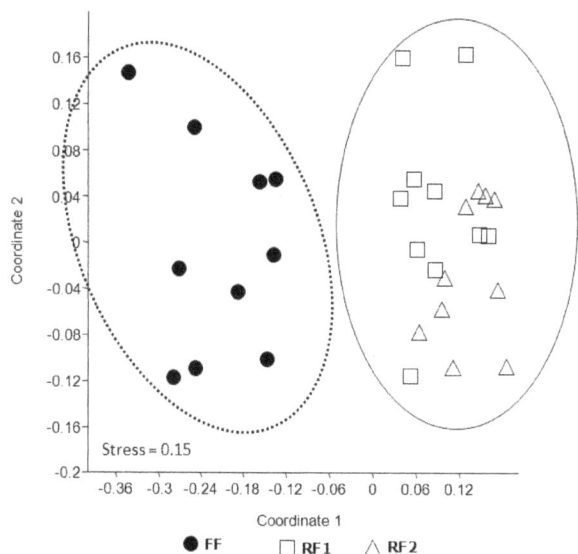

**Figure 2.** Analysis of non-metric multidimensional scaling ordination (NMDS) from the assemblage composition of ants in three sites of different forest regeneration stages: a fragment of secondary forest (FF) one area of reforestation with five years (RF2) and other with seven years of reforestation (RF1).

Changes in the frequency of foraging ants have been observed with environmental seasonality (Wolda, 1988; Kaspari, 2000; Castro et al., 2011; Cook et al., 2011). The availability of resources is reduced during the dry months of the year, and the increased mobility of ants in this period might explain the rise in species richness in the dry season (Andow, 1991; Dantas et al., 2011). These results, however, should be viewed with caution since the data were collected considering just a year and a longer period of collection is needed to establish more secure inferences about effects of seasonality on ant species.

Considering that thirty-four species were restricted to FF site (42% of all species found) our results make it clear that environmental differences among the areas were crucial for determining the composition of species.

The most common species in the FF were *Camponotus renggeri*, *Pheidole* sp.4 and *Pseudomyrmex tenuis*, indicating a preference of these species for less-disturbed environments. Notably, several studies also demonstrate the occurrence of *C. renggeri* in seasonal forest formations such as Cerrado (Del-Claro & Oliveira, 1999; Christianini et al., 2007; Neves et al., 2012) and forests in the semi-arid region of Brazil (Hites et al., 2005). Conceição et al. (2006) also have reported the presence of *P. tenuis* in environments with low disturbance. Besides, predatory species of the genus *Pyramica* (Masuko, 2009) were also found only in the FF, suggesting that this fragment has a more suitable resource availability than the other areas.

The genera *Pheidole* (group *Fallax*) sp.7, *Camponotus* (*Myrmaphaenus*) sp.9, *Crematogaster abstinens*, *Solenopsis globularia* and *Cyphomyrmex transversus*, were found in RF1 and RF2, indicating that these species could have preference

for colonization in degraded areas. Other studies in the Atlantic Forest recorded *S. globularia* in a disturbed mangrove area (Delabie et al., 2006) and *C. transversus* in grasslands (Braga et al., 2010). In fact, the areas RF1 and RF2 have low density of tree species, allowing the establishment of herbaceous species.

Individuals of *Solenopsis saevissima* exhibit aggressive behavior and are also usually associated with disturbed environments (Silvestre et al., 2003). The presence of this species is favored in sites colonized by pioneer plant species, typically found in early sucessional areas (Vasconcelos, 2008; Schmidt et al., 2013). Although *Camponotus vittattus* was sampled in all three areas, this species had similar occurrence to *S. saevissima* being more frequent in samples from RF1 and RF2, thus suggesting its preference for opened sites. In contrast, we also reported the presence of some ant species in sites with late regeneration time (RF2 and FF), such as *Ectatoma edentatum*. Previous studies have associated the occurrence of this species with advanced stages of plant recovery (Ramos et al. 2003; Vasconcelos 2008).

The RF1 site had a similar species richness of epigeic ants compared with RF2 or FF sites. However, the species composition differs considerably among environments with similar historical disturbances *versus* a forest fragment with late regeneration. Our study shows that ant assemblages can vary greatly along a gradient of plant recovery and the conservation of forest fragments with different stages of reforestation is important to sustain more diverse ant fauna. Therefore, reforestation programs that prioritize the conservation and the adoption of native plant species in these areas are a good alternative (Gillespie et al. 2000), enabling the development of scientific studies and especially the maintenance of local biodiversity.

## Acknowledgments

The authors are grateful to Júlio Cezar Mário Chaul, Laboratory of Community Ecology - UFV and Jacques H. C. Delabie, Laboratory of Myrmecology CEPEC / CEPLAC for their help in species identification. This study was supported by the Coordenação de Aperfeiçoamento de Pessoal de Nível Superior (CAPES).

## References

Andow, D.A. (1991). Vegetational diversity and arthropod population responses. Annual Review of Entomology, 36: 561-586.

Barona, E., Ramankutty, N., Hyman, G. & Coomes, O. T. (2010). The role of pastureand soybean in deforestation of the Brazilian Amazon.Environmentl Research Letters, 5: 1-9.

Bolton, B. (1994). Identification guide to ant genera of the world. Harvard University Press, Cambridge, 222p.

Braga, D.L., Louzada, J.N.C., Zanetti, R. & Delabie, J. (2010). Avaliação rápida da diversidade de formigas em sistemas de uso do solo no sul da Bahia. Ecology, Behavior and Bionomics, 39: 464-469.

Calmon, M., Brancalion, P.H.S., Paese, A., Aronson, J., Castro, P., Silva, S.C. & Rodrigues, R.R. (2011). Emerging Threats and Opportunities for Large-Scale Ecological Restoration in the Atlantic Forest of Brazil. Restoration Ecology, 19: 154-158. doi: 10.1111/j.1526-100X.2011.00772.x.

Castro, F.S., Gontijo, A.B., Castro, P.T.A. & Ribeiro, S.P. (2011). Annual and Seasonal Changes in the Structure of Litter-Dwelling Ant Assemblages (Hymenoptera: Formicidae) in Atlantic Semideciduous Forests. Psyche, vol. 2012, Article ID 959715, 12 pages, 2012. doi:10.1155/2012/959715

Christianini, A.V., Nunes, A.J.M. & Oliveira, P.S. (2007). The role of ants in the removal of non-myrmecochorous diasporas and seed germination in a neotropical savanna. Journal of Tropical Ecology, 23: 343-351.

Clarke, K.R. (1993). Non-parametric multivariate analysis of changes in community structure. Austral Ecology, 18: 117-143.

Colombo, A.F. & Joly, C.A. (2010). Brazilian Atlantic Forest lato sensu: the most ancient Brazilian forest, and a biodiversity hotspot, is highly threatened by climate change. Brazilian Journal of Biology, 70: 697-708.

Conceição, E.S., Costa-Neto, A.O., Andrade, F.P., Nascimento, I.C., Martins, L.C.B., Brito, B.N., Mendes, L.F. & Delabie, J. (2006). Assembléias de Formicidae da serapilheira como bioindicadores da conservação de remanescentes de Mata Atlântica no extremo sul do Estado da Bahia. Sitientibus Série Ciências Biológicas, 6: 296-305.

Cook, S.C., Eubanks, M.D., Gold, R.E. & Behmer, S. T. (2011). Seasonality Directs Contrasting Food Collection Behavior and Nutrient Regulation Strategies in Ants. PLoS ONE, 6: 1-8.

Crawley, M. J. (2007). The R Book. John Wiley & Sons Ltd, England, 950p.

Cuenca, M. A. G.& Mandarino, D. C. (2007). Mudança da Atividade Canavieira nos Principais Municípios Produtores do Estado de Sergipe de 1990 a 2005. Documentos 122, Embrapa Tabuleiros Costeiros, 22p.
(Disponível em HTTP://<www.cpatc.embrapa.br>).

Dantas, K.S.Q., Queiroz, A.C.M., Neves, F.S., Júnior, R.R. & Fagundes, M. (2011). Formigas (Hymenoptera: Formicidade) em diferentes estratos numa região de transição entre os biomas do Cerrado e da Caatinga no norte de Minas Gerais. MG Biota, 4: 17-36.

Delabie, J.H.C., Paim, V.R.L.D.M., Nascimento, I.C.D., Campiolo, S. & Mariano, C.D.S.F. (2006). Ants as biological indicators of human impact in mangroves of the southeastern coast of Bahia, Brazil. Neotropical Entomology, 35: 602-615.

Del-Claro, K. & Oliveira, P.S. (1999). Ant-Homoptera Interactions in a Neotropical Savanna: The Honeydew-Producing Treehopper, Guayaquila xiphias (Membracidae), and its Associated Ant Fauna on Didymopanax vinosum (Araliaceae). Biotropica, 31: 135-144.

Dirzo, R. & Raven, P.H. (2003). Global State of Biodiversity and Loss. Annual Review of the Environment and Resources, 28: 137-167.

Fernandez, F. (2003). Introducción a las hormigas de la region Neotropical. Acta Noturna, Bogotá, 398pp.

Floren, A. & Linsenmair, K.E. (1997). Diversity and recolonization dynamics of selected arthropod groups on different tree species in a lowland rainforest in Sabah, with special reference to Formicidae. Canopy Arthropods (eds N.E. Stork, J. Adis & R.K. Didham). pp. 344–381, Chapman & Hall, London.

Folgarait, P.J. (1998). Ant biodiversity and its relationship to ecosystem functioning: a review. Biodiversity and Conservation, 7: 1221-1244.

Gibson, L., Lee, T.M., Koh, L.P., Brook, B.W., Gardner, T.A., Barlow, J., Peres, C.A., Bradshaw, C.J.A., Laurance, W.L., Lovejoy, T.E. & Sodhi, N. (2011). Primary forests are irreplaceable for sustaining tropical biodiversity. Nature, 478: 378-383. doi:10.1038/nature10425.

Gillespie, T.W., Grijalva, A. & Farris, C.N. (2000). Diversity, composition, and structure of tropical dry forests in Central America. Plant Ecology, 147: 37-47.

Hammer, O., Harper, D.A.T. & Ryan, P.D. (2001). PAST: Palaeonthological Statistics Software Package for education and data analysis. Palaeontologia Electronica, 4: 1-9.

Hernández-Ruiz, P. & Castaño-Meneses, G. (2006). Ants (Hymenoptera: Formicidae) diversity in agricultural ecosystems at Mezquital Valley, Hidalgo, Mexico. European Journal of Soil Biology, 42: 208-212.

Hites, N.L., Mourao, M.A.N., Araujo, F.O., Melo, M.V.C., Biseau, J.C. & Quinet, Y. (2005). Diversity of the ground-dwelling ant fauna (Hymenoptera: Formicidae) of a moist, montane Forest of the semi-arid Brazilian "Nordeste". Revista de Biologia Tropical, 53: 165-173.

Hölldobler, B. & Wilson, E.O. (1990). The Ants. Harvard University Press, Massachusetts, Cambridge. 732p.

Holway, D. A. & Suarez, A. V. (2006). Homogenization of ant communities in mediterranean California: The effects of urbanization and invasion. Biological Conservation, 127: 319-326.

Kaspari, M. (2000). A primer of ant ecology. In: Agosti, D. & Alonso, L. (Eds.), Measuring and monitoring biological diversity, standard methods for Ground-living Ants (pp. 9-24). Washington: Smithsonian Institution Press.

Leal, I.R., Filgueiras, B.K.C., Gomes, J.P., Lannuzzi, L. & Andersen, A.N. (2012). Effects of habitat fragmentation on ant richness and functional composition in Brazilian Atlantic forest. Biodiversity Conservation, 21: 1687-1701.

Masuko, K. (2009). Studies on the predatory biology of Oriental dacetine ants (Hymenoptera: Formicidae) II. Novel prey specialization in *Pyramica benten*. Journal of Natural History, 43: 13-14.

Matos, J.A., Yamanaka, C.N., Castellani, T.T. & Lopes B.C. (1994). Comparação da fauna de formigas de solo em áreas de plantio de *Pinus elliottii, com diferentes graus de complexibilidade estrutural (Florianópolis, SC.)*. Biotemas, 7: 57-64.

Melo, F.V., Brown, G.G., Constantino, R., Louzada, J.N.C., Luizão, F.J., Morais, J.W. & Zanetti, R.A. (2009). A importância da meso e macrofauna do solo na fertilidade e como biondicadores. Boletim Informativo da SBCS. (Disponível em http://<sbcs.solos.ufv.br/solos/boletins/biologia%20macrofauna.pdf.>).

Metzger, J.P. (2009). Conservation issues in the Brazilian Atlantic forest. Biological Conservation, 142: 1138-1140.

Myers, N., Mittermeier R.A., Mittermeier C.G, Fonseca G.A.B. & Kent, J. (2000). Biodiversity hotspots for conservation priorities. Nature, 403: 853-858.

Neves, F.S., Braga, R.F., Araujo, L.S., Campos, R.I. & Fagundes, M. (2012). Differential effects of land use on ant and herbivore insect communities associated with *Caryocar brasiliense* (Caryocaraceae). Revista de Biologia Tropical, 60: 1065-1073.

Neves,F.S., Braga,R.F., Espírito-Santo, M.M., Delabie, J.H.C., Fernandes, G.W. & Sanchez-Azofeifa, G. A.(2010). Diversity of arboreal ants an a Brazilian Tropical Dry Forest: Efects of seasonality and successional Stage. Sociobiology, 56: 177-194.

Oliveira MA, Grillo AA, Tabarelli M (2004). Forest edge in the Brazilian Atlantic Forest: drastic changes intree species assemblages. Oryx, 38: 389–394.

Oliveira, M.A., Della-Lucia, T.M.C., Araújo, M.S. & Cruz, A.P. (1995). A fauna de formigas em povoamentos de eucalipto na mata nativa no estado do Amapá. Acta Amazonica, 25: 117-126.

Pais, M.P. & Varanda, E.M. (2010). Arthropod Recolonization in the Restoration of a Semideciduous Forest in Southeastern Brazil. Neotropical Entomology, 39: 198-206.

Primack, R.B. & Corlett, R.T. (2005). Tropical Rain Forests: An Ecological and Biogeographical Comparison. Blackwell Science, Oxford. 336p.

R Development Core Team. (2009). R: A language and environment for statistical computing. R Foundation for Statistical Computing, Vienna, Áustria, 409 pp.

Ramos, L.D., Filho, R.Z.B., Delabie, J.H.C., Lacau, S., Santos, M.F.S., Nascimento, I.C. & Marinho, C.G. (2003). Ant communities (Hymenoptera: Formicidae) of the leaf-litter in cerrado "stricto sensu" areas in Minas Gerais, Brazil. Lundiana, 4: 95-102.

Roth, D.S., Perfecto, I. & Rathcke, B. (1994). The effects of management systems on ground-foraging ant diversity in Costa Rica. Ecological Applications, 4: 423-436.

Schmidt, F.A. & Solar, R. R.C. (2010). Hypogaeic pitfall traps: methodological advances and remarks to improve the sampling of a hidden ant fauna. Insectes Sociaux, 57: 261-266.

Schmidt, F.A., Ribas, C.R. & Schoereder, J.H. (2013). How predictable is the response of ant assemblages to natural forest recovery? Implications for their use as bioindicators. Ecological Indicators, 24: 158-166.

Schoereder, J.H.; Sobrinho, T.G.; Ribas, C. R. & Campos, R.B.F. (2004). Colonization and extinction of ant communities in a fragmented landscape. Austral Ecology, 29: 391-398.

Silva, R.R. & Silvestre, R. (2004). Diversidade de formigas (Hymenoptera: Formicidae) que habita as camadas superficiais do solo em Seara, Oeste de Santa Catarina. Papéis Avulsos de Zoologia, 44: 1-11.

Silvestre, R.C., Brandão, C.R.F. & Silva, R.R. (2003). Grupos funcionales de hormigas: el caso de los gremios del Cerrado. In F. Fernández (Ed.). Introducción a las hormigas de la región neotropical (pp. 113-148). Bogotá: Acta Nocturna.

Sobrinho, T.G., Schoereder, J.H., Sperber, C.F. & Madureira, M.S. (2003). Does fragmentation alter species composition in Ant communities (Hymenoptera: Formicidae)? Sociobiology, 42: 329-342.

Tabarelli, M., Aguiar, A.V., Ribeiro, M.C., Metzger, J.P. & Peres, C.A. (2010). Prospects for biodiversity conservation in the Atlantic Forest: Lessons from aging human-modified landscapes. Biological Conservation, 143: 2328-2340.

Teodoro, A. V., Sousa-Souto, L., Klein, A.M. & Tscharntke, T. (2010). Seasonal contrasts in the response of coffee ants to agroforestryshade-tree management. Environmental Entomology, 39:1744-1750.

Ulyshen, M.D. (2011). Arthropod vertical stratification in temperate deciduous forests: Implications for conservation-oriented management. Forest Ecology and Management, 261: 1479-1489.

Vasconcelos, H. L. (1999). Effects of forest disturbance on the structure of ground-foraging ant communities in central Amazonia. Biodiversity and Conservation, 8: 409-420.

Vasconcelos, H.L. (2008). Formigas do solo nas florestas da Amazônia: padrões de diversidade e respostas aos distúrbios naturais e antrópicos. In: Moreira, F.M., Siqueira, J.O. &

Brussaard, L. Biodiversidade do solo em ecossistemas brasileiros (pp. 323-343). Lavras: Editora UFLA.

Wetterer, J.K. (2012). Worldwide spread of Emery's sneaking ant, *Cardiocondyla emeryi* (Hymenoptera: Formididae). Mymercological News, 17: 13-20.

Wolda, H. (1988). Insect seasonality: why? Annual Review of Ecology and Systematics, 19: 1-18.

# 6

# Evaluation of Insects that Exploit Temporary Protein Resources Emphasizing the Action of Ants (Hymenoptera, Formicidae) in a Neotropical Semi-deciduous Forest

LC Santos-Junior[1,2], JM Saraiva[2], R Silvestre[1], WF Antonialli-Junior[1,2]

1 - Universidade Federal da Grande Dourados, Dourados, Mato Grosso do Sul, Brazil.
2 - Universidade Estadual do Mato Grosso do Sul, Mato Grosso do Sul, Brazil.

**Keywords**
Formicidae, Interspecific Competition, Foraging Activity

**Corresponding author**
Luiz Carlos Santos Junior
Post Graduation Program in Entomology
and Biodiversity Conservation
Universidade Federal da Grande Dourados
Highway Dourados/Itahum, Km 12
Post Office Box 241
79804-970, Dourados, MS, Brazil
E-mail: lc.santosjunior@yahoo.com.br

**Abstract**
The majority of the ants is opportunistic and generalist foragers, commonly feeding on vegetable secretions, seeds, and living or dead animal material. They may be present on any type of substrate even, occasionally on carcasses. This work, then, aimed to evaluate the action of insects, especially ants, in the exploitation of protein resources in forest environment. Monthly collections were made over a year and, in each collection, were made observations during 12 consecutive hours. To simulate exposure of protein resources we used three types of baits, sardines, beef liver and chicken. To evaluate the importance of ants on protein resources for each type of bait there was a control replica with physical barrier to prevent their access. The ants were observed on all baits throughout the collection period. In total, the baits were visited by 34 species of ants. The main genus of ants to visit the baits were: *Pheidole*, *Crematogaster* and *Solenopsis*. These results demonstrate that the presence of ants is important to ecological succession on temporary protein sources in forest environments interfering in the occurrence of other frequent groups in this type of resource, such as flies, for instance. The species that dominated the baits, when presents, were those that regardless of size and aggressiveness, presented mass recruitment and exploited the baits with large flow of individuals Although some species that exhibit certain characteristics can locate the baits faster and eventually dominate them at some point, depending on the ants species that co-occur, the results for the sequence of colonization can be modified.

## Introduction

The insects of the order Hymenoptera have a wide diversity of habits and complex behaviors, culminating in the social organization of wasps, bees and ants (Wilson, 1971; Triplehorn & Jonnson, 2011; Rafael et al., 2012), In the tropics, ants have a strong presence in most terrestrial ecosystems (Fittkau & Klinge, 1973, Erwin, 1989, Stork, 1991, Longino et al., 2002; Ellwood & Foster, 2004). These insects have broad geographic distributions and high species richness, forming one of the most ecologically successful groups (Hölldobler & Wilson, 1990; Longino et al., 2002), and more than 2000 species are estimated to inhabit the Neotropical Region (Fernández, 2000). The evolutionary success of ants is due to several aspects of social life, but especially the strategies for obtaining resources, particularly food, for their colonies.

Some groups have a more specialized feeding mode, such as fungi cultivators (Weber, 1972); others particularly prefer liquid food (Delabie & Fernández, 2003); and, mostly, ants are opportunistic and generalist foragers, commonly feeding on vegetable secretions, seeds, and living or dead animal material (Fowler et al., 1991; Kaspari, 2000). Ants may be present on any type of substrate if conditions are favorable for foraging. According to Clark and Blom (1991), vertebrate or invertebrate carcasses, even if only occasionally, can be a source of additional food for ants that normally feed on seeds, for example.

The decision made by an ant when locating a resource is to maximize the energy balance, in order to obtain a higher gain at low energetic cost for obtaining food, as predicted by the optimal foraging theory (Sih, 1982 a, b, Stephens & Krebs 1986). Due to restrictions on dominating and carrying resources, small ants should recruit other ants to ensure their domination after encountering a resource, avoiding the loss of that resource to a larger ant, or other animals (Pearce-Duvet & Feener Junior, 2010). Additionally, the recruitment speed is directly related to the amount of resources that an ant can carry. Therefore small ants should recruit faster than larger ants, since the smaller body size is satisfied quickly. The speed of food sources recruitment can be an important determinant of ant's communities, since the evolutionary trade-off between exploitative and interference competition may be a key influence to the dominance of resources (Davidson, 1998; Parr & Gibb, 2012).

The intra or interspecific competition during the foraging activity occurs when individuals exploit similar resources (Begon et al., 2006). The competition also occurs when there is an overlap of activity periods and areas for several species of ants that visit the same food source employing similar foraging strategies (Petal, 1978; Brandão et al., 2000; Hölldobler & Wilson, 1990). A species is competitively superior and considered dominant when it presents features that allow the monopoly of the resource, such as aggressive behavior or mass recruitment. The other species that co-occur with the dominant and do not have these characteristics are considered subordinate and usually have alternative strategies for obtaining resources (Andersen, 1992; Brandão et al., 2000). An example of these strategies is the infiltration behavior, in which some individuals of a subordinate species infiltrate among the dominants using a small fraction of the available resources (Brandão et al., 2000; Parr & Gibb, 2010). These aggressive behaviors between individuals can lead in some resource domination cases by workers of one of the species preventing the access of others (Brandão et al., 2000).

The relationship between dominant and subordinate species can also be influenced by environmental factors. This influence can occur by direct physiological effects resulting from the tolerance of each species to microclimatic variations (De Bie & Hewitt, 1990). In particular, when competing species are subjected to a limiting condition, the dominant species may no longer use the resource as a way to reduce the physiological stress. Consequently, subordinate species may take a risk under such adverse conditions and use the resource (Bestelmeyer, 2000). In this sense, especially for ants, temperature is one of the most important factors, since the myrmecofauna is sensitive to desiccation. Thus, high temperatures can influence the foraging strategy of ants and therefore their interactions on resource places (Cerdá et al., 1997, 1998).

Therefore, this study aimed to evaluate the action of ants, especially the interspecific relationships, while visiting temporary protein sources in a forest environment.

**Material and Methods**

Samples were collected monthly between June 2010 and July 2011 in a forest fragment of around 800,000 sqm (square meters). This area is semi-deciduous forest, according to the classification system of IBGE, the Brazilian Institute of Geography and Statistics (Veloso et al., 1991), and is located in Dourados, Mato Grosso do Sul (S 22°12'56.63" – W 54° 54'57.05"). The area was divided into 20 quadrats of 40,000 sqm. The quadrats were numbered from 1 to 20, and in each collection month, each quadrat was randomly assigned a number which then was removed to avoid repetition. Samples were collected for 12 consecutive hours from 06:00 to 18:00.

To evaluate the effect of ants under temporary protein sources in forest environments, 50g of each of three different baits were used at each collection point. The baits consisted of sardines, which are commonly used as bait for ant collections (Benson & Brandão, 1986; Moutinho, 1991; Brandão et al., 2000), as well as beef liver and chicken.

At each collection point, the three types of baits were placed on disposable plates directly on the ground, 10m apart, in a straight line. The use of plates prevented access by some species of ants that can exploit baits beneath the plant litter. In this way, it was possible to monitor the interaction of all the species that occur only on the substrate. In order not to overestimate the occurrence and probable dominance of any species of ant, before the baits were installed at each site, a systematic search was made to avoid installing the bait on or near ant nests.

To evaluate more specifically the action of ants at the baits, isolated control baits were installed, under the same conditions, with a physical barrier (colorless and odorless gel) around the plate, thus preventing access by any insect that was foraging on the soil.

The consumption of baits was determined at the end of each collection, with the aid of an analytical scale, determined by the difference in weight between the beginning and the end of the period of exposure. Occasional weight loss from drying was not taken into account.

To evaluate whether the difference in consumption between the baits with and without barriers was significant, we applied a T test (using a 0.05% confidence level) to compare them. To evaluate whether the change in climatic conditions during the two seasons in the state of Mato Grosso do Sul (Zavatini, 1992) affected the richness and number of interactions between ant species that occurred on the baits, in the first 15 minutes of each hour of observation, temperature and relative humidity were measured with a hygrometer, and these data were evaluated by a Pearson correlation test.

Throughout the observation period, the individual acts of behavioral interactions between different species of ants that co-occurred on the baits were quantified and qualified, according to the following parameters adapted by Brandão et al. (2000): Action: **Going forward** = Going toward an individual of another species with jaws open in an abrupt movement; **Biting** = Clasping body parts of the other individual with the jaws; **Exhibiting the stinger region or sting** = turning the gaster downward from the abdomen; **Lifting the gaster** =

Shaking the gaster to expel pheromones; **Killing** = Attacks that resulted in the death of the individual attacked. Reaction: **Staying on bait** = The individual does not leave the bait even after attacked; **Escaping** = The attacked individual leaves the bait quickly; **Exhibiting the stinger region or stinging the aggressor** = The attacked individual displays the sting, and/or stings the attacker; **Lifting the gaster expelling pheromones** = The attacked individual displays the gaster region, expelling toxic substances; **Fighting** = The attacked individual defends itself, struggling with the attacker by using the jaw or other body parts; **Killing** = The attacked individual, when defending itself, kills the attacker.

The time that the species spent to locate and exploit the bait was quantified, as well as the number of individuals of each species present on the bait, in 1-minute intervals (flow of individuals). The mean flow was categorized as: **Weak flow:** 3 to 10 individuals per minute; **average flow:** 11 to 30 individuals per minute; **intense flow:** more than 30 individuals per minute.

The levels of ant aggression were categorized during interactions in values from 0-2 (0 = not aggressive; 1 = aggressive, 2 = very aggressive), taking into account the following parameters: **Not aggressive:** they always fled and displayed no agonistic behavior; **aggressive:** they moved most of the time, but did not maintain any physical contact; **very aggressive:** they bit and/or killed, or even performed another aggressive act that caused injury to another individual.

The types of interactions that mainly indicate the dominance or exclusion of other ants from the food source, according to Brandão et al. (2000), were categorized as follows: Dominated by being the only individual on the bait; dominated by being abundant; dominated by being aggressive; dominated by being abundant and aggressive; excluded other ants from the bait.

To evaluate whether there was any relationship between the size of the species and the strategy adopted during the interactions between the species, the alitrunk of each individual collected on the bait was measured. According to Brandão et al. (2000), this measurement is not affected by the individual's physiological state, and is traditionally termed in taxonomic articles on ants as the measure of Weber (WL). The sizes of the ants were categorized as: **Small:** from 0.01 to 1.0 mm; **average:** from 1.01 to 2.0 mm; **large:** above 1.2 mm.

All these parameters described above were correlated by a Jaccard cluster analysis, to attempt to identify groups of ant species that adopt the same behavioral strategies during interactions on baits.

After the interactions and/or consuming the baits, some foragers (one or two depending on the species) were collected while they were leaving the bait, and were placed in 70% ethanol for later identification at the genus level, using the keys of Bolton (1994, 1995 and 2003); and at the species level, when possible, by comparing with standard specimens in the Formicidae Reference Collection of the Museum of Myrmecology

CEPEC/CEPLAC - Ilhéus, Bahia. Vouchers from this study were deposited in this collection under number # 5675.

Specimens of other insects were collected with forceps and/or brushes and stored in jars containing 70% ethanol for later identification to family level, with the aid of the dichotomous key of Rafael et al. (2012) and by comparison with standard specimens in the Entomological Collection of the Museum of Biodiversity, UFGD/MS.

**Results and Discussion**

*Occurrence of insects in general*

The average consumption on baits without a barrier was 17.87g ± 5.45, and on baits with a barrier was 15.95g ± 5.86. The T test indicated no significant difference between these values (F= 1.41; p = 0.261). The presence or absence of ants, on these food sources, does not influence their consumption. Baits where ants do not occur must be consumed by other groups of insects.

During the months of collection, the mean temperature and relative humidity were 25.4°C ± 2.86% and 56.58% ± 14.16, respectively. There was no significant correlation between the consumption of baits and the temperature (F = 2.83; p = 0.163), or relative humidity (F= 2.68; p = 0.163), in both seasons.

In general, the assembly of insects varied little on both types of baits (Figs. 1 and 2).

Dipterans occurred most frequently on both baits (Figs. 1 and 2). According to Souza and Linhares (1997), the insects most frequently evaluated in this type of substrate are the dipterans, especially the families Calliphoridae, Sarcophagidae and Muscidae (Oliveira-Costa, 2003, 2008; Gullan & Cranston, 2008; Pujol-Luz et al., 2008). Fly larvae compete intensely for resources on the carcasses in an attempt to consume the largest possible volume of food before the resource is exhausted (Goodbrod & Goff, 1990).

Although coleopterans are also an important group

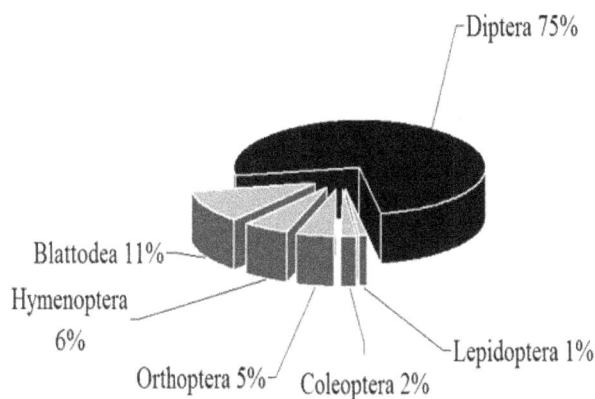

Figure 1: Relative frequency of the different insect orders that visited the 3 types of baits with physical barrier, exposed in forest areas between June/2010 to July/2011.

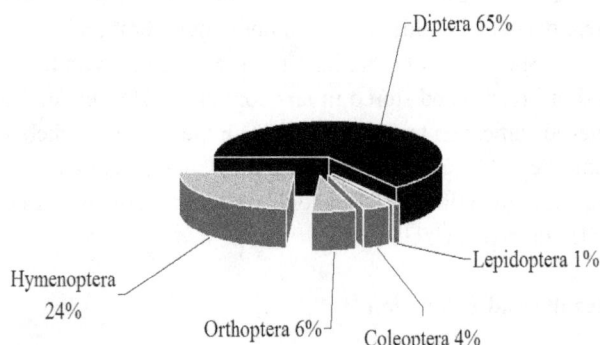

Figure 2: Relative frequency of the different insect orders that visited the 3 types of baits without physical barrier exposed in forest areas between June/2010 to July/2011.

that occurred in this type of resource according to the literature (Oliveira-Costa, 2003; Rafael et al., 2012), they occurred in relatively low frequency compared to the other groups; as Lepidoptera and Orthoptera (Figs. 1 and 2). Souza et al. (2008) reported that insects such as coleopterans, even if frequent, most frequently visit protein sources only at the initial stage of putrefaction.

Cruz and Vasconcelos (2006) discussed that the low occurrence of Blattodea and Orthoptera is associated with their feeding habits. Most representatives of the latter group are phytophagous, occurring almost as accidental visitors (Oliveira-Costa, 2003). Blattodea are opportunistic insects that exploit the most easily available resource. In general they are omnivorous, feeding on organic matter of any kind; however, they are sometimes also predators and attack other insects (Triplehorn & Jonnson, 2011). In this study, Blattodea were present only on baits that were surrounded by a barrier (Fig. 1) and hence without ants, indicating that the presence of ants seems to inhibit the action of this group.

The baits with barriers were visited by insects of the orders Diptera (75%), Blattodea (11%), Orthoptera (5%), Coleoptera (2%) and Lepidoptera (1%); the order Hymenoptera was represented only by wasps (6%) (Fig. 1). These results demonstrate that ants can inhibit the occurrence of wasps. Moretti et al. (2011) demonstrated that these types of substrates may be an additional food source for the wasps; they observed wasps feeding directly on the baits and preying on adult insects.

Also for baits with barriers, Coleoptera of the family Staphylinidae and Diptera of the family Calliphoridae were very common. Calliphorid flies occur in great abundance in this type of substrate, and are very common in manure and carrion (Oliveira-Costa, 2003; Gullan & Cranston, 2008; Triplehorn & Jonnson, 2011).

The baits without barriers were visited by Diptera (65%), Hymenoptera (24%), Orthoptera (6%), Coleoptera (4%) and Lepidoptera (1%) (Fig. 2). On two occasions, baits without barriers were visited by spiders of the family Lycosidae. According to Centeno et al. (2002), and Oliveira-Costa (2008), this group of spiders plays an ecological role as predators of

insects that belong to the cadaverous fauna.

The frequency of occurrence of flies was about 10% lower on bait without barriers, which allowed ants to visit (Table 1). Even so, flies were the most frequent group, as described in trials such as those of Souza and Linhares (1997), Oliveira-Costa (2008), and Rafael et al. (2012). Still, this clear effect on the occurrence of flies should be analyzed with caution because it is a very common group in this type of resources besides being highly important to Forensic Entomology (Oliveira-Costa, 2003; Pujol-Luz et al., 2008).

**Table 1.:** Frequencies relative (%) of occurrences of different orders and families of insects in the three types of baits exposed forested areas between the period June/2010 July/2011.

| Insects | | Baits (%) | | |
|---------|--------|---------|---------|-------|
| Order | Family | Chicken | Sardine | Liver |
| Hymenoptera | Formicidae | 34.73 | 33.22 | 32.04 |
| | Vespidae | 69.56 | 21.73 | 8.69 |
| Diptera | Calliphoridae | 81.25 | 15.62 | 3.12 |
| | Sarcophagidae | 67.98 | 24.02 | 8.00 |
| | Muscidae | 80.01 | 15.96 | 4.03 |
| | Syrphidae | 66.66 | 20.00 | 13.33 |
| Coleoptera | Staphylinidae | 87.09 | 12.9 | 3.22 |
| | Scarabeidae | 75.00 | 20.83 | 4.11 |
| | Histeridae | 96.87 | 3.12 | 0.00 |
| Orthoptera | Gryllidae | 42.85 | 50.00 | 7.14 |
| Blattodea | Blattellidae | 35.71 | 35.71 | 28.57 |

The most frequent insects on the sardine baits without barriers were members of the family Gryllidae (50%); on beef-liver baits, Formicidae (32.04%); and Histeridae (96.87%) were most frequent on chicken baits (Table 1). As seen in Table 1, the chicken bait was the most frequently visited overall. Although subjective, the reason may be the strong odor emitted by the chicken bait on decomposition, compared to the others.

*Occurrence of ants*

Ants occurred on all types of baits throughout the collection period (Table 2), although some species occurred more frequently on a certain type of bait than on another. However, one should take into account that the low occurrence of a species in different areas of collection may explain its low frequency on a specific type of bait.

Throughout the collection period, 34 ant species were observed (Table 2); however, only 27 (80%) interacted with other species of ants. The other species were alone when visiting the baits, with no other ant species at that time. We quantified 194 behavioral acts during interactions between species (Table 3). The most effective act involving action was biting (43.75%).

**Table 2.:** Relative frequency (%) of occurrence of different species of ants in each type of bait attractive exposed in the forest, between the period of the June/2010 July/2011.

| Species | Baits | | |
|---|---|---|---|
| | Chicken | Sardine | Liver |
| **SUBFAMILY: PONERINAE** | | | |
| *Odontomachus meinerti* Forel, 1905 | 1.29 | 0.00 | 0.00 |
| *Odontomachus chelifer* (Latreille, 1802) | 1.29 | 0.00 | 0.00 |
| *Pachycondyla striata* Smith, 1858 | 3.89 | 2.12 | 0.00 |
| *Pachycondyla verenae* (Forel, 1922) | 5.19 | 4.25 | 1.92 |
| *Pachycondyla villosa* (Fabricius, 1804) | 11.68 | 8.51 | 3.84 |
| **SUBFAMILY: ECTATOMMINAE** | | | |
| *Ectatomma brunneum* F. Smith, 1858 | 1.29 | 4.25 | 1.92 |
| *Ectatomma tuberculatum* F. Smith, 1858 | 1.29 | 2.12 | 1.92 |
| *Ectatomma permagnum* Forel, 1908 | 0.00 | 2.12 | 0.00 |
| *Gnamptogenys* sp. | 0.00 | 2.12 | 5.76 |
| **SUBFAMILY: DOLICHODERINAE** | | | |
| *Azteca* sp. | 1.29 | 2.12 | 0.00 |
| *Linepithema iniquum* (Mayr, 1870) | 0.00 | 4.25 | 1.92 |
| *Linepithema pulex* Wild, 2007 | 3.89 | 0.00 | 0.00 |
| **SUBFAMILY: FORMICINAE** | | | |
| *Camponotus crassus* Mayr, 1862 | 5.19 | 2.12 | 3.84 |
| *Camponotus fastigatus* Roger, 1863 | 0.00 | 2.12 | 3.84 |
| *Camponotus melanoticus* Emery, 1894 | 3.89 | 2.12 | 5.76 |
| *Camponotus (myrmaphaenus)* sp | 0.00 | 2.12 | 3.84 |
| *Camponotus sericeiventris* Guérin, 1838 | 2.59 | 0.00 | 0.00 |
| *Nylanderia* sp. | 3.89 | 0.00 | 0.00 |
| *Nylanderia guatemalensis* (Forel, 1885) | 0.00 | 0.00 | 5.76 |
| **SUBFAMILY: ECITONINAE** | | | |
| *Labidus coecus* (Latreille, 1802) | 0.00 | 2.12 | 1.92 |
| **SUBFAMILY: PSEUDOMYRMECINAE** | | | |
| *Pseudomyrmex tenuis* (Fabricius, 1804) | 2.59 | 6.38 | 1.92 |
| **SUBFAMILY: MYRMICINAE** | | | |
| *Acromyrmex coronatus* (Fabricius, 1804) | 2.59 | 0.00 | 0.00 |
| *Atta sexdens rubropilosa* Forel, 1908 | 2.59 | 2.12 | 7.69 |
| *Crematogaster nigropilosa* Mayr, 1870 | 3.89 | 8.51 | 0.00 |
| *Crematogaster limata* Smith, 1858 | 6.49 | 10.63 | 5.76 |
| *Pheidole oxyops* Forel, 1908 | 6.49 | 10.63 | 7.69 |
| *Pheidole pubiventris* Mayr, 1887 | 7.79 | 10.63 | 5.76 |
| *Pheidole radoszkowskii* Mayr, 1884 | 6.49 | 0.00 | 5.76 |
| *Sericomyrmex* sp. | 2.59 | 0.00 | 1.92 |
| *Solenopsis invicta* Buren, 1972 | 6.49 | 0.00 | 10.00 |
| *Solenopsis* sp. | 0.00 | 0.00 | 7.69 |
| *Trachymyrmex iheringi* (Emery, 1888) | 2.59 | 2.12 | 0.00 |
| *Trachymyrmex* sp. | 1.29 | 2.12 | 1.92 |
| *Wasmannia scrobifera* Kempf, 1961 | 1.29 | 4.25 | 0.00 |

This behavior was also the most frequently described by Brandão et al. (2000) and it seems to be one of the most effective behavioral strategies to dominate a resource. The most frequent act involving reaction was lifting the gaster (66.32%). According to Longino (2003), species of the genus *Crematogaster* exhibit their gaster, raising it in order to demonstrate that it is apparently larger than it actually is; or it can be used to eject formic acid or other substances as chemical defenses, depending on the situation and also on the species.

**Table. 3:** Relative frequency of action and reaction behaviors executed by different ant species during interactions in the 3 types of baits exposed in forest areas between June/2010 to July/2011.

| Action | % | Reaction | % |
|---|---|---|---|
| Biting | 43.75 | Lifting the gaster | 66.32 |
| Going Forward | 26.04 | Staying on bait | 20.4 |
| Killing | 13.54 | Escaping | 13.26 |
| Expelling | 17.7 | | |

The results demonstrate that as the frequency of interactions between species on baits increased, the consumption decreased (Fig. 3 A, B and C). It seems that in most cases, the species opted to dominate the resource before exploiting it, which leads them to spend more time interacting with other species than consuming the resource.

The correlations between the number of species of ants that visited the baits (F= 10.88; p= 0.030) and the number of interactions (F= 5.38; p= 0.01) with temperature were significant and positive in the rainy season. That is, the more the temperature increased, the more the number of visitor species increased, and consequently the number of interactions on baits increased as well (Figura 4).

The temperature, especially at the soil surface, is one of the factors that regulate the activity of foraging ground insects. At higher temperatures and favorable relative humidity, ants tend to increase their foraging activities and consequently the number of interactions between species also increases (Hunt, 1974; Levings, 1983; Cerdá et al., 1997, 1998; Dajoz, 2000).

Several species of the Attini group visited and consumed the bait: *Atta sexdens rubropilosa*, *Acromyrmex coronatus*, *Sericomyrmex* sp. and *Trachymyrmex* sp. All of them carried pieces of bait to their colonies. However, these species are known to be restricted to feeding on fungi that grow on a composite substrate consisting mainly of plant material gathered by their workers (Weber 1972; Delabie & Fernández, 2003). On the other hand, Clark and Blom (1991) stated that vertebrate carcasses may be an additional food source for ants that feed on seeds, for example, considering the frequency of availability of carcasses. Conconi and Rodríguez (1977) suggested that species of *Atta* must feed on alternate materials such as meat. Marques and Del-Claro (2006), also observed that ants of the genus *Atta* was one of the most frequent visi-

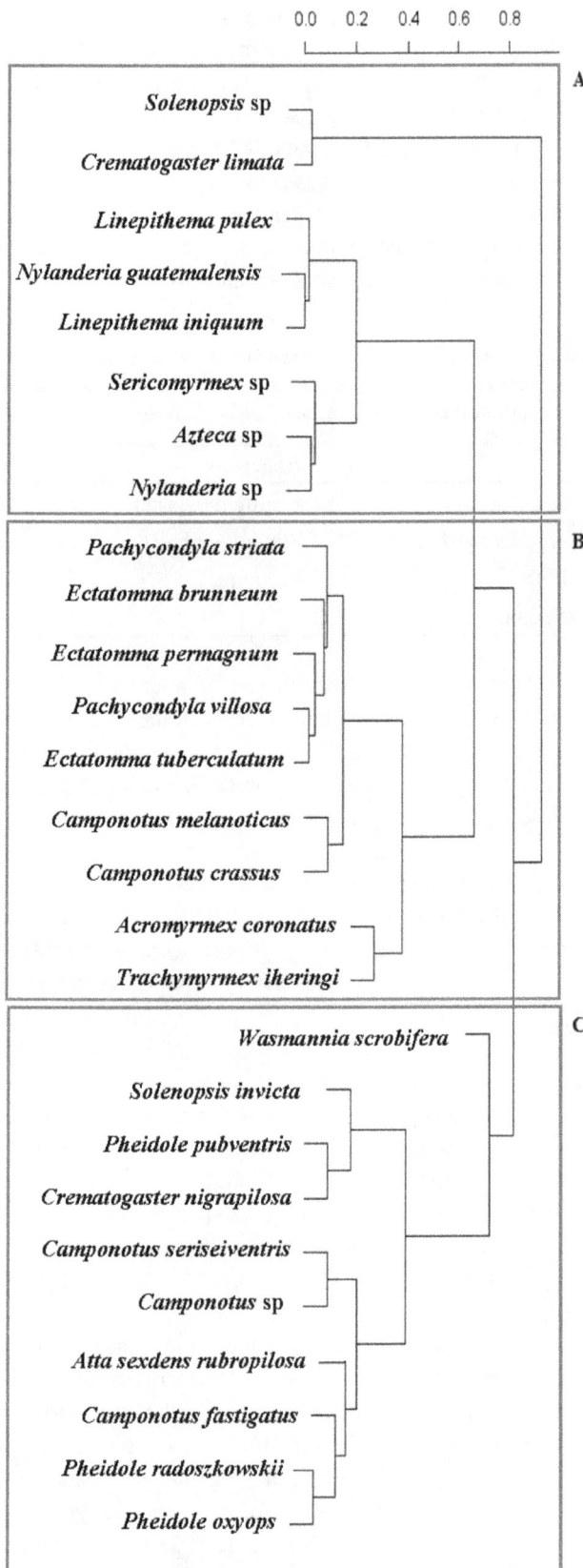

Figure 3: (Jaccard) Grouping evaluating size, aggressiveness and average flow of different ant species during interactions in baits. A: medium-sized species, medium foraging flow, little aggressive and not dominant, B: large species, low foraging flow, very aggressive and not dominant, C: small species, intense foraging flow, not aggressive and dominant.

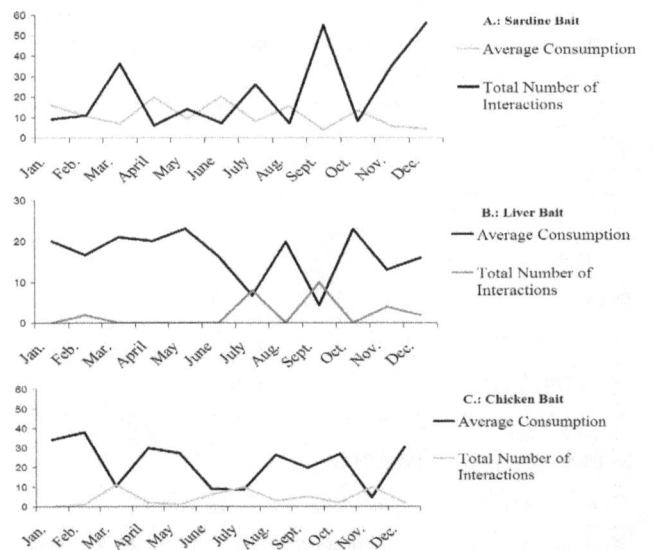

Figure 4: Average consumption and total number of interactions per collection of different ant species in the 3 types of baits without physical barrier, exposed in forest areas between June/2010 to July/2011.

tors in inventory held in a Cerrado area using sardine baits.

Here, the ants, regardless of species, required a mean of $4.1 \pm 1.8$ minutes to locate the bait, which was visited by a mean of 2.11 foragers per minute, regardless of the type of bait. Species of the genera *Pheidole* and *Crematogaster*, and *A. sexdens rubropilosa* were the first to find and exploit resources in 33%, 27.77% and 13.88% of cases respectively, always with a mean flow of over 22 foragers per minute. Species with mass recruitment and that forage in large flows are more likely to detect and numerically dominate food resources more rapidly, as noted by Holldöbler and Wilson (1990).

Brandão et al. (2000) argued that the order of arrival of species on the bait is not necessarily associated with their relative dominance, but rather with other factors such as proximity to the source of the nest, colony size, and foraging strategy.

The Jaccard cluster analysis (J = 0.92) indicated three distinct groups (Fig. 5). Group "A" included ants with a mean size of $1.24 \pm 0.23$ mm and a mean flow of $12 \pm 7.0$ foragers per minute. In this group 83% of the species were categorized as non-aggressive, unable to dominate the bait at any time. An exception in this group was *Azteca* sp. which was considered aggressive; however, its mean flow was $2.1 \pm 2$ foragers per minute, which was weak according to the criteria used here.

Group B (Fig. 5) included ants with a mean size of $3.28 \pm 1.35$ mm and a mean flow of $2 \pm 0.5$ foragers per minute. In this group, 95% of the ants were highly aggressive; however, they dominated the bait in only 5% of cases. In 60% of the cases in which species of this group dominated baits, it was because they were alone, as many species of poneromorphs that comprise this group forage individually (Fig. 5). According to Brandão et al. (2000), ants of this group, although they are large and generally aggressive predators, almost never dominate, and when they do, it is because they are the only ones present on the baits. Their failure to dominate the bait when other species are present is due to their strategy of foraging individually. According to Brandão et al. (2000), ants of the genus *Pachycondyla*, for example, although they are relatively large, cannot monopolize baits and every time that they confront species that use group attack strategies, they are excluded from the baits. However, in 75% of the cases, although they may not dominate baits, they can remove relatively large pieces and carry them off, infiltrating between the dominant species.

Group "C" consisted of small ants (Fig. 5) with a mean size of 0.80 ± 9.5mm and a mean flow of 35.45 ± 9.5 foragers per minute. In this group 81% of the species were not aggressive; however, they dominated baits in 85% of the cases. *Pheidole radoszkowskii*, was a typical species of this group, which, when it occurred, dominated baits in 95% of the cases, maintaining a continuous and intense flow, according to the criteria used here. In particular, in 80% of cases where they were present, *A. sexdens rubropilosa* dominated the bait by being abundant and also by maintaining an intense flow.

Throughout the exposure period of baits it was possible to observe that depending on the time there was one species predominating in number in the bait, demonstrating that there is a succession of dominant species in these food sources, competition occurs more intensively when there is an overlap of activity periods and collection sites by several species of ants visiting the same food source (Brandão et al. 2000). In this case, they can take very aggressive actions that may result in some cases of monopolization of the resource by workers of one species, preventing access by others. A species is considered dominant and competitively superior when it possesses features that allow it to monopolize a resource, such as aggressive behavior or mass recruitment. The other species that co-occur with the dominant species and do not possess these characteristics are considered subordinate and usually have alternative strategies for obtaining resources (Andersen, 1992). The relationship between dominant and subordinate species may also be influenced by environmental factors. This influence may occur through a direct physiological effect resulting from the tolerance of each species to microclimate variations (Bie & Hewitt, 1990), such as temperature, because the myrmecofauna is sensitive to desiccation. High temperatures may influence the foraging strategy of ants and therefore their interactions at sites where resources are present (Hunt, 1974; Levings, 1983; Cerdá et al., 1997 and 1998; Dajoz, 2000).

These results demonstrate that the presence of ants is important to ecological succession on temporary protein sources in forest environments interfering in the occurrence of other frequent groups in this type of resource. Their presence may simply inhibit the presence of other insects, especially those that also forage on the ground, or even flies that avoid landing on the resource when there is an intense flow of ants exploiting it. Another important action in this sense is when they prey on

immature, especially of flies and adults of other species that detect and exploit this type of resource. The results show that there are three distinct groups of ants that can interact in this type of resource according to size, flow and aggressiveness toward other species. However, the ones that dominate the source are always those that arrive with less flow of individuals regardless of whether or not detecting the resource first than other species. Therefore, depending on the ant species that co-occur, the results for the sequence of colonization can be modified.

## Acknowledgments

We thank the post-graduation program in Entomology and Biodiversity Conservation - UFGD, the Coordination of Improvement of Higher Education Personnel (CAPES) for granting the scholarship to the first author, the National Council of Scientific and Technological Development (CNPq) for granting the scholarship to the fourth author. We also thank two anonymous reviewers for valuable comments that improved this article.

## References

Andersen, A.N. (1992). Regulation of "momentary" diversity by dominant species in exceptionally rich ant communities of Australian seasonal tropics. American Naturalist, 140: 401-420. doi: 10.1086/285419.

Begon, M., Townsend, C. R. & Harper, J. L. (2006). Ecology: from individuals to ecosystems. Oxford: Blackwell Publishing.

Benson, W. W. & Brandão, C. R. F. (1986). *Pheidole* Diversity in the humid tropics: a survey from Serra dos Carajas, Pará, Brasil. In International Congress of Iussi X, Chemistry and Biology of Social Insects (pp. 593-594). Munique.

Bestelmeyer, B.T. (2000). The trade-off between thermal tolerance and behavioral dominance in a subtropical South American ant community. Journal of Animal Ecology, 69: 998-1009. doi: 10.1111/j.1365-2656.2000.00455.x

Bie, G. & Hewitt, P.H. (1990). Thermal responses of the semi-arid zone ants *Ocymyrmex weitzeckerii* (Emery) and *Anoplolepis custodiens* (Smith). Journal of the Entomological Society of South Africa, 53: 65-73.

Bolton, B. (1994). Identification guide to the ant genera of the world. Cambridge: Harvard University Press, 222p.

Bolton, B. (1995). A new general catalogue of the ants of the world. Cambridge: Harvard University Press, 504p.

Bolton, B. (2003). Synopsis and classification of Formicidae. Memoirs of the American Entomological Institute, 71: 01-370.

Brandão, C.R.F., Silvestre, R. & Reis-Menezes, A. (2000). Influência das interações comportamentais entre espécie de formigas em levantamentos faunísticos em comunidades de cerrado. In R.P. Martins, T.M. Lewinsohn, & M.S. Barbeitos (Eds.), Ecologia e comportamento de Insetos. (pp. 371-404). Brasil: Oecologia Brasiliensis.

Centeno, N., Maldonado, M. & Oliva, A. (2002). Seasonal patterns of arthropods occurring on sheltered and unsheltered pig carcasses in Buenos Aires Province (Argentina). Forensic Science International, 126: 63-70.

Cerdá, X., Retana, J. & Cross, S. (1997). Thermal disruption of transitive hierarchies in Mediterranean ant communities. Journal of Animal Ecology, 66: 363-374.

Cerdá, X., Retana, J. & Manzaneda, A. (1998). The role of competition by dominants and temperature in the foraging of subordinate species in Mediterranean ant communities. Oecologia. 117: 404-412. doi: 10.1007/s004420050674

Clark, W.H., & Blom, P.E. (1991). Observations of ants (Hymenoptera: Formicidae: Myrmicinae, Formicinae, Dolichoderinae) utilizing carrion. Southwestern Naturalist, 36: 140 - 142.

Conconi, J.R.E., & Rodríguez, H.B. (1977). Valor nutritivo de ciertos insectos comestibles de México y lista de algunos insectos comestibles del Mundo. Anales del Instituto de Biologia de la UNAM Serie Zoologia, 48: 165-186.

Cruz, T. M., & Vasconcelos, S.D. (2006). Entomofauna de solo associada á decomposição de carcaça de suíno em um fragmento de Mata Atlântica de Pernambuco, Brasil. Biociências, 14: 193-201.

Dajoz, R. (2000). Insects and forests: the role and diversity of insects in the forest environment. Paris: Intercept Lavoisier, 620 p. doi: 10.1023/A:1017498600382

Davidson, D.W. (1998). Resource discovery versus domination in ants: a functional mechanism for breaking the trade-off. Ecological Entomology, 23: 484-490. doi: 10.1046/j.1365-2311.1998.00145.x

De Bie, G. & Hewitt, P. H. (1990). Thermal responses of the semi-arid zone ants *Ocymyrmex weitzeckerii* and *Anoplolepis custodiens* (Smith). Journal of the Entomological Society of South Africa, 53: 65-73.

Delabie, J. H. C. & Fernández, F. (2003). Relaciones entre hormigas y "homópteros" (Hemiptera: Sternorrhyncha y Auchenorrhyncha). In F. Fernández (Eds.), Introducción a las hormigas de la región Neotropical (pp. 181-197). Bogotá: Instituto de Investigación de Recursos Biológicos Alexander von Humboldt.

Ellwood, M. D. F. & Foster, W.A. (2004). Doubling the estimate of invertebrate biomass in a rainforest canopy. Nature. 429: 549-551.

Erwin, T.L. (1989). Canopy arthropod biodiversity: a chronology of sampling techniques and results. Revista Peruana de Entomologia, 32: 71-77.

Fernández, F. (2000). Avispas cazadoras de aranãs (Hymenoptera: Pompilidae) de la región neotropical. Biota Colombiana. 1: 3-24.

Fittkau, E.J. & Klinge, H. (1973). On biomass and trophic structure of the central amazonian rain forest ecosystem. Biotropica. 5: 2-14.

Fowler, H. G., Forti, C.L., Brandão, C.R.F., Delabie, J.H.C. & Vasconcelos, H.L. (1991). Ecologia Nutricional de Formigas. In A.R. Panizzi & J.R. Parra (Eds.), Ecologia nutricional de insetos e suas implicações no manejo de pragas (pp. 131-223). São Paulo: Manole.

Goodbrod, J.R., & Goff, M.L. (1990). Effects of larval population density on rates of development and interactions between two species of *Chrysomya* (Diptera: Calliphoridae) in laboratory culture. Journal of Medical Entomology, 27: 338-343.

Gullan, P. J. & Cranston, P.S. (2008). Os insetos: um resumo de entomologia. São Paulo: Roca, 480 p.

Hölldobler, B. & Wilson, E. O. (1990). The Ants. Cambridge: Harvard University Press, 732 p.

Hunt, J. H. (1974). Temporal activity patterns in two competing ant species (Hymenoptera: Formicidae). Psyche, 81: 237-242.

Kaspari, M. (2000). A primer on ant ecology. In D. Agosti, J.D. Majer, L.E. Alonso & T.R. Schultz (Eds.), Ants: Standard methods for measuring and monitoring biodiversity. (pp. 9-24). Washington: Smithsonian Institution Press.

Levings, S.C. (1983). Seasonal, annual and among-site variation in the ground ant community of a deciduous tropical forest: some causes of patchy species distributions. Ecological Monographs, 53: 435-455.

Longino, J.T. (2003). The *Crematogaster* (Hymenoptera, Formicidae, Myrmicinae) of Costa Rica. Zootaxa, 151: 1-150.

Longino, J.T., Coddington, J. & Colwell, R.K. (2002). The ant fauna of a tropical rain forest: estimating species richness three different ways. Ecology, 83: 689-702. doi: org/10.1890/0012-9658(2002)083[0689:TAFOAT]2.0.CO;2

Marques, G.D.V. & Del-Claro, K. (2006). The Ant Fauna in a Cerrado area: The Influence of Vegetation Structure and Seasonality (Hymenoptera: Formicidae). Sociobiology, 47: 235-252.

Moretti. T. C., Giannotti, E., Thyssen, P.J., Solis, D.R. & Godoy, W.A.C. (2011). Bait and habitat preferences, and temporal variability of social wasps (Hymenoptera: Vespidae) attracted to vertebrate carrion. Journal Medical Entomology, 48: 1069-1075. doi: 10.1603/ME11068.

Moutinho, P. R. S. (1991). Note on foraging activity and diet of two *Pheidole* Westwood species (Hymenoptera: Formicidae)

in an area of "shrub canga" vegetation in Amazonian Brazil. Revista Brasileira de Biologia, 51: 403-406.

Oliveira-Costa, J. (2003). Entomologia Forense: quando os insetos são vestígios: São Paulo. Millennium, 257p.

Oliveira-Costa, J. A (2008). Entomologia Forense: quando os insetos são os vestígios. São Paulo. Millenium, 420p.

Parr, C.L. & Gibb, H. (2012). The discovery-dominance trade-off is the exception, rather than the rule. Journal of Animal Ecology, 81: 233-241. doi: 10.1111/j.1365-2656.2011.01899.x

Petal, J. (1978). The role of ants in Ecosystems. In M.V, Brain. (Eds.), Production Ecology of Ants and Termites (pp. 293-325). Cambridge: Cambridge Univ. Press.

Pearce-Duvet, J.M.C. & Feener Junior, D.H. (2010). Resource discovery in ant communities: do food type and quantity matter? Ecological Entomology, 35: 549-556. doi: 10.1111/j.1365-2311.2010.01214.x

Pujol-Luz, J. R., Arantes, L. C. & Constantino, R. (2008). Cem anos da entomologia forense no Brasil (1908-2008). Revista Brasileira de Entomologia, 52: 485-492. doi: org/10.1590/S0085-56262008000400001

Rafael, J.A., Melo, G.A.R., Carvalho, C.J.B., Casari, S.A. & Constantino, R. (2012). Insetos do Brasil: Diversidade e Taxonomia. Ribeirão Preto: Holos, 810 p.

Sih, A. (1982 a) Foraging strategies and the avoidance of predation by an aquatic insect, notonecta-hoffmanni. Ecology. 63: 786-796.

Sih, A. (1982. b) Optimal patch use: variation in selective pressure for efficient foraging. American Naturalist, 120: 666-685.

Souza, A. S. B., Kirst, F.D. & Krüger, R.F. (2008). Insects of forensic importance from Rio Grande do Sul state in southern Brazil. Revista Brasileira de Entomologia, 52: 641-646. doi: 10.1590/S0085-56262008000400016.

Souza, A.M. & Linhares, A.X. (1997). Diptera and Coleoptera of potential forensic importance in southeastern Brazil: relative abundance and seasonality. Medical and Veterinary Entomology, 11: 8-12.

Stephens, D.W. & Krebs, J.R. (1986). Foraging theory. Princeton: Princeton University Press, 262 p.

Stork, N.E. (1991). The composition of the arthropod fauna of Bornean lowland rain forest trees. Journal of Tropical Ecology, 7: 161-180. doi: /10.1017/S0266467400005319

Triplehorn, C.A. & Jonnson, N.F. (2011). Estudo dos Insetos. São Paulo: Cengage Learning, 808 p.

Veloso, H. P., Rangel-Filho, A.L.R. & Lima, J.C.A. (1991). Classificação da vegetação brasileira adaptada a um sistema universal. São Paulo: IBGE, 24 p.

Weber, N.A. (1972). Gardening Ants: The Attines. Philadelphia: American Philosophical Society, 146p.

Wilson, E. O. (1971). The insect societies. Cambridge: Harvard University Press, 562 p.

Zavatini, J.A. (1992). Dinâmica climática no Mato Grosso do Sul. Geografia, 17: 65-91.

# 7

# Susceptibility of *Melipona scutellaris* Latreille, 1811 (Hymenoptera: Apidae) to *Beauveria bassiana* (Bals.) Vuill

P de J Conceição, CM de L Neves, G da S Sodré, CAL de Carvalho, AV Souza, GS Ribeiro & R de C Pereira

*Federal University of Recôncavo da Bahia (UFRB), Cruz das Almas-BA, Brazil.*

**Keywords**
Uruçu bee, entomopathogenic fungi, biological control

**Corresponding author**
Rozimar de Campos Pereira
Center of Agricultural, Environmental and Biological Sciences
Federal University of Recôncavo da Bahia (UFRB)
Rua Rui Barbosa, 710
44380-000, Cruz das Almas-BA, Brazil.
E-Mail: rozimarcp@uol.com.br

**Abstract**

Entomopathogenic fungi are frequently used as an alternative method for insect pest control. However, only a few studies have focused on the effect of these fungi on bees and on the selectivity of fungi to beneficial organisms in agroecosystems. The objective of the present study was to assess the susceptibility of worker bees of the species *Melipona scutellaris* (locally known as "uruçu") to the isolate (Biofungi 1) of the entomopathogenic fungus *Beauveria bassiana*. The experiment was carried through indirect contact between the fungal suspension and newly-emerged bees and topical application of the fungal suspension on the back of newly-emerged bees. The sampling design was completely randomized and comprised five treatments, which included four different concentrations of the fungus: $1 \times 10^5$, $1 \times 10^6$, $1 \times 10^7$, $1 \times 10^8$ conidia/ml, and a control composed of distilled water. Each treatment had five replicates. The mortality data were subjected to an analysis of variance and a probit regression analysis, which provided an estimate of the lethal dose to 50% of the population ($LD_{50}$). The adjustment of the curves to the model was tested with a chi-squared test and differences between curves were tested with a test for parallelism. *Beauveria bassiana* was virulent to uruçu bees, killing the bees at the lowest dose used. These findings may help minimize the impact of this entomopathogen and, therefore, contribute to the maintenance of natural populations of these insects.

## Introduction

The improper use of agricultural chemicals has led to studies on alternative methods of sustainable pest control. Among these methods biological control with entomopathogenic fungi stands out as a broadly used method of pest control in agroecosystems (Messias 1989; Marques et al. 2004).

These fungi are highly viable, because they are able to preserve parasitoid, predator, and pollinator populations, and represent an important factor in integrated pest management (Neves et al. 2001; Oliveira 2008).

However, some authors state that these fungi can be pathogenic to bees. Espinosa-Ortiz et al. (2011) studied the susceptibility of larva and adult honeybees (*Apis mellifera*) to three types of isolates of entomopathogenic fungi and observed high mortality (90-100%) caused by the fungus *Beauveria bassiana*, when applied at the dose of $1 \times 10^7$ conidia/ml. Bee mortality by entomopathogenic fungi was also recorded by Butt et al. (1994, 1998).

Al mazra'awia (2007) studied the impacts of *B. bassiana* on *A. mellifera* and concluded that bees exposed to high concentrations of this fungus show high mortality. However, he also affirmed that the beehives exposed to high densities of inoculum of the same pathogen had low mortality. According to Hokkaner et al. (2003), the temperature is higher inside than outside the beehives, which makes them safer to fungal infection.

When Hokkenen et al. (2003) assessed the impacts of *Metarhizium* and *Beauveria* on bees they observed that different strategies for the application of these fungi should be considered due to the risks they could bring to insects in the natural ecosystem. The conservation of pollinators requires the attention of scientists due to the large number of problems faced by them in natural ecosystems, including death by pesticides (Otterstatter & Thomson 2008; Freitas & Pinheiro 2010; Rocha 2012).

There are few robust data on the effect of entomopathogenic fungi on social bees (Nogueira-Neto 1953; McGregor

1976; Ferraz et al. 2006; Braga et al. 2010) and the impact of these fungi on beneficial insects associated to crops (Espinosa-Ortiz et al. 2011; Kanga et al. 2002).

Ferraz et al. (2006) stated that, in spite of not having unequivocal data at hand, it is possible that the entomopathogenic fungi *Beauveria bassiana*, *Metarhizium anisopliae*, and *Metarhizium flavoviride* cause the death of indigenous bees.

The objective of the present study was to assess the susceptibility of worker bees of the species *Melipona scutellaris* to *B. bassiana* isolates at different concentrations and contact forms.

## Material and Methods

The experiment was conducted at the laboratory of the Center for the Study of Insects (INSECTA), at the Federal University of Recôncavo da Bahia (UFRB), with newly-emerged worker bees of the species *M. scutellaris*, locally known as uruçu. The bees were provided by the rearing facilities of INSECTA/UFRB. In the treatments we used the commercial isolate of the fungus *B. bassiana* (Biofungi 1), produced in the Laboratory of Research and Production of Microorganisms/Biofactory, State University of Southeast Bahia (UESB), Vitória da Conquista, State of Bahia, where this pathogen has been successfully tested for the control of crop pests.

### Collection and sampling

Brood combs of *M. scutellaris* were removed from the colonies of the rearing facilities at INSECTA and maintained in growth chambers of the B.O.D. type (Biologic Oxygen Demand) at a temperature of $28 \pm 2°C$, relative humidity of 70% $\pm$ 2%, and a photoperiod of 12h for a possible emergence of bees (Espinosa-Ortiz et al. 2011).

### Preparation of different concentrations of conidia and application

One-gram samples were randomly removed from the fungal substrate and added to 10 ml of sterilized water containing Tween 80 adhesive spreader at 1% (v/v). To obtain a homogenized suspension, serial dilutions ($10^2$) were made, so that the conidia could be counted in a Newbauer chamber under a microscope (100x). The preparation of the suspensions followed Alves (1998b). The treatments included four fungal concentrations: $1 \times 10^5$, $1 \times 10^6$, $1 \times 10^7$, $1 \times 10^8$ conidia/ml$^{-1}$ and composed of distilled water.

Newly-emerged bees were anaesthetized for 1min in a refrigerator at 16°C to facilitate the handling of worker bees. Bees were exposed to the fungal suspension through topical application on the dorsum and indirect contact. In the topical application, 1µl of each treatment was applied to the dorsum of each bee with a sterile 10µl micro syringe (BD Plastipak). In the exposure by indirect contact, the bees were placed on

filter paper sheets slightly moistened with 1 ml of each treatment for 5 min (Rother et al. 2009). After each treatment, the worker bees were placed in plastic containers (6.0 cm x 8.0 cm) and transferred to an acclimatized chamber.

Five worker bees were placed in each plastic container, totaling 125 individuals. The bees were fed with honey at 10% (100g/1000ml distilled water), placed on a sterilized wad of cotton to prevent contamination (Rother et al. 2009).

### Data analysis

The experimental design was completely randomized. It was composed of five treatments and five replicates, totaling 25 plots. The mortality of worker bees was monitored at 24h intervals for 10 days. The results were corrected considering natural mortality in accordance with the Abbott formula (Alves 1998).

The corrected mortality data were subjected to an analysis of variance. For this analysis, the data were transformed using the formula $X' = \arcsin (X_i / N)$, where $X'$ is the datum after the transformation, $X_i$ is the mortality observed in the replicate $i$ and $N$ is the total number of insects in the experimental plot. Data normality was assessed with a Kolmogorov-Smirnov test and variance homogeneity was assessed with a Levene test.

Mortality data from different treatments were subjected to a probit regression analysis (Sokal 1958) in Statistica (StatSof Inc.). This analysis provided an estimate of the lethal dose to 50% of the population ($LD_{50}$). The adjustment of the curves to the model was assessed with a chi-squared test and differences between curves for the exposure methods were assessed with a test of parallelism (Alves 1998).

The mortality data at highest dose were used to build survival curves for the two exposure methods following the Kaplan-Meier method (Blanford et al. 2005). Based on these curves the time to mortality (or survival) of 50% of the insects ($S_{50}$) was estimated. The curves were compared using a logrank test ($P = 0.95$) and a Gehan-Breslow-Wilcoxon test (GBW) in GraphPad Prism 5.0 (Motulsky 1995).

## Results and Discussion

Under the experimental conditions, there was no significant effect ($p > 0.05$) of the exposure methods on the mortality of uruçu bees (*M. scutellaris*). However, there was a significant interaction between exposure methods and the doses applied (DF = 4.396; F = 17.68; $P < 0.001$). The corrected mortality caused by different doses of the fungal suspension was higher when applied on the dorsum of the bees than when bees got in indirect contact with the fungal suspension (Figure 1). Even at the lowest dose ($10^5$ conidia ml$^{-1}$), the worker bees were affected: they lost mobility and had an average mortality of 56%.

These results differ from those of Butt et al. (1994), who tested the pathogenicity of *Metarhizium anisopliae* to adult bees of *A. mellifera* and observed significant mortali-

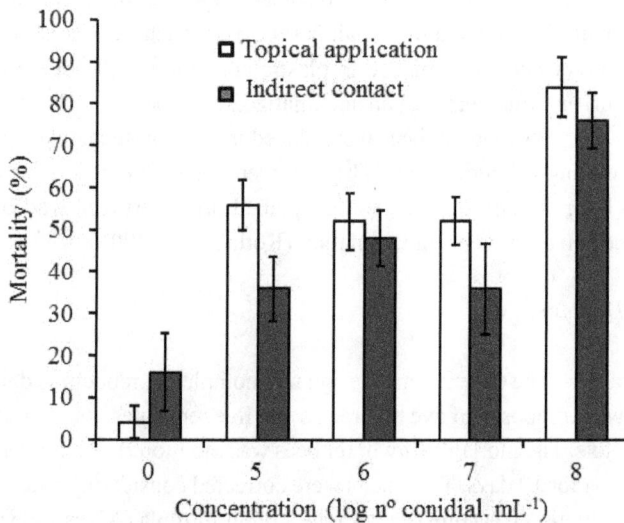

**Fig 1** – Corrected mortality (%) of uruçu bees (*M. scutellaris*) ten days after exposure to fungal suspensions with increasing concentration of *B. bassiana* conidia by the methods of topical application and indirect contact.

ties only at very high doses. In a similar study, Kampongo et al. (2008) reported a mortality rate of 42 - 45% for *Bombus* sp. bees exposed to high doses of *B. bassiana* ($2\times10^{11}$ conidia ml[-1]). Espinosa-Ortiz et al. (2011) tested the virulence of different commercial isolates of *B. bassiana*, *M.anisopliae*, and *P. fumosoroseu* on worker bees of the species *A. mellifera* and obtained a mortality rate of 90% - 100% with a dose of $10^7$ conidia ml[-1] of *B. bassiana*. Therefore, it is possible that the isolates exhibit specificity to bee species.

The topical application of the fungal suspension on the dorsum of bees caused high mortality, with low variability between replicates and uniformity among treatments. The data obtained from this method adjusted well to the probit regres-

sion model ($\chi^2 = 2.897$; DF = 2; $P = 0.235$). The isolate of *B. bassiana* used in the experiments was pathogenic to uruçu bees and caused high mortality even at low doses. Based on the results obtained, a $LD_{50}$ of $2.04\times10^5$ conidia ml[-1] was estimated ($7.95. 10^3$; $3.70.10^5$) (Figure 2).

The data obtained from the exposure method of indirect contact with the fungal suspension at different doses showed high variability among doses, and did not fit the probit regression model ($\chi^2 = 26.811$; DF = 2; $P < 0.001$). With this result, it was not possible to estimate the $LD_{50}$ or make a comparison between exposure methods through the assessment of the parallelism of curves. For the comparison between different exposure methods, the survival curve at the two highest doses was compared with Mantel-Cox and GBW tests.

Data analysis through the construction of survival curves using the Kaplan-Meier method made it possible to assess mortality details over time, and allowed the identification of differences in the survival curve between exposure methods. We observed that the indirect contact method resulted in a lower mortality rate at the lowest doses in the beginning of the observation period, but in the end of the observation period the total mortality was also high.

Figure 3 shows that the topical application of the fungal suspension of *B. bassiana* ($10^8$ ml[-1]) resulted in high mortality in the end of the experiment, with a survival of only 20.4% ($\pm 5.1$) of the bees; whereas the exposure method through indirect contact resulted in higher survival (69.1% $\pm 4.2$). The survival curves obtained for different methods of exposure to the fungus were significantly different from the control and from one another when compared by Mantel-Cox (Log Rank Test) and GBW tests (Table 1).

The mean survival value ($S_{50}$) estimated for the topical application method was 216.0 days (Table 1). For the con-

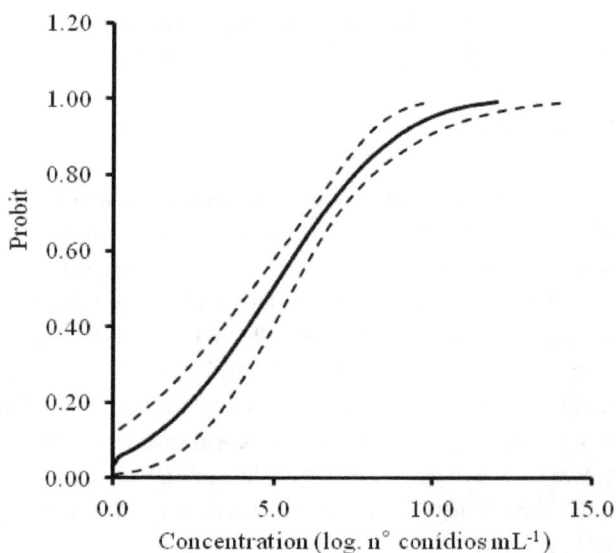

**Fig 2** – Dose-response curve resulting from the methods of topical application of fungal suspensions containing *B. bassiana* conidia on uruçu bees (*M. scutellaris*). Dotted lines represent the fiduciary limits of the estimated doses.

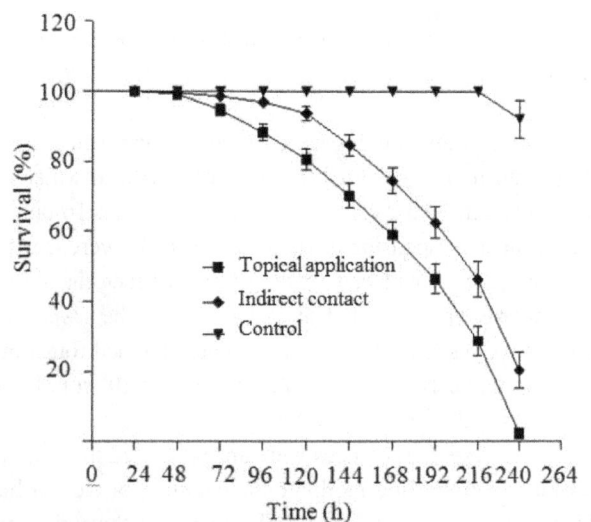

**Fig 3** – Kaplan-Meier Survival curves of uruçu bees (*M. scutellaris*) subjected to two exposure methods to fungal suspensions containing *B. bassiana* conidia ($10^8$ ml[-1]). The fungal suspensions were applied topically on the dorsum of bees or by indirect contact of the bees with a surface previously sprayed with the suspension.

trol treatment or application by indirect contact, the amount of $S_{50}$ could not be estimated. For the control treatment or exposure method by indirect contact, $S_{50}$ could not be estimated because bee mortality did not exceed 50%, and the use of the Kaplan-Meier model to calculate $S_{50}$ is limited by the increased survival time of the individuals studied. The estimated value of hazard ratio (HR) between the two exposure methods was 0.35, between the topical application and the control it was 8.48, and between the indirect contact and the control it was 5.96 (Table 1). The hazard ratio estimates the difference in mortality between treatments based on the slope of the respective survival curves. In this particular case, the average mortality estimated for the topical application exposure method was consistently 35% higher throughout the experiment in comparison with the indirect contact exposure method.

**Table 1** – Comparison of survival curves estimated for uruçu bees (*M. scutellaris*) exposed to topical application or indirect contact with the fungal suspension of *B. bassiana* conidia ($10^8$ ml$^{-1}$).

| Mantel-Cox Test (Logrank) | |
| --- | --- |
| Topical application x Control | 76.96**[1] |
| Indirect contact x Control | 21.23** |
| Topical application x Indirect contact | 26.17** |
| Gehan-Breslow-Wilcoxon Test (GBW) | |
| Topical application x Control | 57.51** |
| Indirect contact x Control | 19.58** |
| Topical application x Indirect contact | 7.69** |
| Median Survival ($S_{50}$) | |
| Topical application | 216 |
| Indirect contact | 192 |
| Control | - |
| Hazard Ratio (HR) | |
| Topical application x Control | 8.48 (5.26 to 13.67)[2] |
| Indirect contact x Control | 5.86 (2.76 to 12.45) |
| Topical application x Indirect contact | 0.35 (0.22 to 0.61) |

[1] Significant at $P < 0.001$; [2] Confidence interval estimated (CI)

Delaplane & Mayer (2005) reported that methods used to apply the compounds may interfere with the results of toxicity assessment of pesticides on non-target insects in the laboratory; there may be an interaction between the active ingredients and exposure methods. Carvalho et al. (2009) tested four methods to assess the toxicity of pesticides to *A. mellifera* and found different responses according to the active ingredient used. Thiamethoxam and methidathion were highly toxic, with low median lethal time ($LT_{50}$) for topical application, supply of contaminated food, and indirect contact of bees to previously sprayed surfaces. Abamectin showed lowest $LT_{50}$ when provided in contaminated food, whereas deltamethrin showed highest toxicity when the insects were exposed to pre-

viously sprayed surfaces (Carvalho et al. 2009).

Pest control with entomopathogenic agents has the advantage of leaving no toxic residues, and therefore can be used for long periods with low environmental impact (Alves 1998). However, the susceptibility of stingless bees to other commercial isolates of entomopathogenic fungi should be tested in future studies, with special attention to the concentration. Only by knowing the effects of entomopathogenic agents on bees, it will be possible to achieve greater efficiency in pest control with minimal impact on these beneficial insects.

*Beauveria bassiana* (Biofungi 1) was highly virulent to uruçu bees (*M. scutellaris*), killing them at the lowest dose used. This information is important because the use of biological products for insect pest control has been growing, and these products require good management to avoid damage to beneficial insects.

## Acknowledgments

The authors thank the Brazilian Council for Scientific and Technological Development (CNPq) and the Brazilian Coordination for the Improvement of Higher Education Personnel (CAPES) for the scholarships granted, the Center for the Study of Insects (INSECTA/UFRB), Dr. Tiyoko Nair Hojo Rebouças, researcher at the Laboratory of Research and Production of Microorganisms/Biofactory, State University of Southeast Bahia (UESB), for supplying the fungal material, and Dr. Carlos Alberto Tuão Gava of Embrapa Semiárido (Brazilian Agricultural Research Corporation) for the help in the statistical analysis. We thank two anonymous reviewers for valuable comments.

## References

Al Mazraawi, M.S. (2007). Impact of the entomopathogenic fungus *Beauveria bassiana* on the honey bees, *Apis mellifera* L (Hymenoptera: Apidae). World Journal of Agricultural Sciences, 3: 7-11.

Alves, S.B. (1998). Fungos entomopatogênicos, 289-381. In: Alves, S.B. (ed.), Controle Microbiano de Insetos. Piracicaba, FEALQ, 1.163p.

Alves, S.B. (1998b). Técnicas de laboratório: vantagens e desvantagens. In: Alves, S.B. (ed.), Controle Microbiano de Insetos. Piracicaba, FEALQ. 22-37.

Blanford, S., Chan, B. H. K., Jenkins, N., Sim, D., Turner, R. J., Read, A. F. & Thomas, M. B. (2005). Fungal pathogen reduces potential for malaria transmission. Science, 308: 1638-1641. doi: 10.1126/science.1108423

Braga, A.S.N.; Silva, M.N. da; Silva, F.O. da; Castro, M.S. de; & Bautista, A.R.P.L. & Viana, B.F. (2010). Avaliação da toxicidade aguda de agrotóxicos em abelhas polinizadoras de fruteiras agrícolas. http://www.labea.ufba.br/polinfrut/produtos/res_adriana.pdf acesso em 04 de julho de 2013.

Butt, T.M., Ibrahima, L., Balla, B.V. & Clarka, S.J. (1994). Pathogenicity of the entomogenous fungi *Metarhizium anisopliae* and *Beauveria bassiana* against crucifer pests and the honey bee. Biocontrol Science and Technology, 4: 207-2014. doi: 10.1080/09583159409355328

Butt, T. M., Carreck, N. L., Ibrahim, L. & Williams, I. H. (1998). Honey-bee-mediated infection of pollen beetle (*Meligethes aeneus* Fab.) by the insect-pathogenic fungus, *Metarhizium anisopliae*. Biocontrol Science and Technology, 4: 533-538. doi: 10.1080/09583159830045

Carvalho, S.M., Carvalho, G.A., Carvalho, C.F., Bueno-Filho, J.S.S. & Baptista, A.P.M. (2009). Toxicidade de acaricidas/inseticidas empregados na citricultura para a abelha africanizada *Apis mellifera*, 1758 (Hymenoptera: Apidae). Arquivos do Instituto Biológico,76: 597-606.

Delaplane, K.S. & Mayer, D.F. (2005). Crop pollination by bees. Oxon: CABI Publishing. 344p.

Espinosa-Ortiz, G.E., Lara-Reyna, J., Otero-Colina, G., Alatorre-Rosas, R. & Valdez-Carrasco, J. (2011). Susceptibilidade de larvas , pupas y abejas adultas a aislamientos de *Beauveria bassiana* (Bals.) Vuill., *Metarhizium anisopliae* (Sorokin) y *Paecilomyces fumosoroseus* (Wize). Interciencia, 36: 148-152.

Ferraz, R.E., Lima, P.M., Pereira, D.S., Alves, N.D, & Feijó, F.M.C. (2006). Microbiota fúngica de abelhas sem ferrão (*Melipona subnitida*) da região semi-árida do nordeste brasileiro. Agropecuária Científica no Semi-árido, 2: 44-47.

Freitas, B.M. & Pinheiro, J.N. (2010). Efeitos sub-letais dos pesticidas agrícolas e seus impactos no manejo de polinizadores dos agroecossistemas brasileiros. Oecologia Australis, 1: 282-298. doi:10.4257/oeco.2010.1401.17

Hokkanen, H.M., Zeng, Q.Q., & Menzler-Hokkanen, I.N.G. E. B. O. R. G. (2003). Assessing the impacts of Metarhizium and Beauveria on bumblebees. Environmental impacts of microbial insecticides. 63-71p.

Kapongo, J.P., Shipp, L., Kevan, P. & Broadbent, B. (2008). Optimal concentration of *Beauveria bassiana* vectored by bumble bees in relation to pest and bee mortality in greenhouse tomato and sweet pepper. Biocontrol, 53: 797-812. doi: 10.1007/s10526-007-9142-9

Kanga LH, James, R.R & Boucias, D. (2002). *Hirsutella thompsonii* and *Metarhizium anisopliae* as potential microbial control agents of *Varroa destructor*, a honey bee parasite. Journal of Invertebrate Pathology, 81: 175-184. doi: 10.1016/S0022-2011(02)00177-5

Marques, R. P.; Monteiro, A. C. Pereira, G. T. (2004). Crescimento, esporulação e viabilidade de fungos entomopatogênicos em meios contendo diferentes concentrações do óleo de Nim (*Azadirachta indica*). Ciência Rural, 34: 1675-1680. doi: 10.1590/S0103-84782004000600002

Mcgregor, S.E . (1976). Insect pollination of cultivated crop plants. Washington: United States Department of Agriculture, 411 p.

Messias, L.C. (1989). Fungos, sua utilização para o controle de insetos de importância médica e agrícola. Memórias do Instituto Oswaldo Cruz, 84: 57-59.

Motulsky, H. (1995). Comparing survival curves. In. Motulsky, H. (Eds.) Intuitive biostatistics (pp. 272-276). New York, Oxford:Oxford University Press.

Neves, P.M.O.J., Hirose, E., Tchujo, P.T. & Moino, J.R.A. (2001). Compatibility of entomopathogenic fungi with neonicotinoid insecticides. Neotropical Entomology, 30: 263-268. doi: 10.1590/S1519-566X2001000200009

Nogueira, N. P. A. (1953). Criação de Abelhas Indígenas sem ferrão. São Paulo: Editora Chácaras e Quintais. 369p.

Oliveira, M.A.P de., Oliveira, M.A.P., Marque, E.J., Wanderley-Teixeira V. & Barros, R. (2008). Efeitos de *Beauveria bassiana* (Bals.) Vuill. e *Metarhizium anisopliae* (Metsch.) Sorok sobre características biológicas de *Diatraea saccharalis* F. (Lepidoptera: Crambidae). Acta Scientarum, Biol. Sci., 30: 219-224. doi: 10.4025/actascibiolsci.v30i2.3627

Otterstatter, M.C., & Thomson, J.D. (2008). Does pathogen spillover from commercially reared bumble bees threaten wild pollinators?. PLoS One, 7: 2771.doi:10.1371/journal.pone.0002771

Rocha, M. C. de L e S. de A. (2012). Efeitos dos agrotóxicos sobre as abelhas silvestres no Brasil: proposta metodológica de acompanhamento – Brasília: Ibama, 88 p

Rother, D.C., Souza, T.F., Malaspina, O., Bueno, O.C., Silva, M.F.G.F., Vieira, P.C. & Fernandes, J.B. (2009). Suscetibilidade de operárias e larvas de abelhas sociais em relação à ricinina. Porto Alegre - RS. Iheringia. Série Zoologia, 99: 61-65. doi: 10.1590/S0073-47212009000100009

Silva, V.C.A., Barros, R., Marques, E.J. & Torres, J.B. (2003). Suscetibilidade de *Plutella xylostella* L. (Lepidoptera: Plutellidae) aos fungos *Beauveria bassiana* (Bals.) Vuill. e *Metarhizium anisopliae* (Metsch.) Sorok. Neotropical Entomology, 32: 653-658. doi: 10.1590/S1519-566X2003000400016

Sokal, R. (1958). Probit analysis. Journal of Entomology, 50: 738-739.

# 8

# Colony performance of *Melipona quadrifasciata* (Hymenoptera, Meliponina) in a Greenhouse of *Lycopersicon esculentum* (Solanaceae)

BF Bartelli, AOR Santos, FH Nogueira-Ferreira

*Universidade Federal de Uberlândia, Instituto de Biologia, Uberlândia, Minas Gerais, Brazil*

**Keywords**

stingless bees, management of bees, external activity, resource collection

**Corresponding Author**

Fernanda Helena Nogueira-Ferreira
Universidade Federal de Uberlândia
Instituto de Biologia
Ceará St., Umuarama
Uberlândia-MG, Brazil, 38400-902
E-Mail: ferferre@inbio.ufu.br

**Abstract**

The use of stingless bees in greenhouses has provided tremendous benefits to diverse crops in terms of productivity and fruit quality. However, knowledge about management techniques in these environments is still scarce. The present study aimed to evaluate colony performance of *Melipona quadrifasciata* Lepeletier, 1836 in a greenhouse of *Lycopersicon esculentum* Mill. and its potential use in pollinating this crop. Six nests of *M. quadrifasciata* were introduced in a greenhouse in Araguari, Minas Gerais state, Brazil. The development of colonies inside the greenhouse was investigated and the foraging behavior of the workers was assessed before introduction, into the greenhouse and after the nests had been removed from the greenhouse. The vital activities of colony maintenance were performed unevenly throughout the day inside and outside the greenhouse, but with confinement the daily period of foraging decreased and bees started collecting pollen from the flowers after approximately six months. The difficulty in orienting to and identifying flowers by the workers was attributed to sunlight diffusion and blockage of ultraviolet radiation caused by the cover on the greenhouse. Structural changes in the greenhouses, as well as improvements in management techniques, are required for better utilization of stingless bees for pollination of plant species grown in greenhouses.

## Introduction

Worldwide, there are approximately 20,000 described bee species (Michener, 2007). Among them, the stingless bees are highly social organisms that constitute an important group due to the ecological and economic roles they play. These bees belong to the subtribe Meliponina, which consists of approximately 400 species grouped in about 50 genera, distributed across tropical and subtropical regions, in America, Southeast Asia, Africa, Madagascar and Australia (Silveira et al., 2002).

Stingless bees are responsible for the pollination of many native (Michener, 2007) and cultivated (Heard, 1999) plant species. As a result, since the 1990s, the number of studies involving the introduction of these bees in greenhouses to evaluate their pollination efficiency on different crops has grown steadily. Particularly noteworthy among these are: eggplant pollination, using *Melipona quadrifasciata* Lepeletier, 1836 (Bispo dos Santos, 2008); basil pollination, tested with *Nannotrigona testaceicornis* Lepeletier,

1836 (Bispo dos Santos, 2008); strawberry pollination, tested with *Tetragonisca angustula* Latreille, 1811 (Malagodi-Braga, 2002), *Scaptotrigona* aff. *depilis* Moure, 1942 (Roselino et al., 2009) and *N. testaceicornis* (Roselino et al., 2009); sweet pepper pollination, using *M. subnitida* Ducke, 1910 (Cruz et al., 2005), *M. quadrifasciata anthidioides* Lepeletier, 1836 (Roselino et al., 2010) and *M. scutellaris* Latreille, 1811 (Roselino et al., 2010); and tomato pollination, tested with *N. perilampoides* Cresson, 1878 (Cauich et al., 2004; Palma et al., 2008b) and *M. quadrifasciata* (Del Sarto et al., 2005; Bispo dos Santos et al., 2009).

Tomato, *Lycopersicon esculentum* Mill. (Solanaceae), is one of the most widespread vegetable crops in the world, being cultivated in almost all parts of the world under different cropping systems and various levels of management. It is a self-fertilizing plant and its flowers are bisexual, do not produce nectar and present poricidal anthers. Therefore, in order to release the pollen, vibration of the anthers with consequent opening thereof is necessary (Buchmann, 1983). Pollination can be performed by the shaking of the anthers by the wind, but cross-pollination is ensured through visits of bees that exhibit vibratory behavior or buzz-pollination (Buchmann & Hurley, 1978; Heard, 1999; Nunes-Silva et al., 2010).

Tomato can be grown in open areas or in greenhouses. When grown in open areas, the released pollen is carried by the wind (McGregor, 1976; Free, 1993) and/or natural pollinators, especially bees, which have free access to the flowers. As for cultivation in greenhouses, pollination is normally performed by the mechanical method of vibrating the flowers, to compensate for the absence of wind and natural pollinators. However, studies involving the management of pollinators in greenhouses have indicated improvements in productivity and fruit quality (Cruz & Campos, 2009). Among the pollinators of tomato in greenhouses, *M. quadrifasciata* proved to be efficient in pollinating the long-lived variety (Del Sarto et al., 2005), but studies testing colony performance and pollination efficiency of this bee on the grape variety, grown mostly in greenhouses, are yet to be performed.

Stingless bees are considered particularly promising for use as commercial pollinators (Cruz & Campos, 2009), given that they do not present a functional sting, are easy to handle (usually low aggressive-nonstinging), have populous and perennial nests, present a marked worker recruitment behavior and stock a large amount of food (Heard, 1999; Malagodi-Braga et al., 2004). Nevertheless, the lack of techniques for management and multiplication of nests has hindered the availability and use of these bees in agriculture on a large-scale (Imperatriz-Fonseca et al., 2006).

The present study sought to advance knowledge about management of stingless bees in greenhouses for use as commercial pollinators. We have specifically evaluated colony performance of *M. quadrifasciata* in a greenhouse of grape tomato, *L. esculentum*.

**Material and Methods**

*Study area*

The study was performed at the "Meliponário da Universidade Federal de Uberlândia (UFU)", located at the "Fazenda Experimental do Glória" (FEG) (18°56'57"S - 48°12'14'"W), Uberlândia, Minas Gerais state, Brazil, and in the "Chácara Paraíso" (CP) (18°39'3.55"S - 48°11'7.51"W), located in the municipality of Araguari, Minas Gerais state, Brazil. At FEG, agricultural and cattle raising activities are developed, but preserved fragments of cerrado and semideciduous forest can be found. CP consists of areas of cultivation and pasture, fragments of cerrado and 12 greenhouses where grape tomatoes are grown. One of these was used for the experiments.

The climate of the regions of Uberlândia and Araguari is marked by two distinct seasons, the rainy season that extends from October to March and a dry season from April to September. Annual rainfall varies between 1,160 and 1,460 mm/year and the average annual temperature is between 23 and 25°C, being uniform throughout the year (Alves & Rosa, 2008).

The greenhouse used in the experiment comprised approximately 1,344 m² (48 m x 28 m), being covered at the top with an Extra Long Life (ELL) Diffuser Antivirus plastic diffuser and fully enclosed on the sides with anti-aphid screens (Fig 1A). The greenhouse presented 24 planting rows and each of these had an average of 112 tomato plants, adding up 2,688 plants.

*Introduction of nests of M. quadrifasciata*

Six nests of *M. quadrifasciata* with similar population sizes were introduced at the onset of flowering, in March 2012. The nests were kept in wooden boxes with an approximate size of 40x25x25cm (Fig 1B). Before introduction, due to the difficulty in bee orientation inside the greenhouse (Bartelli, personal communication; Del Sarto, 2005), the population of old foragers was removed to avoid their loss in the enclosed space, following the methodology used by Cauich et al. (2004). Moreover, for the same reason, the nests were placed in the greenhouse after dark (Cuypers, 1968).

In order to allow a homogeneous distribution of the bees on the flowers, nests were arranged in the central region of the greenhouse (Free, 1993) and supported by plastic boxes installed in eucalyptus logs located within the planting rows. To increase the number of reference points for the bees, since the uniformity of greenhouses may hinder orientation (Dyer, 1994), the entrance to each nest was painted with different color patterns.

Containers with water, mud and cerumen (alternative source for plant resin) of *T. angustula* were placed on the plastic boxes (Fig 1C). Nests were sporadically fed with *A. mellifera* pollen macerated with sugar and water until the onset of flower visitation, and fed weekly with syrup (a mixture of honey of *A. mellifera*, sugar and water in the ratio 1:1:1) over the entire period of confinement. For pesticide application, a common management practice, the entrances of all nests were sealed with paper, the nests themselves were protected with plastic bags and the containers covered with cardboard boxes. These protections were removed only after the pesticide had dried; then, the entrances of the nests were unobstructed after dark. Furthermore, to increase luminosity inside the greenhouse, another common management practice, the ceiling and sides of the greenhouse were washed with soap and water in late August.

In order to facilitate bee orientation, one mercury vapour lamp and two mixed lamps (alternative sources of light and ultraviolet radiation) were installed inside the greenhouse in one of the corridors where some of the nests were placed (Fig 1D). The lamps were connected to a digital timer, which was programmed to switch them on at 6 am and switch them

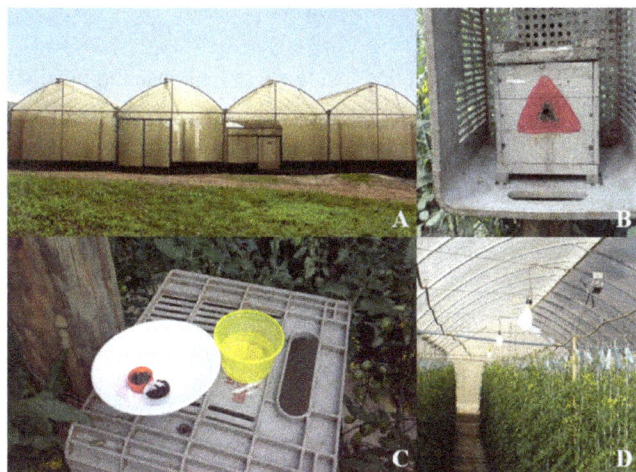

Fig 1. A) Greenhouse used in the experiment at Chácara Paraíso in Araguari; B) Nest of *Melipona quadrifasciata* installed inside the greenhouse; C) Containers with water, mud and cerumen available inside the greenhouse; D) Lamps installed inside the greenhouse.

off at 6 pm. They remained in the greenhouse from early June to mid-July. The lamps' efficiency, as well as the influence of the diffuser film on solar radiation focused on the greenhouse, was evaluated by analysing the pattern of reflectance of the flowers. 10 flowers were photographed between 10 and 11 am in a white background with a Sony Cyber-shot DSC-H20 digital camera. Each flower had its picture taken in three different circumstances: inside the greenhouse and next to the lamps; inside the greenhouse and 12m far from the lamps; and outside the greenhouse. Photoshop software CS3 10.0 was used to quantify patterns of chromatic (RGB - red, green and blue in the light spectrum) and achromatic (luminosity) saturation of flower corollas.

## Colony performance of M. quadrifasciata

To assess colony performance, the internal conditions of nests, behavior of workers in flight and on flowers, foraging activities and temperature inside and outside the greenhouse were analyzed. Figure 2 outlines the arrangement of nests inside the greenhouse. Nests D and F remained in the greenhouse for three months and then they were relocated to the outside area.

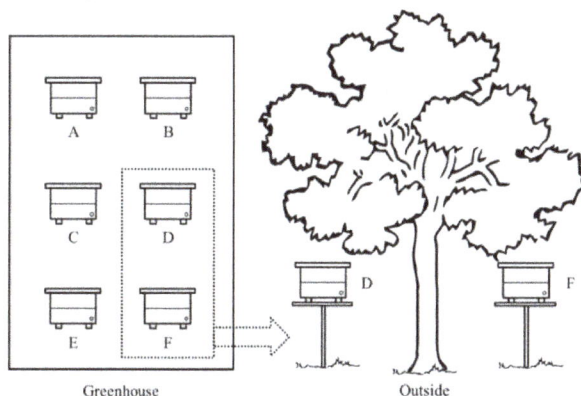

Fig 2. Diagram representing the arrangement of nests inside and outside the greenhouse.

### Internal conditions

Biweekly, from March to October 2012, we assessed qualitative features of nests, such as the presence of cells in construction (yes / no), number of workers (very low / low / medium / high) and presence of guards (yes / no).

### Behavior of workers

Biweekly, from May to October 2012, the flight behavior of bees and the behavior of workers visiting tomato flowers were observed inside the greenhouse. These behaviors were assessed every 15 minutes during each hour (from 6 am to 6 pm).

### Foraging activities

Foraging activities of bees were assessed through direct observations of the flow of workers from nests. The quantity and quality of resources (pollen, nectar/water or water, resin, mud and garbage) that entered and/or left the nests were recorded every 10 minutes during each hour (from 6 am to 6 pm). In order to compare the daily pattern of resource collection of workers inside and outside the greenhouse, the procedures described above were performed: before introduction in three nests (B, D and F, Fig 2), over three non-consecutive days in late February 2012 at FEG; inside the greenhouse in two nests (C and E, Fig 2), biweekly from May to October 2012; and outside the greenhouse in one (nest F, Fig 2) of the nests removed from the greenhouse at CP, biweekly from June to October 2012.

### Temperature

Through the use of a digital thermohygrometer, the temperature inside and outside the greenhouse was measured every day at 8 am, 11 am and 3 pm from April to October 2012.

### Statistical analyses

To verify whether the patterns of reflectance of tomato flowers depended on the presence of the lamps and diffuser film or not, we conducted an analysis of variance (ANOVA) followed by Tukey's test, since there was no difference between blocks (flowers) for RGB ($F = 1.118$; df = 9; $P = 0.399$) and luminosity ($F = 1.037$; df = 9; $P = 0.449$) (Zar, 2010). In order to evaluate the daily pattern of resource collection of bees inside and outside the greenhouse, circular analyses were performed using Oriana 4.1 software (Kovach, 2011), using the Rayleigh test for the calculation of probabilities of distribution of bees throughout the day. To investigate whether the temperature differed inside and outside the greenhouse during the months the nests were confined, we conducted a paired t test (Zar, 2010).

## Results

### Internal conditions

Two nests (A and B, Fig 2) did not survive the conditions of confinement and died after about two months. To avoid their loss, after three months of confinement, two other nests (D and F, Fig 2) were removed from the greenhouse and placed outside due to the absence of cells in construction, the low number of workers and the absence of guard workers. One of these nests (F, Fig 2) survived and increased the number of workers to "high" after two months, but the other died (D, Fig 2), attacked by phorids (Diptera). Nests C and E (Fig 2) survived the conditions of confinement and remained in the greenhouse during the entire experiment, from March to October 2012, with cells in construction, guard workers and a high number of workers.

### Behavior of workers

Inside the greenhouse, flight activities of bees began after 22 days of confinement. However, workers were limited only to the removal of garbage from the colonies and many of them clashed against the ceiling or sides of the greenhouse, where they remained and ended up dying. This flight behavior toward the ceiling and sides of the greenhouse decreased over time during confinement but did not cease, even after the beginning of collection activities in the flowers.

The beginning of pollen collection from the tomato flowers occurred in late July, but there was only one record during that period and none in the following month. Intensive visits of workers to flowers (six records of pollen collection per nest) occurred in early September, after nearly six months of confinement, and the ceiling and sides of the greenhouse had been washed with soap and water. Bees landed on the anthers of a flower and bowed around or at the apex of the anthers cone to grab it (Fig 3A). Thus, they transmitted vibrations to the anthers through their thorax and legs to release the pollen. For the transfer of pollen from the body to the corbicula, some bees remained stuck to the base of the anthers cone by the legs and/or jaws, while others performed the transfer during flight or on other parts of the plant, such as the leaves and fruits (Fig 3B).

Fig 3. A) *Melipona quadrifasciata* worker collecting pollen in tomato flower; B) *M. quadrifasciata* worker transferring pollen from the body to corbicula on a fruit.

### Pattern of reflectance of the flowers

The lamps installed inside the greenhouse did not alter the flight behavior of the bees and did not influence the RGB and luminosity of tomato flower corollas. However, the patterns of chromatic ($F = 10.51$; df = 2; $P < 0.001$) and achromatic ($F = 8.44$; df = 2; $P = 0.001$) saturation of the flowers were significantly different inside and outside the greenhouse (Fig 4).

### Foraging activities

Before introduction, at Fazenda Experimental do Glória (FEG), the bees began to forage at around 5:40 am. The number of forager workers coming in and out of nests was not uniform throughout the day and the period of highest activity occurred at 6-8 am. The relative frequency of collection per resource varied considerably between the nests. Of the workers observed, 7.7-18.2% transported pollen, 17.0-81.4% nectar/water, 3.6-55.7% resin and 0.2-11.1% mud. The resources were not obtained uniformly throughout the day. Pollen collection occurred only in the morning, with a peak at 6-8 am. The highest mean frequency observed for nectar/water occurred between 6 and 7 am and the collection of this resource gradually decreased throughout the day. The peaks of collection of resin and mud were, respectively, at 8-10 am and at 8-11 am (Fig 5, Table 1).

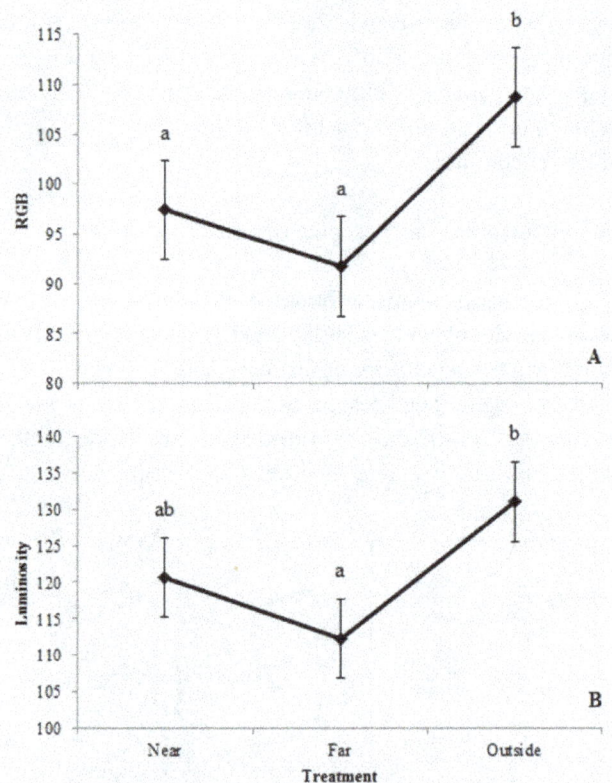

Fig 4. Mean values (± standard error) of patterns of chromatic (or RGB - A) and achromatic (or luminosity - B) saturation of tomato flowers in the different treatments (Near: inside the greenhouse and next to the lamps; Far: inside the greenhouse and far from the lamps; and Outside: outside the greenhouse) at Chácara Paraíso in Araguari. Distinct letters indicate significant differences between treatments.

Table 1. Mean time vector (with the number of observations (n) and probability values (p), according to Rayleigh test) for resources collection and total external activity throughout the day by *Melipona quadrifasciata* at Fazenda Experimental do Glória (FEG), in Uberlândia, and inside and outside the greenhouse at Chácara Paraíso, in Araguari.

| Resource | FEG | | | Inside | | Outside |
|---|---|---|---|---|---|---|
| | Nest B | Nest D | Nest F | Nest C | Nest E | Nest F |
| **Pollen** | 7:56 am (n = 31) ($P < 0.001$) | 6:51 am (n = 60) ($P < 0.001$) | 6:44 am (n = 67) ($P < 0.001$) | 7:40 am (n = 3) ($P = 0.036$) | 8:10 am (n = 6) ($P < 0.001$) | 7:29 am (n = 10) ($P < 0.001$) |
| **Nectar/water or Water** | 7:18 am (n = 329) ($P < 0.001$) | 7:44 am (n = 79) ($P < 0.001$) | 7:52 am (n = 117) ($P < 0.001$) | 4:00 pm (n = 3) ($P = 0.042$) | 2:45 pm (n = 6) ($P = 0.011$) | 7:33 am (n = 19) ($P < 0.001$) |
| **Resin** | 10:41 am (n = 15) ($P < 0.001$) | 9:50 am (n = 259) ($P < 0.001$) | 9:39 am (n = 136) ($P < 0.001$) | ----- | 11:30 am (n = 2) ($P = 0.144$) | 6:30 am (n = 2) ($P = 0.144$) |
| **Mud** | 9:00 am (n = 1) ($P = 0.512$) | 9:42 am (n = 50) ($P < 0.001$) | 10:38 am (n = 23) ($P < 0.001$) | ----- | ----- | ----- |
| **Total external activity** | 7:32 am (n = 801) ($P < 0.001$) | 9:00 am (n = 910) ($P < 0.001$) | 8:36 am (n = 719) ($P < 0.001$) | 11:43 am (n = 17) ($P < 0.001$) | 11:28 am (n = 39) ($P < 0.001$) | 7:45 am (n = 62) ($P < 0.001$) |

(-----) Insufficient data for analysis or resource not collected

Inside the greenhouse, the activity of bees began approximately at 7 am and was not uniform throughout the day, with the greatest movement of workers occurring between 5 and 6 pm for nest C, and between 12 and 1 pm for nest E. The relative frequency of collection per resource varied between the nests. From the workers observed over the entire period of confinement, 14.8-21.4% transported pollen, 37.5-60.7% water and 7.8-23.3% resin (cerumen). There was no collection of mud. Workers obtained pollen and water in a heterogeneous way throughout the day and there was no defined peak of resin collection. Pollen collection was limited to the morning, with a peak between 8 and 9 am. The highest mean frequency observed for water occurred between 5 and 6 pm (Fig 5, Table 1).

After the nests were removed from the greenhouse and placed in the outer area, it was possible to observe workers carrying pollen in their corbiculas on the following day. The movement of bees was not uniform throughout the day and peak activity occurred between 6 and 7 am. Of workers observed over the months, 31.4% carried pollen, 57.5% nectar/water and 7.0% resin. The resources were obtained in a heterogeneous way throughout the day, except resin. Pollen collection, limited again to the morning, and nectar/water collection had peaks between 7 and 8 am, and between 6 and 7 am, respectively. There was no collection of mud (Fig 5, Table 1).

*Temperature*

The temperature inside the greenhouse was significantly higher over the entire period of confinement of the nests of *M. quadrifasciata* ($t = 6.99$; df = 6; $P < 0.001$; Fig 6).

**Discussion**

The results of our experiments showed that acclimation of *Melipona quadrifasciata* to conditions inside greenhouses are colony-dependent and, besides foraging activities varied a little inside and outside the greenhouse, the daily period of foraging into the greenhouse decreased and bees took a long time to visit flowers consistently for pollen collection.

The foraging behavior of stingless bees is related both to factors intrinsic to the nest, including the ability to communicate and population size, and extrinsic factors, such as the abundance and distribution of resources in the environment and susceptibility to abiotic factors (Fidalgo & Kleinert, 2007). However, despite the methodological differences used in the introduction of the nests and the different conditions of confinement (size and structure of the greenhouses, crop type, etc), our results about the foraging behavior of *M. quadrifasciata* were similar to results found in other studies that evaluated the introduction of stingless bees in greenhouses (Del Sarto et al., 2005; Nunes-Silva et al., 2013). The workers concentrated pollen collection in the morning.

Inside the greenhouse, the foraging behavior of *M. quadrifasciata* showed little variation compared to the outside. However, stronger patterns (represented by smaller probability values, Table 1) in the external activities of bees were observed at FEG, before confinement. Additionally, the daily period of foraging in the greenhouse decreased, as well as for *N. perilampoides* (Cauich et al., 2004; Palma et al., 2008a), and bees visited flowers consistently for pollen collection only after approximately six months of confinement. The time required for acclimation to protected environments varies both between species and between colonies of the same species (Malagodi-Braga, 2002). However, this timing was probably determined by the presence of the diffuser film on the coverage of the greenhouse. As evidence of this, pollen collection observed in nests placed in the outer area started immediately after their removal from the greenhouse.

Due to the dispersant effect of plastic films used in greenhouses coverage, solar radiation is one of many environmental factors that can be changed by using protected crops (Schwengber et al., 1996). This could be evidenced via the different reflectance patterns found on tomato flowers inside and outside the studied greenhouse. This dispersant effect is favorable to plants, since the fraction of diffuse solar radiation is more effective for photosynthesis (Farias et al., 1993), but may have hindered bee orientation and identification of flowers by worker bees, affecting foraging activities inside the greenhouse and thereby delaying the beginning of pollen collection.

In relation to the external environment, global solar radiation (measured by luminous flux density) and diffuse solar radiation (multidirectional) are respectively lower and higher inside the greenhouse as a result of reflection and absorption by the material of the plastic coverage. In turn, this reflection and absorption are determined, for example, by the conditions of coverage at the time of use and dust deposition (Farias et al., 1993). This explains why visits to flowers by *M. quadrifasciata* workers intensified when the ceiling and sides of the greenhouse were washed with soap and water as part of the mana-gement of the protected tomato crop.

For orientation, bees use elements like the sun, polarized light, visual cues present in the environment and ultraviolet radiation (Kerr, 1973; Dyer, 1994; Briscoe & Chittka, 2001). The presence of clouds and fog cause dispersion of light rays and reduces the polarization signal (Shashar & Cronin, 1998). Thus, greenhouse cover can produce a similar effect to that promoted by clouds and restrict bee activity to a period in which light rays are less dispersed by the cover (Malagodi-Braga, 2002).

Besides the dispersant effect, the film ELL Diffuser Antivirus eliminates the entry of ultraviolet (UV) to hinder the vision of tomato pest insects (Electro Plastic, 2013). This could also have been an aggravating factor hindering the orientation of *M. quadrifasciata* forager workers inside the greenhouse. Although this species present a wide distribution throughout Brazil (Camargo & Pedro, 2012), foraging in places relatively poor in UV, such as under the canopies of dense tropical forests (Briscoe & Chittka, 2001), the nests were in an open environment at FEG and were accustomed to high UV exposure before introduction to the greenhouse.

In temperate areas, where bumble bees are largely used for greenhouse tomato pollination, showing high efficiency (Banda & Paxton, 1991; Morandin et al., 2001), the structure of the greenhouses are similar to those used in this study. However, in the case of bumblebees, it seems that the reduction in UV light can be compensated and as a result their visit to flowers is not affected by the type of greenhouse coverage (Dyer & Chittka, 2004).

In the present study, as in Del Sarto et al. (2005), the performance of *M. quadrifasciata* to conditions of confine-

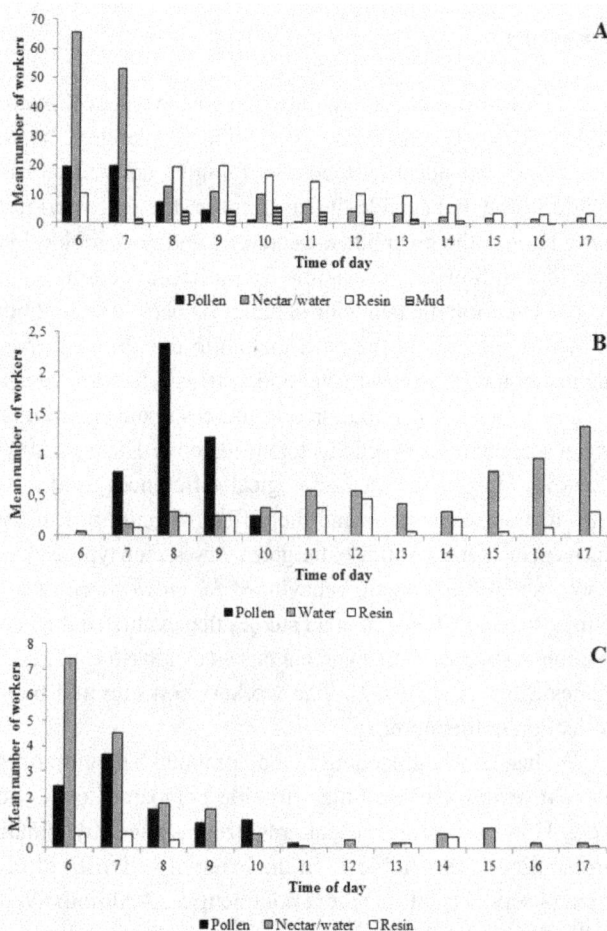

Fig 5. Mean number of workers for each resource collected throughout the day by *Melipona quadrifasciata* before introduction (A), at Fazenda Experimental do Glória, in Uberlândia, and inside (B) and outside (C) the greenhouse at Chácara Paraíso, in Araguari.

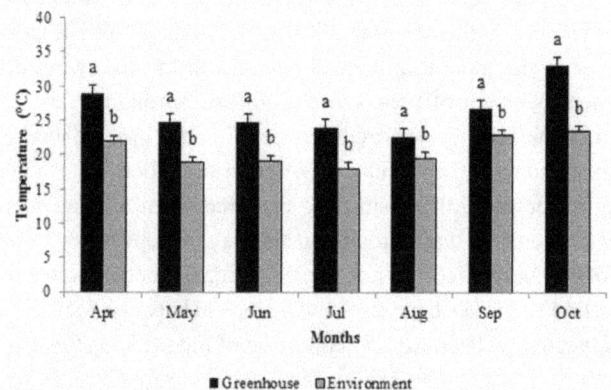

Fig 6. Mean values (± standard error) of temperature (°C) inside and outside (environment) the greenhouse during the period of confinement of nests of *Melipona quadrifasciata* at Chácara Paraíso in Araguari. Distinct letters indicate significant differences for each month.

ment proved to be colony-dependent. Studies have shown that stingless are tolerant to high temperatures and capable of cooling the inside of their colony by ventilation generated by beating their wings in the nest entrance. However, we believe that the high temperatures inside the greenhouse may have been an aggravating factor for colony development, since the intense heat decreases the density of larval food, causing the eggs to sink and death of larvae by drowning (Amano et al., 2000).

From results obtained in this study and considering the benefits that the use of stingless bees has provided to diverse crops grown in greenhouses (Cruz & Campos, 2009), the control of temperature and humidity inside greenhouses would be an important step. Another idea would be to exchange the plastic greenhouse coverings for materials that interfere less significantly in solar radiation, or, alternatively, a less drastic alternative would be to intercalate the plastic cover with such materials, allowing at least partial diffusion of solar radiation to occur in the greenhouse. For the farmer, taking into account the duration of the cycle of grape tomato in greenhouses, which is eight months, using stingless bees is not practical if these bees require a great deal of time to acclimate and initiate foraging activities. However, studies asses-sing the cost-benefit relationship of such structural changes in greenhouses must still be made and more information that will permit improvements in management techniques for stingless bees in greenhouses is needed.

## Acknowledgements

The authors are grateful to Conselho Nacional de Desenvolvimento Científico e Tecnológico (CNPq), to Fundação de Amparo à Pesquisa do estado de Minas Gerais (FAPEMIG) and to Coordenação de Aperfeiçoamento de Pessoal de Nível Superior (CAPES) for financial support, to the farmers Edson, Clóvis and João, who allowed and gave full support to this study, to Dr. Paulo Eugênio Alves Macedo de Oliveira for suggestions given to this study and to the biologists Isabel Farias Aidar and Jaqueline Eterna Batista for contributions in the field.

## References

Alves, K.A. & Rosa, R. (2008). Espacialização de dados climáticos do Cerrado mineiro. Horizonte Científico, 8: 1-28.

Amano, K., Nemoto, T. & Heard, T. (2000). What are stingless bees and why and how to use them as crop pollinators? A review. Japan Agricultural Research Quarterly, 34: 183-190.

Banda, H.J. & Paxton, R.J.(1991). Pollination of greenhouse tomatoes by bees. Acta Horticulturae, 288: 194-198.

Bispo dos Santos, S.A. (2008). Polinização em culturas de manjericão, Ocimum basilicum L. (Lamiaceae), berinjela, Solanum melongena L. (Solanaceae) e tomate Lycopersicon esculentum (Solanaceae) por espécies de abelhas sem ferrão (Hymenoptera, Apidae, Meliponini). Dissertation (MSc in Entomology), Universidade de São Paulo, Ribeirão Preto.

Bispo dos Santos, S.A., Roselino, A.C., Hrncir, M. & Bego, L.R. (2009). Pollination of tomatoes by the stingless bee Melipona quadrifasciata and the honey bee Apis mellifera (Hymenoptera, Apidae). Genetics and Molecular Research, 8: 751-757.

Briscoe, A.D. & Chittka, L. (2001).The evolution of color vision in insects. Annual Review of Entomology, 46: 471-510.

Buchmann, S.L. (1983). Buzz pollination in angiosperms. In: Jones, C.E. & Little, R.J. (Eds.), Handbook of Experimental Pollination Biology (pp. 73-113). New York: Scientific and Academic Editions.

Buchmann, S.L. & Hurley, J.P. (1978). A biophysical model for buzz pollination in angiosperms. Journal of Theoretical Biology, 72: 639-657. doi: 10.1016/0022-5193(78)90277-1.

Camargo, J. M. F. & Pedro, S. R. M. (2012). Meliponini Lepeletier, 1836. In: Moure, J. S., Urban, D. & Melo, G. A. R. (Eds.), Catalogue of bees (Hymenoptera, Apoidea) in the neotropical region – online version (Available in: http://www.moure.cria.org.br/catalogue).

Cauich, O., Quezada-Euán, J. J. G., Macias-Macias, J. O., Reyes-Oregel, V., Medina-Peralta, S. & Parra-Tabla, V. (2004). Behavior and pollination efficiency of Nannotrigona perilampoides (Hymenoptera: Meliponini) on greenhouse tomatoes (Lycopersicon esculentum) in subtropical México. Journal of Economic Entomology, 97: 475-481.

Cruz, D.O. & Campos, L.A.O. (2009). Polinização por abelhas em cultivos protegidos. Revista Brasileira de Agrociência, 15: 5-10.

Cruz, D.O., Freitas, B.M., Silva, L.A., Silva, E.M.S. & Bomfim, I.G.A. (2005). Pollination efficiency of the stingless bee Melipona subnitida on greenhouse sweet pepper. Pesquisa Agropecuária Brasileira, 40: 1197-1201.

Cuypers, J. (1968). Using honeybees for pollinating crops under glass. Bee World, 49: 72-76.

Del Sarto, M.C.L. (2005). Avaliação de Melipona quadrifasciata Lepeletier (Hymenoptera: Apidae) como polinizador da cultura do tomateiro em cultivo protegido. Dissertation (MSc in Entomology), Universidade Federal de Viçosa, Viçosa.

Del Sarto, M.C.L., Peruquetti, R.C. & Campos, L.A. O. (2005). Evaluation of the neotropical stingless bee Melipona quadrifasciata (Hymenoptera: Apidae) as pollinator of greenhouse tomatoes. Journal of Economic Entomology, 98: 260-266. doi: 10.1603/0022-0493-98.2.260.

Dyer, F.C. (1994). Spatial cognition and navigation in insects. In: Real, L.A. (Ed.), Behavioral mechanisms in evolutionary ecology (pp. 66-98). Chicago: University of Chicago Press.

Dyer, A.G. & Chittka, L.(2004). Bumblebee search timewithout ultraviolet light. Journal of Experimental Biology, 207: 1683-1688. doi:10.1242/jeb.00941.

Electro Plastic. (2013). E.L.V – Difusor Antivírus. Available at: <http://www.electroplastic.com.br/categorias/produto/42/e-l-v-difusor-antivirus.html>

Farias, J.R.B., Bergamaschi, H., Martins, S.R. & Berlato, M.A. (1993). Efeito da cobertura plástica de estufa sobre a radiação solar. Revista Brasileira de Agrometeorologia, 1: 31-36.

Fidalgo, A.O. & Kleinert, A.M.P. (2007). Foraging behavior of Melipona rufiventris Lepeletier (Apinae, Meliponini) in Ubatuba/SP, Brazil. Brazilian Journal of Biology, 67: 137-144.

Free, J.B. (1993). Insect pollination of crops. San Diego: Academic.

Heard, T.A. (1999). The role of stingless bees in crop pollination. Annual Review of Entomology, 44: 183-206. doi: 10.1146/annurev.ento.44.1.183.

Imperatriz-Fonseca, V.L., Saraiva, A.M., De Jong, D. (2006). Information technology and pollinators iniciatives. In: Imperatriz-Fonseca, V.L., Saraiva, A.M. & De Jong, D. (Eds.), Bees as pollinators in Brazil: assessing the status and suggesting best practices (pp. 20-20). Ribeirão Preto: Holos Editora.

Kerr, W.E. (1973). Sun compass orientation in the stingless bees Trigona (Trigona) spinipes (Fabricius, 1793) (Apidae). Anais da Academia Brasileira de Ciências, 45: 301-308.

Kovach, W.L. (2011). Oriana – Circular Statistics for Windows, ver. 4. Kovach Computing Services, Pentraeth, Wales, U.K.

Malagodi-Braga, K.S. (2002). Estudo de agentes polinizadores em cultura de morango (Fragaria x ananassa Duchesne – Rosaceae). Thesis (Doctorate in Sciences - Ecology Area), Universidade de São Paulo, São Paulo.

Malagodi-Braga, K.S., Kleinert, A.M.P. & Imperatriz-Fonseca, V.L. (2004). Abelhas sem ferrão e polinização. Revista Tecnologia e Ambiente, 10: 59-70.

McGregor, S.E. (1976). Insect pollination of cultivated crop plants. Washington: United States Department of Agriculture.

Michener, C.D. (2007). The bees of the world. Baltimore: The Johns Hopkins University Press.

Morandin, L.A., Laverty, T.M. & Kevan, P.G. (2001). Bumble bee (Hymenoptera-Apidae) activity and pollination levels in commercial tomato greenhouses. Journal of Economic Entomology, 94: 462-467.

Nunes-Silva, P., Hrncir, M. & Imperatriz-Fonseca, V.L. (2010). A polinização por vibração. Oecologia Australis, 14: 140-151. doi: 10.4257/oeco.2010.1401.07.

Nunes-Silva, P., Hrncir, M., Silva, C.I., Roldão, Y.S. & Imperatriz-

Fonseca, V.L. (2013). Stingless bees, Melipona fasciculata, as efficient pollinators of egg plant (Solanum melongena) in greenhouses. Apidologie, 44: 537-546. doi: 10.1007/s13592-013-0204-y.

Palma, G., Quezada-Euán, J.J.G., Meléndez-Ramirez, V., Irigoyen, J., Valdovinos-Nuñez, G.R. & Rejón. M. (2008a). Comparative Efficiency of Nannotrigona perilampoides, Bombus impatiens (Hymenoptera: Apoidea), and Mechanical Vibration on Fruit Production of Enclosed Habanero Pepper. Journal of Economic Entomology, 101: 132-138. doi: 10.1603/0022-0493(2008)101[132:CEONPB]2.0.CO;2.

Palma, G., Quezada-Euán, J.J.G., Reyes-Oregel, V., Meléndez, V. & Moo-Valle, H. (2008b). Production of greenhouse tomatoes (Lycopersicon esculentum) using Nannotrigona perilampoides, Bombus impatiens and mechanical vibration (Hym.:Apoidea). Journal of Applied Entomology, 132: 79-85.

Roselino, A.C., Santos, S.B., Hrncir, M. & Bego, L.R. (2009). Differences between the quality of strawberries (Fragaria x ananassa) pollinated by the stingless bees Scaptotrigona aff. depilis and Nannotrigona testaceicornis. Genetics and Molecular Research, 8: 539-545.

Roselino, A.C., Bispo dos Santos, S.A. & Bego, L.R. (2010). Qualidade dos frutos de pimentão (Capsicum annuum L.) a partir de flores polinizadas por abelhas sem ferrão (Melipona quadrifasciata anthidioides Lepeletier 1836 e Melipona scutellaris Latreille 1811) sob cultivo protegido. Revista Brasileira de Biociências, 8: 154-158.

Schwengber, F.E., Peil, R.M.N., Martins, S.R. & Assis, F.N. (1996). Comportamento de duas cultivares de morangueiro em estufa plástica em Pelotas – RS. Horticultura Brasileira, 14: 143-147.

Shashar, N. & Cronin, T.W. (1998).The polarization of light in a Tropical Rain Forest. Biotropica, 30: 275-285.

Silveira, F.A., Melo, G.A.R. & Almeida, E.A.B. (2002). Abelhas brasileiras: sistemática e identificação. Belo Horizonte: Fundação Araucária.

Zar, J.H. (2010). Biostatistical Analysis - Fifth Edition. New Jersey: Prentice-Hall/Pearson.

# 9

# Ant Communities along a Gradient of Plant Succession in Mexican Tropical Coastal Dunes

P Rojas[1], C Fragoso[1], WP Mackay[2]

1 - Instituto de Ecología A.C. (INECOL). Xalapa, México.

2 - University of Texas at El Paso, El Paso, USA.

**Keywords**
Ant assemblages, species richness, diversity, La Mancha, México

**Corresponding author**
Patricia Rojas
Instituto de Ecología A.C. (INECOL)
Carretera Antigua a Coatepec 351
El Haya, Xalapa 91070
Veracruz, México.
E-Mail: patricia.rojas@inecol.mx

**Abstract**

Most of Mexican coastal dunes from the Gulf of Mexico have been severely disturbed by human activities. In the state of Veracruz, the La Mancha Reserve is a very well preserved coastal community of sand dunes, where plant successional gradients are determined by topography. In this study we assessed species richness, diversity and faunal composition of ant assemblages in four plant physiognomies along a gradient of plant succession: grassland, shrub, deciduous forest and subdeciduous forest. Using standardized and non-standardized sampling methods we found a total of 121 ant species distributed in 41 genera and seven subfamilies. Grassland was the poorest site (21 species) and subdeciduous forest the richest (102 species). Seven species, with records in ≥10% of samples, accounted 40.8% of total species occurrences: *Solenopsis molesta* (21.6%), *S. geminata* (19.5%), *Azteca velox* (14%), *Brachymyrmex* sp. 1LM (11.7%), *Dorymyrmex bicolor* (11.2%), *Camponotus planatus* (11%) and *Pheidole susannae* (10.7%). Faunal composition between sites was highly different. Nearly 40% of all species were found in a single site. In all sites but grassland we found high abundances of several species typical of disturbed ecosystems, indicating high levels of disturbance. A species similarity analysis clustered forests in one group and grassland and shrub in another, both groups separated by more than 60% of dissimilarity. Similarity of ant assemblages suggests that deciduous and subdeciduous forests represent advanced stages of two different and independent successional paths.

## Introduction

Ants are social insects with important ecological functions. They influence ecosystems through soil bioturbation (Lobry de Bruyn & Conacher, 1990), predation of invertebrates (Gotwald, 1995) and mutualistic interactions with hundreds of plant species (Jolivet, 1996). Due to their high diversity, numerical and biomass dominance (Fittkau & Klinge, 1973, Brown, 2000) and sensitivity to environmental changes (Andersen, 1995), ants constitute an ideal group to inquire about patterns in community characterization. Coastal dunes are complex and very dynamic environments that have been shaped by biological and physical processes like water and wind action (McLachlan, 1991). Its high environmental heterogeneity is determined by distinct landforms and different plant communities (Martínez et al., 2004).

Ant communities from coastal environments have been poorly studied. However, the importance of these ecosystems for conservation of ants has been recently recognized (Howe et al., 2010). Communities from temperate and tropical coastal ecosystems are markedly different. In general, ant communities from temperate marine coasts have a low number of species and, independently of the number of sites included in a given locality, the number of species never surpass two dozen. For example, Boomsma and De Vries (1980) in The Netherlands recorded only three species in sparsely vegetated sand dunes and grasslands, whereas studies in successional gradients from pioneer vegetation to mature forests carried out in Finland (Gallé, 1991) and Spain (Ruano et al.,1995), recorded 19 and 24 species respectively. In temperate dunes, species richness, abundance and equitability increases along vegetation succession, with higher values being found in sites with a more dense plant cover (Boomsma & Van Loon, 1982; Ruano et al., 1995). Positive correlations have also been observed between composition of ant assemblages and a more complex habitat (Boomsma & Van Loon, 1982). These attributes of ant assemblages, however, are relatively independent of diversity and composition of vegetation, suggesting that in temperate coastal systems, plant succession stages are not coordinated with successional stages of ant communities (Gallé, 1991).

In tropical coastal dunes the number of ant species is more variable. Studies in comparable vegetation mosaics, recorded from 22 to 92 species in Cuba (Fontenla, 1993, 1994), Mexico (Durou et al., 2002), and Brazil (Bonnet & Lopes, 1993; Texeira et al., 2005; Vargas et al., 2007; Cardoso et al., 2010). In these systems species richness varies between habitats, with higher values being found in more complex and heterogeneous habitats (Fontenla, 1993; Durou et al., 2002; Vargas et al., 2007). Faunal composition is another attribute of tropical coast ant assemblages that strongly varies with the type of vegetation (Fontenla, 1993; Cardoso et al., 2010) and plant cover (Dorou et al., 2002).

Most of natural undisturbed coastal dunes of the Mexican littoral zone of the Gulf of Mexico have disappeared due to extensive farming management, human settlements and touristic activities (Moreno-Casasola, 2006). However, the state of Veracruz still harbors some well-preserved sites of coastal dunes that can enter inland up to 3 km. One of these sites is found within the Ecological Reserve of the Centro de Investigaciones Costeras La Mancha (CICOLMA). In the dunes of La Mancha environmental gradients related to the force of the wind, sand movement, and depth of water table (ultimate mediated by topography), determine the establishment of different plant communities (Moreno-Casasola & Vázquez, 2006). In this zone the vegetation follows a successional gradient, from pioneer plants and grasslands growing on the beach and young dunes, to deciduous and subdeciduous tropical forests established on older dunes. This last community constitutes the last remnant of this kind of forest in the Gulf of Mexico growing in sandy soils (Moreno-Casasola & Travieso, 2006).

Several studies of ants have been conducted at La Mancha, including numerous aspects of plant-ant interactions (Rico-Gray, 1989, 1993; Mehltreter et al., 2003), evaluation of some invasive ants (Fragoso & Rojas, 2009), records in checklists (Rojas 2001, 2011) and taxonomical studies (Mackay et al., 2004). So far, no studies characterize the complete ant community in any vegetation type of this site have been published.

The main objective of our study was to describe the ant communities of La Mancha in four types of vegetation that represent a gradient of plant succession on coastal dunes. Communities were characterized considering species richness, diversity, abundance and species composition. We hypothesized that richness and diversity would increase along the vegetation successional gradient; that each stage will have different faunal composition, and that differences will be higher between early and late successional stages.

## Material and methods

### Study area

The study area is located at the Centro de Investigaciones Costeras La Mancha (CICOLMA) in the coast central region of Veracruz State (96°22'40" W; 19°36'00" N) with an altitudinal range of 0-80 m elevation. The CICOLMA field station covers a total area of 83 ha. The zone is geologically young, shaped by Miocene volcanic activity and by Quaternary deposits (Geissert, 1999). The weather is characterized by an annual average temperature of 25°C and an annual precipitation of 1500 mm, with the large amount of rains (78%) occurring during the rainy season of June-September. Soils are unstructured luvic and calcaric arenosols (Travieso & Campos, 2006).

Sampling sites were established on four different successional stages of coastal dune vegetation: grassland, shrub, deciduous forest and subdeciduous forest (Fig 1). Grasslands and shrubs are found in semi stabilized dunes, with grasslands being found in upper dry parts of the dunes and shrubs on humid depressions, where water table is higher. Tropical forests grow over stabilized dunes with subdeciduous forests being established on flat sites or sites with level relief, and deciduous forests located on more steep sites (Moreno-Casasola & Travieso, 2006).

Grassland (G), (Fig 1A) - With a plant cover of nearly 40%, this community includes grasses and short shrubs that alternate with open spaces of bare sand. Common species are *Trachypogon plumosus* (Humb. & Bonpl. ex Willd.) Nees, *Andropogon glomeratus* (Walter) B.S.P. and *Chamaecrista chamaecristoides* (Colladon) Greene (Moreno-Casasola & Travieso, 2006).

Fig 1. La Mancha plant physiognomies sampled in this study. A – Grassland, B – Shrub, C – Deciduous forest, D – Subdeciduous forest.

Shrub (S), (Fig 1B) - Medium size (2-3 m high) closed canopy shrub with a plant cover of 80% and with some isolated tress. Predominant species are shrubs *Randia laetevirens* Standl., *Pluchea odorata* (L.) Cass. and *Mimosa chaetocarpa* Brandeg. (Moreno-Casasola & Travieso, 2006).

Deciduous forest (DF), (Fig 1C) - Trees reach 12m height, with many deciduous species. Most common species are *Bursera simaruba* (L.) Sarg., *Coccoloba barbadensis* Jacq. and *Ocotea cernua* (Nees) Mez. The understory stratum is dominated by *Crossopetalum uragoga* (Jacq.) Kuntze, *Chiococca alba* (L.) Hitchc. and *Randia aculeata* L.; grasses and herbs are almost absent (Castillo, 2006).

Subdeciduous forest (SF), (Fig 1D) – The canopy (>20m height) is dominated by *Brosimum alicastrum* Sw., *Ficus cotinifolia* Kunth, *Cedrela odorata* L. and *Enterolobium cyclocarpum* (Jacq.) Griseb., whereas the lower canopy (6-15 m) is characterized by *Erithroxylum havanense* Jacq., *Nectandra salicifolia* (Kunth) Nees and *Ocotea cernua*. The species *Crossopetalum uragoga, Schaefferia frutescens* Jacq. and *Hippocratea celastroides* Kunth predominate in the shrub stratum (Castillo, 2006).

*Ant sampling*

Our sampling was performed only in one plot per vegetation type (no replicates); however we consider that the amount of traps and eight different sampling methods made this study valid for site comparisons.

Standardized sampling (SS). Sampling of grassland, deciduous forest and subdeciduous forest was performed during 1992 whereas shrub was sampled two years later (1994). In each site, a 20x20m (400m²) plot was delimited and located at least 30m inside the respective vegetation type; the plot was a grid each five meters. Sampling was made in the dry (May) and rainy (October) seasons. A total of 588 samples were obtained, using the following standardized sampling methods:

Pitfall traps (based on Greenslade, 1964) (200 traps) - In each site and plot 25 plastic container traps were set at distances of five meter intervals in a grid pattern. Each container, with a volume of 250 ml and a diameter of 8 cm, was buried at ground level and filled 3/4 with 70% ethanol and a small quantity of commercial detergent. Traps remained in the field five days.

Subterranean baits (200 baits) - Following the same grid pattern, 25 subterranean baits were set in each site and plot. Each bait was buried at 20-30 cm depth, inside a plastic vial of 5 ml volume and with several holes in the walls (Mackay & Vinson, 1989); tuna fish and a mixture of honey and oatmeal were used as bait. Traps were left buried for 48 hours.

Surface baits (72 baits) (see Bestelmeyer et al., 2000) - In each site and plot nine surface (three per row) baits were placed along three parallel rows separated 10 meters. In each row baits were separated 10 m each. A teaspoon of bait (the same type used in subterranean sampling) was placed within a

10cm diameter plastic Petri dish, deposited aboveground and left for 5 hours (from 10:00 to 15:00 h). Attracted ants were collected at one hour intervals.

Arboreal baits (70 baits) - This method was used only in the two forests, because trees were nonexistent or very scarce in G and S plots, respectively. Baits (same type used in former methods) were placed inside plastic containers of 150ml and tied at 1.5m height in different tree species with a minimal diameter of 20cm. Baits were left for 5 hours (from 10:00 to 15:00h); attracted ants were collected at one hour intervals. Twenty arboreal baits were placed in DF and 15 in SF.

Leaf-litter samples (30 samples) (see Bestelmeyer et al., 2000) - This method was used in all sites except G, where no litter stratum was found. In each site five samples of litter (1m²) were collected and processed with Berlese funnels until the litter was dry. Samples were taken from the center and two meters inside the four corners of each plot.

Hand sampling (16 samples) - In all sites, search and capture of ants was performed by two persons during 4h (8 hours per site). Sampling included vegetation (epiphytes, foliage and hollow stems) and soil (litter, first centimeters of soil and within and under decaying logs).

Non standardized sampling (NSS). It refers to any sample of ants obtained out of the SS protocol by hand sampling.

Ants were identified to genus level using Bolton (1994); species were determined using specialized publications and revisions or by comparison with reference material from the ant collections of W.P. Mackay and the Laboratorio de Invertebrados del Suelo (INECOL, Xalapa). Voucher specimens of all species were deposited in the latter Collection. Nomenclature follows Bolton et al. (2007).

*Data analysis*

Seasonality - Differences in the number of species captured in rainy and dry season were analyzed with parametric t-test for dependent samples (when normality was fitted) and nonparametric Wilcoxon matched pair test. Analyses were performed for each site and the following SS methods: pitfall traps, subterranean baits, surface baits and arboreal baits. All these tests were performed using STATISTICA (Statsoft, 1999). Seasonality differences for a given site (SS + NSS samples) were compared considering 95% confidence intervals derived from rarefaction curves (Mao Tau, EstimateS program version 9, Colwell, 2013).

Abundance - In order to avoid over estimation of those species with large foraging areas or legionary habits, abundance was calculated as presence/absence data (occurrence) from SS samples. For a given species, abundance was calculated as frequency of occurrence (FO=$n_i$/N x100, where $n_i$ is the number of samples where species i was found, and N is the total number of samples). The number of occurrences in samples was considered as an indirect measure of the relative abundance of each species in each site (Gottelli & Colwell, 2011) and can be used to estimate abundance distribution of species along gradients (Andersen, 1997).

Species richness and diversity - Considering SS data, species accumulation curves were generated as rarefaction curves (Mao Tau) and compared to estimated richness (Chao 2 estimator) (Gotelli & Colwell, 2011) using EstimateS program (version 9, Colwell, 2013). Diversity was calculated using Shannon and Simpson indices; differences between vegetation types were evaluated using the t-tests of Hutcheson (1970) and Brower et al. (1998) for, respectively, Shannon and Simpson indices. These analyses were performed using PAST program (version 3, Hammer et al., 2001).

Similarity - Similarity of the four sites was calculated following two approaches. In the first one a hierarchical similarity analysis of sites was made with the presence/absence data (SS and NSS sampling) of each species in each site (121 species per four sites). Variables were associated and clustered using Kulczinski index (incidence-based) and UPGMA, respectively and represented as a similarity dendrogram; these analyses were performed using PATN software (Belbin, 1989). Incidence-based Sorensen index was also calculated for comparative purposes. In the second approach we used FO of each species obtained from SS sampling. The resultant species per site matrix (97 x 4) was analyzed using the Morisita-Horn index (abundance-based), from EstimateS (version 9, Colwell, 2013).

Vertical distribution - Each species was assigned to one of the following vertical categories: soil only, vegetation only, both. Assignment was made considering the more frequent localization of nests (personal observations) and by consulting natural history information available in current literature.

## Results

Considering that no significant differences in number of species were found between dry and rainy seasons for all sites and sampling methods, except for pitfall trap data from G and SF (see Table 1), and that similar numbers of species were collected during dry and rainy seasons (78 and 77 species, respectively) our results correspond to clumped data from both seasons. In the four sites 121 ant species of 41 genera and seven subfamilies were found. With the standardized sampling 97 species (80.2 % of total) were obtained from 1440 species occurrences from all sites; the remaining 24 species were captured using non standardized sampling. Subfamilies with more species were Myrmicinae (57 species) and Formicinae (23); the genera with more species were *Camponotus* (16 species), *Pheidole*, *Pseudomyrmex* (14 each), *Solenopsis* and *Crematogaster* (5 each) (Appendix 1).

The following seven species, with records in ≥10% of samples, accounted for 40.8% of total species occurrences: *Solenopsis molesta* (21.6%), *S. geminata* (19.5%), *Azteca velox* (14%), *Brachymyrmex* sp. 1LM (11.7%), *Dorymyrmex bicolor* (11.2%), *Camponotus planatus* (11%) and *Pheidole susannae* (10.7%). Twenty eight species (23%) were repre-

**Table 1**. Average values and standard deviations (in brackets) in the number of ant species found in dry ($x_d$) and rainy ($x_r$) seasons at La Mancha with four standardized sampling methods (SS). The last column shows values (and 95% confidence intervals) including all species in dry ($N_d$) and rainy ($N_r$) seasons collected by SS and non-standardized methods, NSS. Significant differences between seasons are in bolds (ns = no significant differences).

| | Pitfall (t-test) | Subterranean (Wilcoxon) | Superficial (t-test) | Arboreal (Wilcoxon) | SS + NSS |
|---|---|---|---|---|---|
| Grassland | $X_d$=3 (1.6) $X_r$=1.6 (1.3) n=25 *P*=**0.0007** | $X_d$=0.52 (0.59) $X_r$=0.6 (0.65) n=25 ns | $X_d$=2.3 (1) $X_r$=2.1 (1.3) n=9 ns | Not used | $N_d$=14 ± 5.1 $N_r$=10 ± 3.6 |
| Shrub | $X_d$=6.2 (1.7) $X_r$=6.4 (2.5) n=25 ns | $X_d$=0.16 (0.37) $X_r$=0.2 (0.41) n=25 ns | $X_d$=3.8 (1.5) $X_r$=3.8 (1.3) n=9 ns | Not used | $N_d$=29 ± 6.3 $N_r$=38 ± 8.4 |
| Deciduous forest | $X_d$=2.7 (1.6) $X_r$=2.5 (1.3) n=25 ns | $X_d$=0.88 (0.53) $X_r$=0.80 (0.41) n=25 ns | $X_d$=4.1 (1.7) $X_r$=4.7 (1.7) n=9 ns | $X_d$=1.9(1.1) $X_r$=2(1) n=20 ns | $N_d$=26 ± 8.5 $N_r$=34 ± 8.6 |
| Subdeciduous forest | $X_d$=2.5 (1.7) $X_r$=4.2 (1.5) n=25 *P*=**0.001** | $X_d$=0.36 (0.76) $X_r$=0.44 (0.58) n=25 ns | $X_d$=3.8 (1.9) $X_r$=3.2 (1.6) n=9 ns | $X_d$=1(0.8) $X_r$=2(1.6) n=15 ns | $N_d$=50 ± 8.8 $N_r$=44 ± 9.5 |

sented only by a single record.

Species richness - When considering only standardized sampling (SS), G was the site with the lowest species richness (15 species) whereas SF forest was the richest (62). DF and S presented an intermediate value of richness (42 species each). Estimates of expected species richness for each site (Chao2 richness estimator) showed that G was the best sampled site (77% of expected number of species), followed by SF and DF (64 and 59%, respectively). S turned to be the worst sampled site (56%) (Fig 2 and Table 2). When species captured with non-standardized sampling (NSS) are added, sites followed the same order of species richness, although percentages of increase varied for each site. Species richness values of G and SF were closer to those predicted by Chao2 index, even exceeding the estimated total number of species (Table 2).

Abundance - The lowest number of occurrences was found in G (197, 13.7% of total). S was the site with the highest number of occurrences (454, 31.5% of total); however, only with pitfall traps data was this value significantly different from the other sites (ANOVA, $F_{3,196}$=52.88, $P < 0.0001$, Tukey HSD test; $X_S$= 6.30, $SE_S$= 0.31; $X_G$= 2.34, $SE_G$= 0.22; $X_{DF}$= 2.60, $SE_{DF}$= 0.21; $X_{SF}$= 3.32, $SE_{SF}$= 0.25). The two forests showed intermediate occurrences values (deciduous: 393, 27.3%; subdeciduous: 396, 27.5%).

On the basis of the five most abundant species from each site (Table 3) communities were very different. Noteworthy in G these five species, characterized by the presence of three *Dorymyrmex* species, were not found in the two forests. Al-

though *Dorymyrmex* sp. aff. *flavus* and *D. bicolor* appear also in S, they have lower abundances. In G and S these five species accounted for 82.7% and 45.8% of species occurrences, respectively. In DF this group represented 56.5%, with the arboreal *Azteca velox* the more abundant species. In SF forest this group includes only ground ants which accounted 47.7%, with four of these species characteristic of disturbed places. *Solenopsis geminata* and *S. molesta* always appear in S, DF and SF among the five more abundant species.

Diversity - As expected, lower and higher diversity values were respectively observed in G and SF; however, diversity in S was higher than in DF (Shannon and Simpson indices, Table 4). In paired comparisons, all sites significantly differ in Shannon and Simpson indices (Table 5), excepting differences between S and SF with Simpson index. Rank abundance plots (Fig 3) also indicate that G was the less diverse assemblage (higher slope and shorter line) and that SF was the more diverse. Even that S and DF forest had the same number of species (SS sampling) the curve of S had a less steep slope indicating a higher evenness in the abundance of their species. Long curves observed in SF, DF and S reflect presence of many species with very low abundances.

Faunal composition - Appendix 1 (SS and NSS data) show the list of all species found at La Mancha, suggesting at first glance strong differences in faunal composition between sites. A similarity analysis of species presence/absence (Kulczinski index) clustered forests in one group and G and S in another, both groups being separated by more than 60%

**Table 2**. Observed (SS and SS+NSS) and estimated (Chao2) species richness of ants in the four sites. SS= standardized sampling; NSS= non standardized sampling.

|  | SS | SS+NSS | Chao2 |
|---|---|---|---|
| Grassland | 15 | 21 | 19.5 |
| Shrub | 42 | 47 | 74.7 |
| Deciduous forest | 42 | 50 | 70.9 |
| Subdeciduous forest | 62 | 102 | 96.3 |
| Total | 97 | 121 |  |

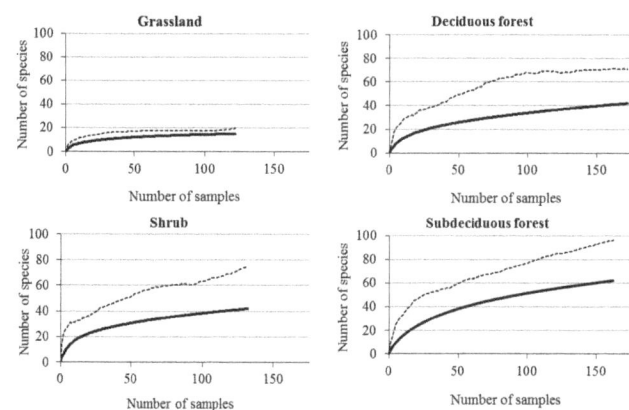

**Fig 2**. Species accumulation curves of ant species for each site. Solid lines correspond to rarefaction curves (Mao Tau) of observed species richness; dashed lines correspond to Chao2 predicted species richness.

**Table 3**. Abundance expressed as percentage of occurrences in samples (FO) of the five most important ant species in each site (in bold). FO values of these species in the other sites are shown for comparison; a dash means absence of species. G = Grassland; S = Shrub; DF = Deciduous forest; SF = Subdeciduous forest.

|  | G (N=122) | S (N=132) | DF (N=172) | SF (N=162) |
|---|---|---|---|---|
| *Azteca velox* | - | 5.3 | **40.7** | 3.7 |
| *Dorymyrmex bicolor* | **42.6** | 10.6 | - | - |
| *Dorymyrmex smithi* | **22.9** | - | - | - |
| *Dorymyrmex* sp. aff. *flavus* | **10.6** | 0.7 | - | - |
| *Forelius pruinosus* | **19.7** | 15.1 | - | - |
| *Brachymyrmex* sp. 1 | **37.7** | 9.1 | 5.1 | 0.6 |
| *Camponotus planatus* | 1.6 | 13.6 | **22.7** | 3.7 |
| *Monomorium ebeninum* | - | **43.9** | - | - |
| *Pheidole punctatissima* | - | - | 1.2 | **23.4** |
| *Pheidole* sp. 11LM | - | **23.5** | - | 5.5 |
| *Pheidole* sp. 5LM | - | - | **18.6** | 1.8 |
| *Pheidole susannae* | - | 5.3 | - | **34.6** |
| *Solenopsis molesta* | 0.8 | **29.5** | **30.8** | **21.0** |
| *Solenopsis geminata* | 8.2 | **39.4** | **16.3** | 15.4 |
| *Tetramorium spinosum* | - | **21.2** | - | - |
| *Wasmannia auropunctata* | 0.8 | - | 7.0 | **22.2** |

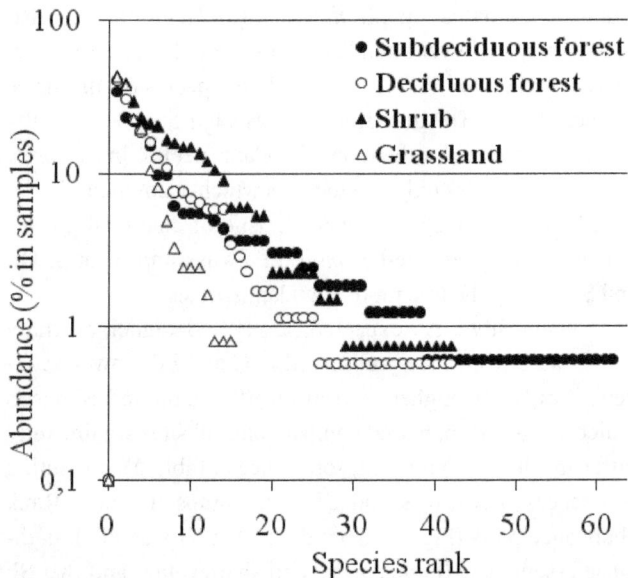

**Fig 3**. Rank-abundance plot of ant species in each site. Abundance is expressed on a log scale.

of dissimilarity (Fig 4). Similar results were obtained when similarity was calculated with Sorensen index (Table 5). Conversely, the Morisita-Horn index indicated a higher similarity of DF and SF with S, leaving G as an isolated site. The number of species considered in Morisita analysis was lower, because abundance data was obtained from standardized sampling only (Table 5).

**Table 4**. Ant diversity of the four sites studied obtained with two diversity indices. Data from standardized sampling (SS).

|  | Shannon (H′) Mean (SD) | Simpson (1/D) Mean (SD) |
|---|---|---|
| Grassland | 2.06 (0.06) | 6.09 (0.04) |
| Shrub | 3.06 (0.04) | 16.30 (0.09) |
| Deciduous forest | 2.89 (0.05) | 12.29 (0.09) |
| Subdeciduous forest | 3.35 (0.06) | 17.92 (0.17) |

Site specificity - At La Mancha 48 ant species (ca. 40% of total) were found in a single site. SF had the largest number of exclusive species (39, 32.2%), whereas DF and S had four exclusive species each (3.3%) and G only one species (0.8%). In contrast number of ubiquitous species (those found in all sites) was very low. Only six species (5%) were found in the four vegetation types: *Brachymyrmex* sp. 1LM, *Camponotus planatus*, *C. atriceps*, *Cyphomyrmex rimosus*, *Solenopsis geminata* and *S. molesta* (Appendix 1).

Vertical distribution - Soil-nesting species (69, 57%) dominated over plant-nesting species (48, 40%), with only four species nesting in both strata (3%) (Appendix 1). Five species in G were plant-nesting ants, but were low in abundance, and always captured with pitfall traps and soil baits; the remaining species (17, 76%) were soil dwellers. In S, very similar numbers of species nesting in soil (23, 49%) and vegetation (22, 47 %) were found, with two species nesting in both strata (4%). A similar situation was observed in DF, 23

species (46%) nesting in soil, 25 (50%) in vegetation and two species (4%) in both strata. Finally SF forest had, respectively, 56 (55%) and 42 (41%) soil and plant-nesting species, with four species (4%) inhabiting both strata.

**Discussion**

In order to capture as many species as possible, ant communities were sampled in two seasons. Sampling was not designed with the aim to compare seasonal patterns.

Accordingly, we conducted some statistical tests to confirm that all the SS data can be grouped. In general no significant differences were observed between rainy and dry season in the amount of species per sampling method and by site, excepting pitfall traps of G and SF. This can be explained by the dependence of this method on the foraging activity of ants (Bestelmeyer, 2000).

Total species richness – The number of ant species en-

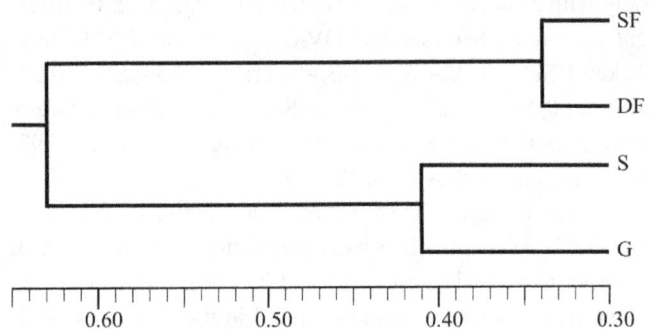

**Fig 4**. Similarity between sites in function of presence/absence of ant species. Scale shows values of dissimilarity (Kulczinski index). G = grassland, S = shrub, DF = deciduous forest, SF = subdeciduous forest.

countered in the four vegetation types studied at La Mancha was notably high, considering the relatively small area of the reserve (83ha, Moreno-Casasola & Monroy, 2006). Moreover, this number of species (121) make up nearly 34% of the total number of species recorded in the state of Veracruz (Rojas, 2011) and 13.7 % of all Mexican ant species (Vázquez-Bolaños, 2011). Comparatively, similar studies completed in tropical coastal ecosystems recorded lower values of species richness. For example in the Brazilian "restinga" Cardoso et al. (2010) reported 71 species along one transect of 6.5 km length, whereas Vargas et al. (2007) recorded 92 species; similarly in Mexico Durou et al. (2002) found 96 species. The highest values found at La Mancha can be explained because our study included two well-developed tropical forests. In spite that Vargas et al. (2007) also sampled a tropical forest, their ant richness values were still lower than in La Mancha.

Species richness and diversity along the vegetation gradient - The four studied sites represent a successional plant gradient established on sand dunes, which also entails the stabilization of dunes. Accordingly, the grassland with 21 ant species and low diversity values, represent the first stage in dune succession and it is the less complex environment with

**Table 5**. Species similarity values between grassland (G), shrub (S), deciduous forest (DF) and subdeciduous forest (SF), calculated with three indices. Diversity differences between sites are also indicated: †= significant differences ($P < 0.005$) comparing Simpson index (Brower et al., 1998). *= Significant differences ($P < 0.01$) comparing Shannon index (Hutcheson, 1970).

|            | Kulczinski | Sorensen | Morisita-Horn |
|------------|------------|----------|---------------|
| G vs S†*   | 0.41       | 0.50     | 0.25          |
| G vs DF†*  | 0.76       | 0.20     | 0.09          |
| G vs SF†*  | 0.65       | 0.19     | 0.05          |
| S vs DF†*  | 0.63       | 0.37     | 0.39          |
| S vs SF*   | 0.49       | 0.44     | 0.34          |
| DF vs SF†* | 0.34       | 0.58     | 0.37          |

only herbaceous vegetation alternating with areas of bare sand. In addition no litter layer exists and temperature differences between high and low coverture sites can be up to 11°C (Moreno-Casasola & Travieso, 2006); thus scarcity of nesting sites for ants is not only restricted to vegetation, but also occurs in the soil. Non structured sandy soils represent an unstable substrate for ants, and they would need to frequently reconstruct nest chambers and galleries (Lubertazzi &Tchinkel, 2003). The higher diversity and number of species (42) in the shrub zone, the second successional stage, can be explained by the presence of both herbs and bushes. These strata produce a litter layer that, even scarce as it was, offers suitable microsites for the arrival and settlement of more ant species.

The two forests correspond to last successional stages, and although they have a well-developed arboreal, bush and litter strata, their plant diversity and environmental conditions are not the same (Castillo, 2006) as was indicated by ant diversity and species richness. By being located in upper and steep places of stabilized dunes, deciduous forest is more exposed to wind, with a higher water runoff, and is consequently dryer. This translates into less plant cover, smaller tree height and fewer epiphytes. Conversely the subdeciduous forest is located in a more humid flat and wind protected place; this causes a higher plant cover, taller trees and more epiphytes (Novelo, 1978). Correspondingly, the more stable subdeciduous forest harbor twofold ant species richness (102 species) than the stressed deciduous forest (50).

It is widely recognized that in most of habitats, plant communities determine the physical structure of environment and therefore have a strong influence over the distribution and interactions of animal species (Lawton, 1983; Rosenzweig, 1995; Tews et al., 2004). Several studies completed in temperate coastal dunes (Boomsma & Van Loon, 1982; Dauber and Wolters, 2005) and in tropical Brazilian "restinga" (Vargas et al., 2007) have shown that more heterogeneous environments correspond to late successional stages and harbor a more diverse fauna and a higher species rich ant communities. In general all of these studies conclude that this pattern is due to the higher amount of microhabitats and microclimates that in turn produces more availability of food and nest sites. The results

of this study partially support our original hypothesis related to an increase of ant species richness along the plant successional gradient (G→S→DF→SF). Sites with the lower and higher species richness were, respectively, G and SF; however no significant differences in species richness were observed between the intermediate gradient stages, with estimated richness (Chao 2) very similar in S and DF, and even higher in the former. Diversity values showed also the same pattern.

On the basis of these results we propose that plant succession at La Mancha has not been unidirectional, but that two independent paths have occurred after the development of shrubs from grasslands: the first path would be the successional change of shrubs into deciduous forest (S→DF) in more stressed environmental conditions; alternatively the second path would be the change of shrubs towards subdeciduous forest (S→SF) in more stable environments.

Faunal composition - Through coastal plant succession, faunal composition of ant assemblages varies between temperate and tropical ecosystems. In temperate sites, species assemblages are very similar across vegetation physiognomies, as it has been shown by studies in coastal dunes (Gallé, 1991; Ruano et al., 1995) and grasslands (Zorilla et al., 1986; Dauber & Wolters, 2005). Independently of the successional stage, ant assemblages from these sites have not shown different number of species; instead the abundances of each species changes across the gradient. On the other hand, ant communities from tropical ecosystems show larger differences across vegetation successions. At La Mancha, the four studied vegetation types presented different species assemblages as has been observed in other coastal dune vegetation studies (Fontenla, 1993; Durou et al., 2002; Cardoso et al., 2010).

Considering its faunal composition the studied sites were divided into two separated groups. The first group included sites of earlier stages of succession (G and S) whereas the second one grouped the two forests (DF and SF) corresponding to late successional stages. From the point of view of vegetation, at La Mancha grasslands and shrubs also comprises a well differentiated group from deciduous and subdeciduous forests, sharing only 14% of plant species (Castillo & Travieso, 2006). Ant membership in the first group was defined by the share of eight species typical of dry environments, in spite of differences in the number of species. Among them *Dorymyrmex bicolor* and *Forelius pruinosus* are well adapted to xeric conditions and nest and forage on the soil of sunny sites, especially in grasslands (Shattuck, 1992). *Tetramorium spinosum* was another shared species widely distributed in arid zones of Mexico, that also nest in exposed soil (Rojas & Fragoso, 1994). Although they nest only in S, other shared species such as *Atta cephalotes* and *Crematogaster crinosa* were found foraging in both G and S habitats; this foraging strategy has been also recorded in other coastal environments, where ants living in adjacent forests use dunes and beaches as foraging areas (Ruano et al., 1995). The remaining two species shared by G and S were *Pseudomyrmex brunneus* and *P. ejectus* which have been reported occurring sympatrically and nesting in dead twigs of woody and herbaceous plants (Ward, 1985). We were unable to find nests of these species, but we observed them foraging in soil and plants at both sites. It remains to be demonstrated whether or not these species are adapted to live in harsh environments. Each site, nevertheless, has some exclusive species. Whereas G has only one unique species, the abundant ant *Dorymyrmex smithii* which nests under dead stems of grasses, S had four exclusive species (*Neivamyrmex rugulosus, Nesomyrmex wilda* and two unidentified *Pheidole*), all very low in abundance.

The second group clustered both forests on the basis of 27 shared species (see Appendix 1). The presence of a well-developed arboreal stratum determines that more than 50% of these species nest in trees.

In spite of their faunal similarities, both forests were separated by species richness and by the amount of exclusive species. While SF contained 39 exclusive species (38% of their 102 species), in DF only four species (all low abundant) were exclusive (8% of their 50 species). Thus, ant fauna of DF can be considered as an impoverished subset of SF with more microhabitats available to ants.

Interestingly in all sites, but G, ants typical of disturbed environments were found in high abundances. This was the case of *Solenopsis molesta*, a generalist soil-nesting species (Mackay & Mackay, 2002), *S. geminata*, commonly found in disturbed ecosystems of the Neotropics, and which have been found even penetrating tropical forests (Risch & Carroll, 1982; Taber, 2000), and *Pheidole susannae* which has been reported in disturbed habitats throughout the Neotropics (Wilson, 2003).

Unexpectedly, the higher number of ant species associated to disturbance was recorded in SF, currently considered as a functional forest in the last stages of succession (Castillo, 2006). An explanation of this finding could be related that in the past botanists recognized this site as a strongly disturbed secondary forest (Novelo, 1978); moreover Gomez-Pompa (as cited in Paradowska & Moreno-Casasola, 2006) suggests that presence of useful trees is an indication that this forest underwent high disturbances in the past, even being used as orchard by prehispanic people.

Vertical distribution - Our results showed that 97% of species were found associated with a single vertical stratum (soil or vegetation) in agreement with other studies which show a high vertical segregation in ants (Bruhl et al., 1998; Yanoviak & Kaspari, 2000). In the three sites with developed plant strata, vertical segregation was nearly 50%. Considering that in our sampled forests canopy ants were under-sampled, this proportion should change once a detailed sampling of canopy is undertaken. Remarkably, the only four species found nesting both in soil and vegetation were tramp and/or invasive species: *Paratrechina longicornis, Pheidole punctattissima, S. geminata* and *Tetramorium bicarinatum* (Kempf, 1972; McGlynn, 1999; Wetterer, 2009).

## Conclusions

Ant communities of tropical coastal environments have been, compared to other ecosystems, poorly studied in spite of their fragility and high risk of change due to climate change. In Mexico this is the first study that characterized ant communities in this kind of ecosystem. Considering that this country will probably be greatly affected in the future by climate change (International Panel for Climate Change [IPCC], 2013), we expect that patterns obtained in this study will constitute a base line to evaluate future changes. Although no continuous plant studies have been conducted in forests and shrubs, information is available on the changes that have occurred in dune grasslands over the last 20 years (Alvarez-Molina et al., 2012). This period of time corresponds to the time elapsed since we sampled these ant communities. Considering that after 20 years dune grasslands have a higher coverture and more plant species typical of shrubs (Alvarez-Molina et al., 2012), we expect to find a similar trend in ant communities.

We can also anticipate that no large changes will be observed in the two forests, as far as it seems that both ecosystems represent advanced stages in a successional process (Castillo, 2006). Changes in abundance of invasive ants however, could significantly influence species richness as it has been observed in other ecosystems (McGlynn, 1999). In this regard the recent record of tramp species *Monomorium pharaonis* (pers. obs.) should be monitored. Considering that SF harbor more than 84% of ant species richness of La Mancha, and that it constitutes the last remnant of subdeciduous forest in the Mexican gulf coasts (Moreno-Casasola & Travieso, 2006) monitoring should be focused mainly at this site.

## Acknowledgments

To all the staff of the Centro de Investigaciones Costeras La Mancha (CICOLMA) for all its support. To Araceli Cartas, Julián Bueno and Griselda Camacho for their invaluable aid in the field. To María Luisa Castillo for donating ant specimens. Antonio Angeles and Martín de los Santos help with the elaboration of data sheets and preparation of figures. Finally we acknowledge two anonymous reviewers for their comments and suggestions that highly improved the manuscript.

## References

Alvarez-Molina, L.L., Martínez, M.L., Pérez-Maqueo, O., Gallego-Fernández, J.B. &. Flores, P. (2012). Richness, diversity, and rate of primary succession over 20 years in tropical coastal dunes. Plant Ecology, 213: 1597-1608. doi: 10.1007/s11258-012-0114-5.

Andersen, A.N. (1995). A classification of Australian ant communities, based on functional groups which parallel plant-life forms in relation to stress and disturbance. Journal of Biogeography, 22: 15-29.

Andersen, A.N. (1997). Functional groups and patterns of organization in North American ant communities: a comparison with Australia. Journal of Biogeography, 24: 433-460.

Belbin, L. (1989). PATN, Technical Reference. CSIRO, Division of Wildlife and Ecology, P.O. Box 84, Lyneham, ACT, 2602. 167 p.

Bestelmeyer, B., Agosti, D., Alonso, L.E., Brandão, C.R.F., Brown Jr.W.L., Delabie, J.H. & Silvestre R. (2000). Field techniques for the study of ground-dwelling ants: an overview, description, and evaluation. In D. Agosti, J.D. Majer, L.E. Alonso & T.R. Schultz (eds.), Ants. Standard Methods for Measuring and Monitoring Biodiversity (pp. 122-144). Washington: Smithsonian Institution Press.

Bolton, B. (1994). Identification guide to the ant genera of the world. Cambridge: Harvard University Press, 22 p.

Bolton, B., Alpert G., Ward P.S. & Naskrecki, P. (2007). Bolton's catalogue of ants of the world 1758-2005. (Compact Disc Edition). Cambridge: Harvard University Press.

Bonnet, A. & Lopes, B.C. (1993). Formigas de dunas e restingas da Praia da Joaquina, Ilha de Santa Catarina, SC (Insecta: Hymenoptera). Biotemas, 6: 107-114.

Boomsma, J.J. & de Vries, A. (1980). Ant species distribution in a sandy coastal plain. Ecological Entomology, 5: 189-204.

Boomsma, J.J. & Van Loon, A.J. (1982). Structure and diversity of ant communities in successive coastal dune valleys. Journal of Animal Ecology, 51: 957-974.

Brower, J.E., Zar, J.H., & von Ende, C.N. (1998). Field and Laboratory Methods for General Ecology. McGraw-Hill, Boston, 237 p.

Brown, W.L. (2000). Diversity of ants. In D. Agosti, J.D. Majer, L.E. Alonso & T.R. Schultz (eds.), Ants. Standard Methods for Measuring and Monitoring Biodiversity (pp. 45-79). Washington: Smithsonian Institution Press.

Bruhl, C., Gunsalam, G. & Linsenmair, E. (1998). Stratification of ants (Hymenoptera, Formicidae) in a primary rain forest in Sabah, Borneo, Journal of Tropical Ecology, 14: 285-297.

Cardoso D.C., Sobrinho, T.G. & Schoereder,J. H. (2010). Ant community composition and its relationship with phytophysiognomies in a Brazilian Restinga. Insectes Sociaux, 57: 293-301. doi: 10.1007/s00040-010-0084-3.

Castillo, G. (2006). Las selvas. In P. Moreno-Casasola (Ed.). Entornos veracruzanos: la costa de La Mancha (pp. 221-229). Instituto de Ecología, A.C., Xalapa, Ver. México.

Castillo, G. & Travieso, A.C. (2006). La flora. In P. Moreno-Casasola (Ed.). Entornos veracruzanos: la costa de La Mancha (pp. 171-204). Instituto de Ecología, A.C., Xalapa, Ver. México.

Colwell, R.K. (2013). EstimateS: Statistical estimation of species richness and shared species from samples. Version 9. http://purl.oclc.org/estimates.

Dauber, J. & Wolters, V. (2005). Colonization of temperate grassland by ants. Basic and Applied Ecology, 6: 83-91. doi:10.1016/j.baae.2004.09.011.

Durou, S., Dejean, A., Olmsted, I. & Snelling, R.R. (2002). Ant diversity in coastal zones of Quintana Roo, Mexico, with special reference to army ants. Sociobiology, 40: 385-402.

Fittkau, E.J. & Klinge, H. (1973). On biomass and trophic structure of the central Amazonian rain forest ecosystem. Biotropica, 5: 1-14.

Fontenla, J.L. (1993). Composición y estructura de comunidades de hormigas en un sistema de formaciones vegetales costeras. Poeyana, 441: 1-19.

Fontenla, J.L. (1994). Mirmecofauna de la Península de Hicacos, Cuba. Avicennia, 1: 79-85.

Fragoso, C. & Rojas, P. (2009). Invasiones en el suelo: la lombriz de tierra *Pontoscolex corethrurus* y la hormiga *Solenopsis geminata* en los ecosistemas tropicales de México. In: G.A. Aragón, M.A. Damián & J.F. López-Olguín (Eds.). Manejo Agroecológico de Sistemas. Vol. I. (pp. 81-107). Publicación especial de la Benemérita Universidad Autónoma de Puebla. México.

Gallé, L. (1991). Structure and succession of ant assemblages in a north European sand dune area. Holarctic Ecology (Ecography), 14: 31-37.

Geissert, D. (1999). Regionalización geomorfológica del estado de Veracruz. Investigaciones Geográficas: Boletín del Instituto de Geografía de la UNAM, 40: 23-47.

Gotelli N.J. & Colwell, R.K. (2011). Estimating species richness. In A.E. Magurran & B.J. McGill (Eds.), Biological diversity: frontiers in measurement and assessment (pp. 39-54). Oxford: Oxford University Press.

Gotwald, W.H. Jr. (1995). Army ants: the biology of social predation. Ithaca: Cornell University Press, 302 p.

Greenslade, P.J.M. (1964). Pitfall trapping as a method for studying populations of Carabidae (Coleoptera). Journal of Animal Ecology, 33: 301-310.

Hammer, Ø., Harper, D.A.T. & P.D. Ryan (2001). PAST: Paleontological Statistics Software Package for Education and Data Analysis. Palaeontologia Electronica, 4: 1-9.

Holway D.A., Lach, L., Suarez, A.V, Tsutsui, N.D & Case, T.J. (2002). The causes and consequences of ant invasions. Annual Review of Ecology and Systematics, 33: 181-233. doi: 10.1146/annurev.ecolsys.33.010802.150444.

Howe, M.A., Knight, G.T. & Clee, C. (2010). The importance of coastal sand dunes for terrestrial invertebrates in Wales and the UK, with particular reference to aculeate Hymenoptera (bees, wasps & ants). Journal of Coastal Conservation, 14: 91-102. doi: 10.1007/s11852-009-0055-x.

Hutcheson, K. (1970). A test for comparing diversities based on the Shannon formula. Journal of Theoretical Biology, 29: 151-154. doi: 10.1016/0022-5193(70)90124-4.

International Panel for Climate Change. IPCC (2013). Fifth report. http://www.ipcc.ch/report/ar5/wg1/#.UnrCrlOwG_I

Jolivet, P. (1996). Ants and plants. An example of coevolution. Leiden: Backhuys Publishers, 303 p.

Kempf, W.W. (1972). Catálogo abreviado das formigas da região Neotropical (Hymenoptera: Formicidae). Studia Entomologica, 15: 2-345.

Lawton, J.H. (1983). Plant architecture and the diversity of phytophagous insects. Annual Review of Entomology, 28: 23-39. doi: 10.1146/annurev.en.28.010183.000323.

Lobry de Bruyn, L.A. & Conacher, A.J. (1990). The role of termites and ants in soil modification: A review. Austral Journal of Soil Research, 28: 55-93.

Lubertazzi, D. & Tschinkel, W.R. (2003). Ant community change across a ground vegetation gradient in north Florida's longleaf pine flatwoods. Journal of Insect Science, 3(21): 1-17. doi: 10.1672/1536-2442(2003)003[0001:ACCAAG]2.0.CO;2.

Mackay, W.P. & Mackay, E. (2002). The ants of New Mexico (Hymenoptera: Formicidae. Lewiston: The Edwin Mellen Press, 400 pp.

MacKay, W.P. & Vinson, S.B. (1989). A versatile bait trap for sampling ant populations. Notes from Underground, 3: 14.

Mackay, W.P., Maes, J.M., Rojas, P. & Luna, G. (2004). The ants of North and Central America: the genus *Mycocepurus* (Hymenoptera: Formicidae). Journal of Insect Science, 4: 1-7.

Martínez, M.L., Psuty, N.P. & Lubke, R.A. (2004). A perspective on coastal dunes. In M.L. Martínez & N.P. Psuty (Eds.). Coastal Dunes. Ecology and Conservation (pp. 3-10). Berlin: Springer.

McGlynn, T.P. (1999). The worldwide transfer of ants: geographical distribution and ecological invasions. Journal of Biogeography, 26: 535-548.

McLachlan, A. (1991). Ecology of coastal dune fauna. Journal of Arid Environments, 21: 229-243.

Mehltreter, K., Rojas, P. & Palacios-Ríos, M. (2003). Moth larvae-damaged giant leather-fern *Acrostichum danaeifolium* as host for secondary colonization by ants. American Fern Journal, 93: 49-55. doi: 10.1640/0002-8444(2003)093[0049:MLGLAD]2.0.CO;2.

Moreno-Casasola, P. (Ed.). (2006). Entornos veracruzanos: la costa de La Mancha. Instituto de Ecología, A.C., Xalapa, Ver. México, 576 p.

Moreno-Casasola, P. & Monroy, R. (2006). Introducción. In P. Moreno-Casasola (Ed.). Entornos veracruzanos: la costa de La Mancha (pp. 17-22). Instituto de Ecología, A.C., Xalapa, Ver. México.

Moreno-Casasola, P. & Travieso, A.C. (2006). Las playas y las dunas. In P. Moreno-Casasola (Ed.). Entornos veracruzanos: la costa de La Mancha (pp. 205-220). Instituto de Ecología, A.C., Xalapa, Ver. México.

Moreno-Casasola, P. & Vázquez, G. (2006). Las comunidades de las dunas. In P. Moreno-Casasola (Ed.). Entornos veracruzanos: la costa de La Mancha (pp. 285-310). Instituto de Ecología, A.C., Xalapa, Ver. México.

Novelo, R.A. (1978). La vegetación de la estación biológica El Morro de La Mancha, Veracruz. Biotica, 3: 9-23.

Paradowska, K. & Moreno-Casasola, P. (2006). La caminera. In P. Moreno-Casasola (Ed.). Entornos veracruzanos: la costa de La Mancha (pp. 539-574). Instituto de Ecología, A.C., Xalapa, Ver. México.

Rico-Gray, V. (1989). The importance of floral and circumfloral nectar to ants inhabiting dry tropical lowland. Biological Journal of the Linnean Society, 38: 173-181.

Rico-Gray, V. (1993). Use of plant-derived food resources by ants in the dry tropical lowlands of coastal Veracruz, México. Biotropica, 25: 301-315.

Risch S.J. & Carroll, C.R. (1982). Effect of a keystone predaceous ant, *Solenopsis geminata*, on arthropods in a tropical agroecosystem. Ecology, 63:1979-1983.

Rojas, P. (2001). Las hormigas del suelo en México: diversidad, distribución e importancia (Hymenoptera: Formicidae). Acta Zoologica Mexicana (n.s.), Número especial 1: 189-238.

Rojas, P. (2011). Hormigas (Insecta: Hymenoptera: Formicidae). In La biodiversidad en Veracruz. Estudio de Estado. Vol. II. (pp. 431-439). Comisión Nacional para el Conocimiento y Uso de la Biodiversidad, Gobierno del Estado de Veracruz, Universidad Veracruzana, Instituto de Ecología, A.C. México.

Rojas, P. & Fragoso, C. (1994). The ant fauna (Hymenoptera: Formicidae) of the Mapimi Biosphere Reserve, Durango, México. Sociobiology, 24: 48-75.

Rosenzweig, M.L. (1995). Species Diversity in Space and Time. New York: Cambridge University Press.

Ruano, F., Ballesta, M., Hidalgo, J. & Tinaut, A. (1995). Mirmecocenosis del Paraje Natural Punta Entinas-El Sabinar (Almería) (Hymenoptera: Formicidae). Aspectos Ecológicos. Boletin de la Asociación Espanola de Entomologia, 19: 89-107.

Shattuck, S.O. (1992). Generic revision of the ant subfamily Dolichoderinae (Hymenoptera: Formicidae). Sociobiology, 21: 1-181.

Statsoft, Inc. Statistica for Windows. Tulsa, OK. 1999.

Taber, S.W. (2000). Fire Ants. College Station, TX: Texas A&M University Press. 308 p.

Teixeira, M.C., Schoereder, J.H., Nascimento, J.T. & Louzada, J.N.C. (2005). Response of ant communities to sand dune vegetation burning in Brazil (Hymenoptera : Formicidae). Sociobiology, 45: 631-641.

Tews, J., Brose, U., Grimm, V., Tielbo, K., Wichmann, M.C., Schwager, M. & Jeltsch, F. (2004). Animal species diversity driven by habitat heterogeneity/diversity: the importance of keystone structures. Journal of Biogeography, 31: 79-92. doi: 10.1046/j.0305-0270.2003.00994.x

Travieso, A.C. & Campos, A. (2006). Los componentes del paisaje. In P. Moreno-Casasola (Ed.). Entornos veracruzanos: la costa de La Mancha (pp. 139-150). Instituto de Ecología, A.C., Xalapa, Ver. México.

Vargas, A.B., Mayhé-Nunes, A.J., Queiroz, J.M., Souza, G.O. & Ramos, E.F. (2007). Efeitos de fatores ambientais sobre a mirmecofauna em comunidade de restinga no Rio de Janeiro, RJ. Neotropical Entomology, 36: 28-37. doi 10.1590/S1519-566X2007000100004.

Vázquez-Bolaños, M. (2011). Lista de especies de hormigas (Hymenoptera: Formicidae) para México. Dugesiana, 18: 95-133.

Ward, P.S. (1985). The Nearctic species of the genus *Pseudomyrmex* (Hymenoptera: Formicidae). Quaestiones Entomologicae, 21: 209-246.

Wetterer, J.K. (2009). Worldwide spread of the penny ant, *Tetramorium bicarinatum* (Hymenoptera: Formicidae). Sociobiology, 54: 811-830.

Wilson, E.O. (2003). *Pheidole* in the New World. A Dominant, Hyperdiverse Ant Genus. Cambridge: Harvard University Press. 794 p.

Yanoviak, S. P. & Kaspari, M. (2000). Community structure and the habitat templet: ants in the tropical forest canopy and litter. Oikos, 89: 259-266. doi: 10.1034/j.1600-0706.2000.890206.x.

Zorilla, J.M., Serrano, J.M., Casado, M.A., Acosta, F.J. & Pineda, F.D. (1986). Structural characteristics of an ant community during succession. Oikos, 47: 346-354.

**Appendix 1.** Number of records in samples of each ant species in the four sites studied at La Mancha. The total number of samples are included in brackets. *= species captured with non-standardized sampling (NSS). v= nesting in vegetation; s= nesting in soil; vs= nesting in vegetation and soil.

| Species list | Grassland (N=122) | Shrub (N=132) | Deciduous forest (N=172) | Subdeciduous forest (N=162) |
|---|---|---|---|---|
| **Dolichoderinae** | | | | |
| *Azteca forelii* Emery, 1893 v | 0 | 0 | * | * |
| *Azteca velox* Forel, 1899 v | 0 | 7 | 70 | 6 |
| *Dolichoderus diversus* Emery, 1894 v | 0 | 0 | 1 | * |
| *Dolichoderus lutosus* (Smith, 1858) v | 0 | 3 | 2 | 3 |
| *Dorymyrmex bicolor* Wheeler, 1906 s | 52 | 14 | 0 | 0 |
| *Dorymyrmex smithi* Cole, 1936 s | 28 | 0 | 0 | 0 |
| *Dorymyrmex* sp. aff. *flavus* s | 13 | 1 | 0 | 0 |
| *Forelius pruinosus* (Roger, 1863) s | 24 | 20 | 0 | 0 |
| **Ectatomminae** | | | | |
| *Ectatomma ruidum* (Roger, 1860) s | 0 | 0 | 0 | 2 |
| **Ecitoninae** | | | | |
| *Eciton burchelli parvispinum* Forel, 1899 s | 0 | 0 | 0 | * |
| *Labidus coecus* (Latreille, 1802) s | 0 | 0 | 0 | 1 |
| *Labidus praedator* (Smith, 1858) s | 0 | 0 | 0 | * |
| *Neivamyrmex opacithorax* (Emery, 1894) s | 0 | 0 | 3 | 0 |
| *Neivamyrmex pilosus* (Smith, 1858) s | 0 | 0 | 0 | * |
| *Neivamyrmex rugulosus* Borgmeier, 1953 s | 0 | * | 0 | 0 |
| *Neivamyrmex swainsoni* (Shuckard, 1840) s | 0 | 0 | 0 | 1 |
| *Nomamyrmex esenbeckii wilsoni* (Santschi, 1920) s | 0 | * | 0 | 1 |
| **Formicinae** | | | | |
| *Acropyga smithii* Forel, 1893 s | 0 | 0 | 0 | * |
| *Brachymyrmex heeri* Forel, 1874 s | 0 | * | * | 0 |
| *Brachymyrmex* sp. 1LM s | 46 | 12 | 10 | 1 |
| *Brachymyrmex* sp. 2LM s | 0 | 0 | 0 | 2 |
| *Camponotus atriceps* (Smith, 1858) v | 3 | 8 | 1 | * |
| *Camponotus cerberulus* Emery, 1920 v | 0 | 0 | 0 | * |
| *Camponotus claviscapus* Forel, 1899 v | 0 | 0 | 0 | * |
| *Camponotus coloratus* Forel, 1904 v | 0 | 1 | 0 | * |
| *Camponotus coruscus* (Smith, 1862) v | 0 | 0 | 0 | 1 |
| *Camponotus etiolatus* Wheeler, 1904 v | 0 | 0 | 1 | * |
| *Camponotus excisus* Mayr, 1870 v | 0 | 0 | 0 | 1 |
| *Camponotus linnaei* Forel, 1886 v | 0 | 3 | 2 | 3 |
| *Camponotus mucronatus hirsutinasus* Wheeler, 1934 v | 0 | 3 | 13 | * |
| *Camponotus novogranadensis* Mayr, 1870 v | 0 | 0 | 19 | 4 |
| *Camponotus planatus* Roger, 1863 v | 2 | 18 | 39 | 6 |
| *Camponotus sericeiventris* (Guerin-Meneville, 1838) v | 0 | 0 | 1 | 7 |
| *Camponotus conspicuus sharpi* Forel, 1893 v | 0 | 0 | 0 | * |
| *Camponotus zoc* Forel, 1879 v | 0 | 0 | 0 | * |
| *Camponotus* sp. 1LM v | 0 | 1 | 1 | 0 |
| *Camponotus* sp. 2 LM v | 0 | 1 | 0 | * |
| *Myrmelachista skwarrae* Wheeler, 1934 v | 0 | 0 | 1 | * |
| *Nylanderia steinheili* (Forel, 1893) s | 0 | 8 | 22 | 1 |
| *Paratrechina longicornis* (Latreille, 1802) sv | 1 | 20 | 0 | * |
| **Myrmicinae** | | | | |
| *Atta cephalotes* (Linnaeus, 1758) s | 4 | 21 | 0 | 0 |
| *Atta mexicana* (Smith, 1858) s | 0 | 0 | 0 | 4 |
| *Cephalotes minutus* (Fabricius, 1804) v | 0 | 3 | 0 | 5 |
| *Cephalotes scutulatus* (Smith, 1867) v | 0 | * | 10 | * |
| *Cephalotes umbraculatus* (Fabricius, 1804) v | 0 | 0 | 10 | 1 |
| *Crematogaster corvina* Mayr, 1870 v | 0 | 0 | 1 | 0 |
| *Crematogaster crinosa* Mayr, 1862 v | * | 27 | 0 | 0 |
| *Crematogaster curvispinosa* Mayr, 1862 v | 0 | 0 | 4 | 1 |
| *Crematogaster torosa* Mayr, 1870 v | 0 | 0 | 1 | 3 |
| *Crematogaster* sp. aff. *curvispinosa* v | 0 | 0 | 0 | 1 |
| *Cyphomyrmex costatus* Mann, 1922 s | 0 | 0 | 0 | * |

| | | | | |
|---|---|---|---|---|
| *Cyphomyrmex rimosus* (Spinola, 1851) s | * | 8 | 13 | 6 |
| *Megalomyrmex silvestri* Wheeler, 1909 s | 0 | 0 | 0 | * |
| *Monomorium ebeninum* Forel, 1891 s | * | 58 | 0 | * |
| *Monomorium floricola* (Jerdon, 1851) v | 0 | 0 | 0 | 16 |
| *Mycetosoritis hartmanni* (Wheeler, 1907) s | 0 | 0 | 1 | 0 |
| *Mycocepurus curvispinosus* Mackay, 1998 s | 0 | 0 | 0 | 3 |
| *Mycocepurus smithii* (Forel, 1893) s | 0 | 0 | * | 6 |
| *Myrmicocrypta* sp. s | 0 | 0 | 0 | 1 |
| *Nesomyrmex echinatinodis* (Forel, 1886) v | 0 | 2 | 0 | * |
| *Nesomyrmex wilda* (Smith, 1943) v | 0 | 1 | 0 | 0 |
| *Pheidole punctatissima* Mayr, 1870 sv | 0 | 0 | 2 | 38 |
| *Pheidole susannae* Forel, 1886 s | 0 | 7 | * | 56 |
| *Pheidole* sp. 1LM s | 0 | 0 | 1 | 9 |
| *Pheidole* sp. 2LM s | 0 | 0 | 2 | 2 |
| *Pheidole* sp. 3LM s | 0 | 0 | 11 | 2 |
| *Pheidole* sp. 4LM s | 0 | 0 | 32 | 3 |
| *Pheidole* sp. 5LM s | 0 | 0 | 0 | 1 |
| *Pheidole* sp. 6LM s | 0 | 2 | 0 | 1 |
| *Pheidole* sp. 7LM s | 0 | 16 | 0 | 0 |
| *Pheidole* sp. 8LM s | 0 | 0 | * | * |
| *Pheidole* sp. 9LM s | 0 | 1 | 0 | 0 |
| *Pheidole* sp. 10LM s | 0 | 31 | 0 | 9 |
| *Pheidole* sp. 11LM s | 0 | 0 | 0 | 9 |
| *Pheidole* sp. 12LM s | 0 | 0 | 2 | 0 |
| *Rogeria belti* Mann, 1922 s | 0 | 0 | * | 1 |
| *Rogeria cuneola* Kugler, 1994 s | 0 | 0 | 0 | * |
| *Sericomyrmex aztecus* Forel, 1855 s | 0 | 0 | 0 | 9 |
| *Solenopsis molesta* (Say, 1836) s | 1 | 39 | 53 | 34 |
| *Solenopsis geminata* (Fabricius, 1804) sv | 10 | 52 | 28 | 25 |
| *Solenopsis isopilis* Pacheco & Mackay, 2013 s | 0 | 0 | 1 | 10 |
| *Solenopsis* sp. aff. *azteca* s | 0 | 0 | 0 | * |
| *Solenopsis* sp. s | 0 | 0 | 0 | * |
| *Strumigenys boneti* Brown, 1959 s | 0 | 0 | 6 | * |
| *Strumigenys eggersi* Emery, 1890 s | 0 | 0 | 0 | * |
| *Strumigenys elongata* Roger, 1863 s | 0 | 0 | 0 | 1 |
| *Strumigenys louisianae* Roger, 1863 s | 0 | 1 | 0 | 1 |
| *Strumigenys ludia* Mann, 1922 s | 0 | 0 | 0 | 2 |
| *Strumigenys nigrescens* Wheeler, 1911 s | 0 | 0 | 0 | * |
| *Temnothorax subditivus* (Wheeler, 1903) s | 0 | 2 | 3 | * |
| *Tetramorium bicarinatum* (Nylander, 1846) sv | * | 0 | 0 | * |
| *Tetramorium simillimum* (Smith, 1851) s | * | 0 | 0 | * |
| *Tetramorium spinosum* (Pergande, 1896) s | * | 28 | 0 | 0 |
| *Trachymyrmex intermedius* (Forel, 1909) s | 0 | 0 | 0 | 1 |
| *Trachymyrmex* sp. aff. *saussurei* s | 6 | 22 | 0 | 6 |
| *Wasmannia auropunctata* (Roger, 1863) s | 1 | 0 | 12 | 36 |
| *Xenomyrmex panamanus* (Wheeler, 1922) v | 0 | 0 | 1 | 1 |
| **Ponerinae** | | | | |
| *Hypoponera nitidula* (Emery, 1890) s | 0 | 0 | 1 | 5 |
| *Hypoponera opacior* (Forel, 1893) s | 0 | 1 | 0 | 5 |
| *Hypoponera* sp. aff. *vana* s | 0 | 0 | * | 5 |
| *Odontomachus brunneus* (Patton, 1894) s | 0 | 0 | 0 | * |
| *Odontomachus laticeps* Roger, 1861 s | 0 | 0 | 0 | 1 |
| *Pachycondyla crenata* (Roger, 1861) v | 0 | 0 | 0 | 1 |
| *Pachycondyla harpax* (Fabricius, 1804) s | 0 | 0 | 3 | 8 |
| *Pachycondyla stigma* (Fabricius, 1804) s | 0 | 1 | 0 | * |
| *Pachycondyla villosa* (Fabricius, 1804) v | 0 | 0 | * | 16 |
| *Platythyrea punctata* (Smith, 1858) s | 0 | 0 | 0 | * |
| **Pseudomyrmeciinae** | | | | |
| *Pseudomyrmex boopis* (Roger, 1863) s | 0 | 0 | 0 | 1 |
| *Pseudomyrmex brunneus* (Smith, 1877) v | 3 | 3 | 0 | 0 |
| *Pseudomyrmex cubaensis* (Forel, 1901) v | 0 | 1 | 0 | 3 |
| *Pseudomyrmex ejectus* (Smith, 1858) v | 3 | 3 | 0 | 0 |

| | | | |
|---|---|---|---|
| *Pseudomyrmex elongatulus* Dalla Torre, 1892 v | 0 | 1 | 1 | 1 |
| *Pseudomyrmex ferrugineus* (Smith, 1877) v | 0 | 0 | 1 | 2 |
| *Pseudomyrmex gracilis* (Fabricius, 1804) v | 0 | 0 | 1 | 1 |
| *Pseudomyrmex ita* (Forel, 1906) v | 0 | 1 | 0 | * |
| *Pseudomyrmex oculatus* (Smith, 1855) v | 0 | 0 | 0 | * |
| *Pseudomyrmex seminole* Ward, 1985 v | 0 | 0 | 0 | 1 |
| *Pseudomyrmex simplex* (Smith, 1877) v | 0 | 1 | 0 | * |
| *Pseudomyrmex spiculus* Ward, 1989 v | 0 | 0 | 1 | * |
| *Pseudomyrmex tenuissimus* (Emery, 1906) v | 0 | 1 | 5 | * |
| *Pseudomyrmex* sp. (*pallens* group) v | 0 | * | 0 | 2 |

# 10

# Food competition mechanism between *Solenopsis invicta* Buren and *Tapinoma melanocephalum* Fabricus

Biqiu Wu[1,2], Lei Wang[1], Guangwen Liang[1], Yongyue Lu[1], Ling Zeng[1]

1 - *South China Agricultural University, Guangzhou, China.*
2 - *Guangxi Province Academy of Agricultural Sciences, Guangxi, China.*

**Keywords**
Invasive species; interspecific coexistence; interspecific competition; experiments

**Corresponding author**
Ling Zeng
Red Imported Fire Ant Research Center
South China Agricultural University
Guangzhou 510642, China
E-Mail: zengling@scau.edu.cn

**Abstract**

This study compared the amount of food resource depletion and interference competition at the individual and colony levels between *Solenopsis invicta* and *Tapinoma melanocephalum* in laboratory. The consumption of sausage, honey water, and mealworm by *S. invicta* colonies were of equal worker number was higher than that by *T. melanocephalum* colonies. However, the amounts of sausage, honey water, and mealworm depleted by *S. invicta* colonies were of equal worker biomass were lower than those by *T. melanocephalum* colonies. The consumption of sausage and mealworm by *S. invicta* colonies were of equal worker biomass were also significantly lower than that by *T. melanocephalum* colonies. Individual-level interference competition between *S. invicta* and *T. melanocephalum* colonies in small space was intense. Competition intensity and the mortality rate reached their maximum when the worker numbers of both colonies were equal. In any proportion, the mortality rate of *T. melanocephalum* reached over 80%, higher than that of *S. invicta*. *S. invicta* colonies were of equal worker biomass and number recruited more workers for colony-level interference competition and used more resources. But the death rates of *S. invicta* colonies were higher than those of *T. melanocephalum* colonies. The highly exploitative and interference-competitive of *S. invicta* in trails had restricted the foraging behavior and active region of *T. melanocephalum*.

## Introduction

*Solenopsis invicta* Buren is a dangerous quarantine pest that significantly affects public safety, human health, husbandry, and the ecological environment in its invasion regions. *S. invicta* builds its nests in soil, workers may infiltrate and destroy infrastructure, especially underground wires, cables, and electrical equipment, at cost of millions of dollars; meanwhile, because of its aggressive, death of creatures also leads by fire ant (Vinson, 2013). *S. invicta* has spread to and propagated in China, and occasionally threaten the people's health (Wang et al., 2013). *S. invicta* directly feeds on crops, livestock, and poultry and have thus caused huge economic losses in husbandry (Lofgren et al., 1975; Stewart & Vinson, 1991; Jetter et al., 2002). *S. invicta* also attacks young birds, spawn, calves of sea turtles and other reptiles, and small rodents (Drees, 1994; Giuliano et al., 1996; Allen et al., 1997; Parris et al., 2002; Pascoe, 2002; Allen et al., 2004).

Invasive ants significantly influence local ant populations, especially ant species with similar ecological characteristics (Holway et al., 2002). For instance, *S. invicta* has replaced *Solenopsis geminata* (F.) and *Solenopsis xyloni* McCook in the North American ecological system through competitive exclusion after invasion (Wilson and Brown Jr, 1958; Porter et al., 1988; Porter, 1992; Porter and Savignano, 1990). Such replacement decreased the abundance and diversity of local ant colonies (Porter & Savignano, 1990) and even changed the coexistence mode of surviving local ant colonies in the biogeographic balance (Gotelli & Arnett, 2000). In 2004, *S. invicta* was found in Wuchuan, Guangdong Province, indicating that the species had invaded the continental China (Wang et al., 2013).

The invasion of South China by *S. invicta* has significantly decreased the diversity of local ant communities

(Shen et al., 2007; Wu et al., 2008). However, few studies have investigated the influence of *S. invicta* invasion on the dominant local species in Guangdong, *Tapinoma melanocephalum* (Wu et al., 2008), as well as the coexistence and competition mechanisms between these species. This study examined the food resource depletion and interference competition at the individual and colony levels between *S. invicta* and *T. melanocephalum* laboratory populations to explore the competition and coexistence mechanisms of *T. melanocephalum* in response to *S. invicta* invasion in Guangdong.

## Materials and methods

### Ant samples

*S. invicta* and *T. melanocephalum* colonies were collected from polygyne colonies at wild grass ground or litchi orchard in Southern China and maintained at the South China Agricultural University and Red Imported Fire Ant Research Center, Guangdong province. Colonies of *S. invicta* and *T. melanocephalum* contained queens, workers and included immature at all developmental stages. These colonies were separated from the soil and placed in an open plastic box (23.5 cm length × 15.5 cm width × 9.0 cm hight) with small plastic boxes (8.5 cm length × 6.0 cm width × 5.0 cm hight) with wet plaster to serve as nest chambers, and then placed in floors at 28 °C, on a diet of 20% honey water (a test tube half full of honey water and plugged with cotton) and fresh mealworms (Coleoptera: Tenebrionidae) until needed for experiments. The inside walls of the bigger boxes were coated with the fluoropolymer resin, Fluon (polytetrafluoroethylene, ICI Fluoropolymers, Exon, PA) to prevent them from escaping.

### Comparison between amounts of food resource depletion of S. invicta and T. melanocephalum

#### Test food

The following were the food used in the tests: honey - a high carbohydrate food resource (Guangzhou Baoshengyuan Corporation); sausage - a protein- and lipid- rich artificial food resource (Guangdong Shuanghui Food Corporation); and mealworm (Coleoptera: Tenebrionidae) - a high protein natural food resource (purchased from market).

#### Test method

Experimental colonies of *S. invicta* and *T. melanocephalum* were set up on both an equal worker biomass and equal worker number basis for average worker size varied among their forms (Morrison et al., 2000). Each experimental equal worker biomass colony contained 0.5 g of workers, 0.25 g of brood (included eggs, larvae and pupae) and two queens.

Each experimental equal worker number colony contained 1000 workers, 0.25g of brood and two queens.

Because counting live worker ants was not practical and the workers of *T. melanocephalum* were activity quickly, the method of workers added to colonies was operated like Morrison (2000). But both ants were anaesthetized with ethyl ether before weigh with an electronic balance. We found that ethyl ether had no effect on ants in preliminary trials.

Each experimental colony occupied a plastic box (25 cm length × 18 cm width × 7 cm hight), equipped with a small plaster plastic box (8.5 cm length ×6 cm width × 5 cm hight) to serve as nest chambers, the sides of which were coated with Fluon to prevent escapes. The mother colonies from which the experimental colonies were kept in the laboratory at 28°C for at least two weeks before experimental colony formation, on a diet of 20% honey water (a test tube half full of honey water and plugged with cotton) and fresh mealworms every day. The experimental colonies were placed in the laboratory at 28°C and 60% 75% RH, starved for 24 h before the beginning of the trials to produce a uniform state of hunger. Each box containing an experimental colony was connected via a test tubing (1 cm inside diameter) to an adjacent (empty) box of the same dimensions. Holes were drilled in the side of the box near the bottom to allow insertion of the tubing. 2.0g of sausage, mealworm, or 20% honey water (drop on the cotton) was placed in the test tube, the ants were allowed to forage for 24h, and then weighed the remaining food with an electronic balance. We conducted ten replicate trails for each food item, for each equivalent colony. Test was stopped when the workers carried larvae and spawn to food, but this situation was rarely observed.

Portions of 2.0g of sausage, mealworm, and honey water were placed in a clean plastic tube, kept at the same conditions, but protected from the ants as evaporation loss control.

### Individual level interference competition

#### Testing method

Healthy and similar size workers of *S. invicta* and *T. melanocephalum* were selected to carry out the interference competitive abilities at the individual level, and the agonistic interactions between *S. invicta* and *T. melanocephalum* were staged in small arenas. Arenas consisted of a dry plastic Petri dish (12 cm inside diameter) which was sterilized with 75% alcohol, washed with distilled water, with inner sides coated with Fluon. Ants were counted by allowing them to climb onto a small paint brush and then placing them into separate plastic boxes (23.5 cm length × 15.5 cm width × 9.0 cm hight) with inside walls coated with Fluon. *T. melanocephalum* workers were placed into the plastic Petri dish first, *S. invicta* workers were placed into the arena later. We chose five encounter ratios of *S. invicta* to *T. melanocephalum*, 1) 50 *S. invicta* workers to 10 *T. melanocephalum* workers (5:1); 2) 45 *S. invicta* workers to 15 *T. melanocephalum* workers (3:1); 3) 30 *S. invicta* workers to 30 *T. melanocephalum* workers (1:1); 4) 15 *S. invicta* workers to 45 *T. melanocephalum* workers (1:3); and 5) 10 *S. invicta* workers to 50 *T. melanocephalum* workers (1:5). The number of major *S. invicta* workers used in areas interaction in each encounter ratio accounted for 10%. We chose the *T. melanocephalum* workers in agonistic interactions were at the uniform size. The ants were observed for 3 h at 26°C and 65% RH. Interaction behavior was recorded, and the number of dead or mortally wounded workers was noted at the end of the trials. We conducted ten replicate trials for each encounter ratio. Healthy individual ants (40 workers) of *S. invicta* and *T. melanocephalum* were independently placed in a clean and dry plastic petri dish; the number of dead workers was recorded to adjust the control death rate. This trail was conducted in ten replicates.

*Calculation formula of adjusted death rate*

Adjusted death rate (%) = (control survival rate % − treatment survival rate %) / control survival rate % × 100

*Colony level interference competition*

*Testing method*

The experimental colonies of both ants which deal with like 1.2.2 were evaluated in two pairwise comparisons. An equal worker biomass or an equal worker number colony of *S. invicta* was connected to an equal 'size' colony of *T. melanocephalum* via three intervening (empty) boxes by 10 cm length of Tygon tubing (Fig. 1). In the experimental set-up, the boxes containing the ant colonies on each end are

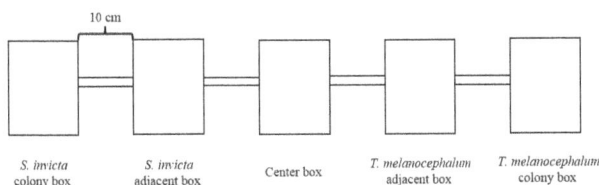

**Fig. 1**. Design of colony level interference competition experiments.

referred to as 'colony' boxes; the empty boxes adjacent to each colony boxes are referred to as 'adjacent' boxes; and the connecting box in the middle is referred to as the 'center' box (Morrison et al., 2000).

1.0 g of fresh mealworm and 0.5 g of sausage were placed in the center box and two adjacent boxes before the experiment, and then recorded the number of active and dead workers in the three empty boxes (center and both adjacent) when the ants foraging foods at 0.5 h, 24 h, 48 h, and 72 h, the number of dead workers and colonies, which invaded by another ants, in colony boxes when the ants foraging foods at 72 h were also noted. To statistically compare the survival rate of *S. invicta* and *T. melanocephalum* workers in the three empty boxes after 0.5 h min, 24 h, 48 h, and 72 h, the mortality of both ants in five boxes after 0.5 h, 24 h, and 72 h, the number of invaded colonies at the end of trail. The interference competition between *S. invicta* and *T. melanocephalum* in each equal 'size' colony was conducted ten replicate trails.

*Data processing*

The mean amount of resource acquired by each equal *S. invicta* and *T. melanocephalum* colonies was compared by T-test, the mean amount of resource acquired by *S. invicta* or *T. melanocephalum* was compared by one-way ANOVA, for each type of resource. All pairwise comparisons of means were made by Tukey method of multiple comparisons. The data were processed by SPSS 18.0.

**Results**

*Comparison amounts of food resource depletion between S. invicta and T. melanocephalum*

Resource consumption of the different food types varied in *S. invicta* and *T. melanocephalum* (Table 1). When colonies were of equal worker number, *S. invicta* consumed significantly more sausage and honey water than did *T. melanocephalum* ($p < 0.05$). When colonies were of equal worker biomass, *S. invicta* consumed significantly less sausage and mealworm than did *T. melanocephalum* ($p < 0.01$).

**Table 1**. Amount of food resource depletion of *S. invicta* and *T. melanocephalum* after 24 h

| Food resource | Colonies were of equal worker number | | Colonies were of equal worker biomass | |
| --- | --- | --- | --- | --- |
| | *T. melanocephalum* | *S. invicta* | *T. melanocephalum* | *S. invicta* |
| Sausage | 0.020±0.001 | 0.027±0.003[*] | 0.045±0.005 | 0.016±0.002[**] |
| Honey water | 0.044±0.002 | 0.127±0.028[*] | 0.065±0.004 | 0.054±0.007[ns] |
| Mealworm | 0.018±0.001 | 0.019±0.002[ns] | 0.033±0.003 | 0.015±0.001[**] |

Note: "[ns]" indicates lack of significance. "[*]" and "[**]" represent significance between the quantities of food transferred by *T. melanocephalum* and *S. invicta* with equal worker number or biomass at the 0.05 and 0.01 probability levels, respectively (independent T-test).

**Table 2**. Mortality rates of *S. invicta* and *T. melanocephalum* workers in individual level interference competition after 3 h.

| Ant species | Confrontation ratio (*S. invicta*: *T. melanocephalum*) | | | | |
|---|---|---|---|---|---|
| | 1:5 | 1:3 | 1:1 | 3:1 | 5:1 |
| *S. invicta* | 0.15 | 0.22 | 0.57 | 0 | 0 |
| *T. melanocephalum* | 80.72 | 87.14 | 100 | 99.33 | 100 |

*Individual level interference competition*

In individual level competition, four interference conditions were observed: 1) *T. melanocephalum* workers voluntarily evading *S. invicta* to avoid fighting are referred to as 'ignore'; 2) When both ants encounter, they touch each other by antennae then avoiding rapidly are referred to as 'contact', we can see this condition frequent when *T. melanocephalum* workers less than *S. invicta*; 3) two or more native ants touch a invader ant by antennae, turn-back to eject defensive compounds from pygidial glands and avoiding quickly are referred to as 'chemical defense'; 4) Ants engaging in "fights" often in prolonged grappling with frequent biting and flexing their gaster in the direction of opponents are referred to as 'physical aggression', fighting usually occurred "group" attack (2 or more workers) by *T. melanocephalum* ants on single *S. invicta* ant, although this condition rarely saw in trails. There was no incidence of multiple invaders fighting lone native opponent.

The individual-level interference competition between *S. invicta* and *T. melanocephalum* after 3 h were shown in Table 2. The mortality rates of both ant workers were the lowest when the ratio of *S. invicta* to *T. melanocephalum* was 1:5. The mortality rates of both ant workers increased when the number of *T. melanocephalum* workers equal to *S. invicta*. When both ant workers were 30 (*S. invicta* to *T. melanocephalum*=1:1), the death rate of *T. melanocephalum* was 100%, and the mortality of *S. invicta* was more than 50%. The mortality rates of *T. melanocephalum* were over 99% in the ratio of 3:1 and 5:1 (*S. invicta* to *T. melanocephalum*), but no dead workers of *S. invicta* were observed.

*Colony level interference competition*

*Proportion of active S. invicta workers in two adjacent boxes and center box*

To count the survival workers of *S. invicta* in each kind of food foraging in adjacent and center boxes, the number of survival workers appearing in each box accounts for the total workers (including dead and alive ants) which foraging in adjacent and center boxes as proportion of active *S. invicta* workers (Table 3). When colonies were of equal worker biomass, sausage was used as the food resource, the active *S. invicta* workers in *S. invicta* colony adjacent box accounted for the maximum proportion (14.97%) at 0.5 h, and no workers were found in the center box and *T. melanocephalum* colony adjacent box. At 24 and 48 h, active *S. invicta* workers in the *S. invicta* colony adjacent box reached their maximum proportions (12.24% and 11.11%, respectively), which were significantly higher than those in *T. melanocephalum* colony adjacent box. At 72 h, active *S. invicta* workers in center box

**Table 3**. Proportion of active *S. invicta* workers in the two adjacent boxes and center box.

| | | Colonies were of equal worker biomass | | Colonies were of equal worker number | |
|---|---|---|---|---|---|
| | | Sausage | Mealworm | Sausage | Mealworm |
| 0.5 h | *S. invicta* colony adjacent box | 14.97±13.56 a | 16.54±8.87 a | 14.98±11.94 a | 16.22±9.65 a |
| | Center box | 0.00±0.00 b | 0.85±2.68 b | 0.00±0.00 b | 0.00±0.00 b |
| | *T. melanocephalum* colony adjacent box | 0.00±0.00 b | 0.00±0.00 b | 0.00±0.00 b | 0.00±0.00 b |
| 24 h | *S. invicta* colony adjacent box | 12.24±4.08 a | 13.37±4.28 a | 13.61±6.02 a | 10.15±4.24 a |
| | Center box | 7.63±3.21 b | 9.18±4.32 b | 4.74±5.68 b | 10.91±2.91 a |
| | *T. melanocephalum* colony adjacent box | 4.75±5.66 b | 4.05±5.06 c | 5.72±6.34 b | 7.89±5.35 a |
| 48 h | *S. invicta* colony adjacent box | 11.11±3.84 a | 12.39±5.35 a | 13.66±2.47 a | 11.46±5.56 a |
| | Center box | 10.83±5.23 a | 9.81±3.39 a | 7.00±2.73 b | 9.68±4.99 a |
| | *T. melanocephalum* colony adjacent box | 6.07±4.63 b | 5.25±4.97 b | 7.86±5.20 b | 7.04±3.28 a |
| 72 h | *S. invicta* colony adjacent box | 11.49±3.18 a | 12.25±7.36 a | 12.74±4.73 a | 12.93±3.32 a |
| | Center box | 11.89±2.84 a | 5.32±5.95 b | 7.62±6.26 ab | 9.76±4.19 a |
| | *T. melanocephalum* colony adjacent box | 4.65±5.01 b | 7.31±5.28 b | 5.45±5.79 b | 5.19±4.60 b |

Note: Same-column means followed by the same letter are not significantly different at the 0.05 and 0.01 levels, respectively, as in Table 4.

**Table 4.** Proportion of active *T. melanocephalum* workers in two adjacent boxes and center box.

| | | Colonies were of equal worker biomass | | Colonies were of equal worker number | |
|---|---|---|---|---|---|
| | | Sausage | Mealworm | Sausage | Mealworm |
| 0.5h | *S. invicta* colony adjacent box | 0.00±0.00 b | 0.00±0.00 b | 0.00±0.00 b | 0.00±0.00 b |
| | Center box | 0.00±0.00 b | 0.00±0.00 b | 0.00±0.00 b | 0.00±0.00 b |
| | *T. melanocephalum* colony adjacent box | 16.98±7.91 a | 17.17±7.34 a | 15.10±11.67 a | 15.29±11.33 a |
| 24h | *S. invicta* colony adjacent box | 3.08±4.50 b | 2.87±5.30 b | 2.91±5.04 b | 0.60±1.88 b |
| | Center box | 6.40±5.03 b | 5.09±6.01 b | 6.48±7.04 b | 10.03±4.88 a |
| | *T. melanocephalum* colony adjacent box | 14.05±7.62 a | 14.13±7.54 a | 13.91±5.38 a | 13.23±6.47 a |
| 48h | *S. invicta* colony adjacent box | 3.31±3.71 b | 5.28±7.51 a | 4.24±5.74 b | 1.07±3.38 b |
| | Center box | 9.14±6.82 a | 7.85±7.29 a | 3.63±4.79 b | 8.55±6.37 a |
| | *T. melanocephalum* colony adjacent box | 11.33±8.23 a | 8.92±8.88 a | 14.23±8.03 a | 12.87±7.65 a |
| 72h | *S. invicta* colony adjacent box | 3.38±4.62 b | 7.21±11.66 a | 3.11±5.12 b | 2.05±4.41 b |
| | Center box | 7.29±6.62 ab | 4.86±8.28 a | 4.80±7.04 b | 8.27±6.08 a |
| | *T. melanocephalum* colony adjacent box | 12.06±8.89 a | 4.86±8.28 a | 14.61±5.19 a | 12.48±8.19 a |

reached their maximum proportion (11.89%), which was significantly higher than that in *T. melanocephalum* colony adjacent box. When mealworm was used as the food resource, active *S. invicta* workers in *S. invicta* colony adjacent box reached their maximum proportions and significantly higher than those in *T. melanocephalum* colony adjacent box at 0.5, 24, and 48 h. At 72 h, active *S. invicta* in workers in center box reached their maximum proportion (12.25%) and significantly higher than those in center box and *T. melanocephalum* colony adjacent box.

When colonies were of equal worker number, sausage as the food resource, active *S. invicta* workers in *S. invicta* colony adjacent box had their maximum proportion at 0.5 h, and no workers were found in center box and *T. melanocephalum* colony adjacent box. At 24 and 48 h, active *S. invicta* workers in *S. invicta* colony adjacent box reached their maximum proportions (13.61% and 13.66%, respectively), significantly higher than those in center box and *T. melanocephalum* colony adjacent box. At 72 h, active *S. invicta* workers in *S. invicta* colony adjacent box reached their maximum proportion (12.74%), significantly higher than that in *T. melanocephalum* colony adjacent box. When mealworm was the food resource, active *S. invicta* workers in *S. invicta* colony adjacent box reached their maximum proportion (16.22 %) at 0.5 h, and no workers were found in the center box and *T. melanocephalum* colony adjacent box. At 24 h, active *S. invicta* workers in center box reached their maximum proportion (10.91%), indistinctively with those in center box and *T. melanocephalum* colony adjacent box. At 48 and 72 h, active *S. invicta* workers in *S. invicta* colony adjacent box reached their maximum proportions (11.46% and 12.93%, respectively), significantly higher than that in *T. melanocephalum* colony adjacent box.

*Proportion of active T. melanocephalum workers in two adjacent boxes and center box*

The proportion of the active *T. melanocephalum* workers in which colonies were equivalent by worker biomass or number was counted and shown in table 4. When colonies were of equal worker biomass, sausage was used as the food resource, active *T. melanocephalum* workers were found only foraging in *T. melanocephalum* colony adjacent box at 0.5 h. Active *T. melanocephalum* workers in *T. melanocephalum* colony adjacent box reached their maximum proportions (14.05%, 11.33%, and 12.06%, respectively) at 24, 48, and 72 h, therefore the minimum proportion of native ants foraging in *S. invicta* colony adjacent box was observed at the same time. At 24 and 48 h, the proportions of active *T. melanocephalum* workers in *T. melanocephalum* colony adjacent box were significantly higher than those in *S. invicta* colony adjacent box; at 72 h, the proportion of active *T. melanocephalum* workers in *T. melanocephalum* adjacent box was significantly higher than that in *S. invicta* colony adjacent box. When mealworm was used as the food resource, active *T. melanocephalum* workers were found only in *T. melanocephalum* colony adjacent box at 0.5 h. At 24 and 48 h, active *T. melanocephalum* workers in *T. melanocephalum* colony adjacent box reached their maximum proportions (14.13% and 8.92%, respectively), and the proportions of active *T. melanocephalum* workers in *T. melanocephalum* adjacent box were significantly higher than those in *S. invicta* colony adjacent box at 24 h. At 72 h, the proportions of active *T. melanocephalum* workers foraging in two adjacent and center boxes was non-significant.

When colonies were of equal worker number, sausage was used as food resource, active *T. melanocephalum* workers

**Table 5**. Comparison of workers mortality rate in five boxes after 72 h

| Test box | Colonies were of equal worker biomass | | | | Colonies were of equal worker number | | | |
| | Sausage | | Mealworm | | Sausage | | Mealworm | |
| | Si | Tm | Si | Tm | Si | Tm | Si | Tm |
| --- | --- | --- | --- | --- | --- | --- | --- | --- |
| *S. invicta* colony box | 31.29 | 0 | 21.55 | 0 | 27.88 | 0.96 | 28.64 | 0 |
| *S. invicta* colony adjacent box | 12.93 | 0 | 11.21 | 0 | 7.69 | 2.40 | 8.04 | 0 |
| Center box | 2.72 | 2.04 | 5.60 | 0 | 3.85 | 1.44 | 4.02 | 2.01 |
| *T. melanocephalum* colony adjacent box | 5.44 | 2.72 | 3.88 | 9.05 | 1.92 | 3.37 | 2.01 | 4.52 |
| *T. melanocephalum* colony box | 21.09 | 21.77 | 28.88 | 19.83 | 22.12 | 28.37 | 24.12 | 26.63 |

Note: Si =*S. invicta* and Tm = *T. melanocephalum.*

were found only in *T. melanocephalum* colony adjacent box at 0.5 h. At 24, 48, and 72 h, active *T. melanocephalum* workers in *T. melanocephalum* colony adjacent box reached their maximum proportions (13.91%, 14.23%, and 14.6%, respectively), significantly higher than that in center box and *S. invicta* colony adjacent box. When mealworm was used as food resource, active *T. melanocephalum* workers were found only in the *T. melanocephalum* colony adjacent box at 0.5 h. At 24, 48, and 72 h, active *T. melanocephalum* workers in *T. melanocephalum* colony adjacent box and center box reached their maximum proportions (13.23%, 12.87%, and 12.48%, respectively), significantly higher than that in *S. invicta* colony adjacent box.

*Comparison of workers mortality rate in five boxes after 72 h*

Although the case of invaders presenting to each other nest chamber can be saw in trails, intense fighting between *S. invicta* and *T. melanocephalum* was rarely observed, and ants usually establish their territories after 72 h (Morrison et al., 2000). Therefore, we recorded the workers mortality rate foraging in two colony boxes, two adjacent boxes, and center box after 72 h (Table 5). When colonies were of equal worker biomass, in the case of sausage depletion by two ants, the mortality rates of both ants in *T. melanocephalum* colony box were the highest (>21%), followed by those in *T. melanocephalum* colony adjacent box and center box. The death rates of *T. melanocephalum* in *S. invicta* colony box and adjacent boxes were zero. In the case of mealworm depletion, the death rates of both ants in *T. melanocephalum* colony box were the highest, followed by those in *T. melanocephalum* colony adjacent box. The death rates of *T. melanocephalum* in center box, *S. invicta* colony adjacent box, and *S. invicta* colony box were zero.

When colonies were of equal worker number, in the case of sausage depletion, the death rates of both ants in *T. melanocephalum* colony box were the highest (>22%). The death rate of *T. melanocephalum* workers in *S. invicta* colony box and *S. invicta* workers in *T. melanocephalum* colony adjacent box were the lowest. In the case of mealworm depletion, the death rates of both ants in *T. melanocephalum* colony box were the highest (>24%), followed by those in

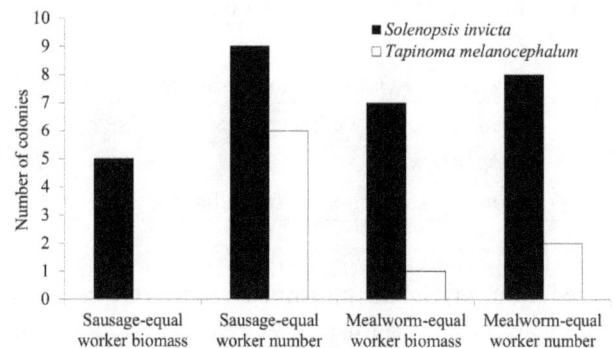

**Fig. 2**. Colonies of *S. invicta* and *T. melanocephalum* invading to each other nest chamber after 72 h.

*T. melanocephalum* colony adjacent box and center box. The death rates of *T. melanocephalum* in *S. invicta* colony and adjacent boxes were zero.

*Colonies of S. invicta and T. melanocephalum invading to each other nest chamber after 72 h*

We recorded and counted the colony number of *S. invicta* and *T. melanocephalum* workers invaded to opponent colony boxes after 72 h (Fig. 2). When sausage was used as food resource, five *S. invicta* colonies (50% of all colonies in trails) which colonies were of equal worker biomass invaded to *T. melanocephalum* colony box, and no-one *T. melanocephalum* colony invaded to *S. invicta* colony box. Nine *S. invicta* colonies (90% of all colonies in trails) which colonies were of equal worker number invaded to *T. melanocephalum* colony box, and six *T. melanocephalum* colonies (60% of all colonies in trails) invaded to *S. invicta* colony box. When mealworm was used as food resource, seven *S. invicta* colonies (70% of all colonies in trails) which colonies were of equal worker biomass, invaded *T. melanocephalum* colony box, while only one *T. melanocephalum* colony invaded *S. invicta* colony box. Eight *S. invicta* colonies (80% of all colonies in trails) which colonies were of equal worker number, invaded *T. melanocephalum* colony box, while only two *T. melanocephalum* colonies invaded *S. invicta* colony box. These results indicate that *S. invicta* was more aggressive.

## Discussion

Interspecific competition refers to the mutual interference or inhibition between two or more species. The essence of interspecific competition lies in the efficiency reduction of the reproduction, survival, growth, and other aspects of individuals of one species because of the exploitation or interference of common resources by individuals of another species. Interspecific competition primarily refers to resource competition, namely, the mutually unfavorable effects of commonly exploiting scarce resources on biological individuals. Resource competition can be classified into exploitation and interference competition. In exploitation competition, individuals of one species obtain more common resources than those of another species; in interference competition, individuals of one species limit or prevent individuals of another species from using the resources (Reitz and Trumble, 2002). Interspecific competition is considered key in structuring local ant communities, and it has been described as the "hallmark of ant ecology" (Cerda et al., 2013).

Local ants have been replaced by *S. invicta* because this species is highly exploitative and interference-competitive (Porter and Savignano, 1990; Bhatkar et al., 1972; Obin and Vander Meer, 1985; Jones and Phillips Jr, 1987; Hook and Porter, 1990; Jones and Phillips Jr, 1990; Morrison, 2000, 1999, 2002). In this study, we found that food consumption of *T. melanocephalum* colonies which colonies were of equal worker number, was less than did *S. invicta*, *T. melanocephalum* workers (monomorphic, one-sized) are extremely small, 1.3 to 1.5 mm long (Scheurer et al., 1998), only *S. invicta* worker (involving major and minor ants) was 3.24 times the average weight of *T. melanocephalum* may be responsible for the results. Higher amount of food resource depleted by equal worker biomass *T. melanocephalum* colonies may be due to the number of *T. melanocephalum* workers more than *S. invicta*. In other words, one *T. melanocephalum* worker need less food than *S. invicta* to maintain its daily activities.

Intense fighting between *S. invicta* and *T. melanocephalum* was found in individual level interference competition in a small arena when worker number in both ants was equivalent or there were more workers of *S. invicta* than *T. melanocephalum*. *S. invicta* workers attacking *T. melanocephalum* usually by 'physical aggression', but *T. melanocephalum* worker would prefer to use 'chemical defense' to repel invader ants, which *T. melanocephalum* displayed alerting, alarm behavior, and the daubing of pygidial gland secretions (Tomalski et al., 1987). *T. melanocephalum* ants would initiative attacking *S. invicta* when its workers were more than invaders, and a single *S. invicta* worker kept far away from *T. melanocephalum* ants to avoid attacked for application of the pygidial gland secretion to the legs or antennae of a foreign ant often resulted in a hindrance of movement or in the limbs adhering together (Tomalski et al.,

1987). However, mortality rate of *T. melanocephalum* was more than 80% in all treatments, *S. invicta* workers were more aggressive and its size (involving major and minor ants) bigger than *T. melanocephalum* may be the main reasons. In colony level interference competition, both ant colonies of equal worker biomass or number foraging in colony boxes and each colony adjacent box were more, and then foraging in father distance food resources. Intense fighting in both ants was usually saw in *T. melanocephalum* colony box for native ant nest chamber intruded by *S. invicta* workers, but *S. invicta* paying its stronger aggressiveness for the higher mortality rate in this interference competition.

In the indoor interference competition between *S. invicta* and *T. melanocephalum*, *S. invicta* recruited larger workers on food resources, intruded to *T. melanocephalum* colony box, indicated that *S. invicta* was highly exploitative and interference-competitive, which restricts the activity of *T. melanocephalum*. Chemical defense used by *T. melanocephalum* repelled the exotic ant was a major reason to explain the highly mortality rate in *S. invicta*, and implied that *T. melanocephalum* was the native ant against with *S. invicta* in invaded region.

The reasons can explain the phenomenon of coexistence between *T. melanocephalum* and *S. invicta* in invaded region in South China are as follow: 1) *T. melanocephalum* is opportunistic nester in places that sometimes remain habitable for only a few days or weeks (Hölldobler and Wilson, 1990), and highly adaptable in its nesting habits outdoors or indoors, the colonies occupy local sites include tufts of dead but temporarily moist grass, plant stems, and cavities beneath detritus in open, rapidly changing habitats (Oster and Wilson, 1978). Indoors, the ant colonizes wall void or spaces between cabinetry and baseboards. It will also nest in potted plants (Smith and Whitman 1992) and the ant nest which abandon by *S. invicta* populations (we found in wild grass ground or litchi orchard in South China); 2) Multiple queens may be spread out in multiple subcolonies, new colonies are probably formed by budding and there does not appear to be any infighting between members of different colonies or nests (Smith and Whitman 1992); 3) *T. melanocephalum* workers are favor many food resources, they are fond of honeydew and tend honeydew-excreting insects, and foraging on honeydew more efficiently than *S. invicta* (Zheng and Zhang, 2010) and *T. melanocephalum* workers are extremely small, a little food can maintain its daily activities; 5) *T. melanocephalum* has the habit of running rapidly and erratically when disturbed (Li et al., 2008), using pygidial gland secretions to repel invaders; in addition, this native ant has higher tolerance to high temperature than *S. invicta* (Zheng et al., 2007).

*S. invicta*, the invaders, is foraging all year in Guangdong province, has great capacity for plundering food resources, more aggressive, and a larger population, therefore they exhibits intense competitiveness than *T. melanocephalum* in our research. Short-term invasions by *S. invicta* also significantly affect *T. melanocephalum* in simple habitats (unpublished). If *S. invicta* invaded in a shortage resource and vegetation over simplified habitat, for example in lawn (*T. melanocephalum* workers only can be saw foraging in the border), may overcome *T. melanocephalum* and eventually replace it as the only dominant species would be further research.

## Acknowledgments

This study was supported by the National Basic Research Program, China (No. 2009CB119200), the National Natural Science Foundation of China (No. 305712427).

## References

Allen, C., Epperson, D. & Garmestani, A. (2004). Red imported fire ant impacts on wildlife: a decade of research. American Midland Nataturalist, 152: 88-103. doi:10.1674/0003-0031(2004)152[0088:RIFAIO]2.0.CO;2

Allen, C.R., Demarais, S. & Lutz, R.S. (1997). Effects of red imported fire ants on recruitment of white-tailed deer fawns. Journal of Wildlife Management, 61(3): 911-916.

Bhatkar, A., Whitcomb, W., Buren, W., Callahan, P. & Carlysle, T. (1972). Confrontation behavior between *Lasius neoniger* (Hymenoptera: Formicidae) and the imported fire ant. Environmental Entomology, 1: 274-279.

Cerdá, X., Arnan, X. & Retana, J (2013). Is competition a significant hallmark of ant (Hymenoptera: Formicidae) ecology? Myrmecological News, 18: 131-147.

Drees, B.M. (1994). Red imported fire ant predation on nestlings of colonial waterbirds. Southwestern Entomology, 19: 355-360.

Giuliano, W.M., Allen, C.R., Lutz, R.S., & Demarais, S. (1996). Effects of red imported fire ants on northern bobwhite chicks. Journal of Wildlife Management, 60: 309-313.

Gotelli, N. & Arnett, A. (2000). Biogeographic effects of red fire ant invasion. Ecology Letters, 3: 257-261. doi:10.1046/j.1461-0248.2000.00138.x

Hölldobler B. & Wilson EO. 1990. The Ants. Belknap Press of Harvard University Press. Cambridge, MA. 732 pp.

Holway, D.A., Lach, L., Suarez, A.V., Tsutsui, N.D. &Box, T.J. (2002). The causes and consequences of ant invasions. Annual Review of Ecology and Systematics, 33: 181-233. doi:10.1146/annurev.ecolsys.33.010802.150444

Hook, A.W. & Porter, S.D. (1990). Destruction of harvester ant colonies by invading fire ants in south-central Texas (Hymenoptera: Formicidae). Southwestern Naturalist, 35: 477-478. doi:10.2307/3672056

Jetter, K.M., Sausageilton, J. & Klotz, J.H. (2002). Eradication costs calculated: Red imported fire ants threaten agriculture, wildlife and homes. California Agriculture, 56: 26-34. doi:10.3733/ca.v056n01p26

Jones, S. & Phillips Jr., S. (1987). Aggressive and defensive propensities of *Solenopsis invicta* (Hymenoptera: Formicidae) and three indigenous ant species in Texas. Texas Journal of Science, 39: 107-115.

Jones, S.R., & Phillips Jr., S.A. (1990). Resource collecting abilities of *Solenopsis invicta* (Hymenoptera: Formicidae) compared with those of three sympatric Texas ants. Southwestern Naturalist, 35: 416-422.

Li, J., Han, S.C., Li, Z.G., & Zhang, B.S. (2008). The behavior observes of *Tapinoma melanocephalum* native competitive species of *Solenopsis invicta* [in Chinese, English abstract]. Plant Quarantine, 22: 19-21.

Morrison, L.W. (1999). Indirect effects of phorid fly parasitoids on the mechanisms of interspecific competition among ants. Oecologia, 121: 113-122. doi:10.1007/s004420050912

Morrison, L.W. (2000). Mechanisms of interspecific competition among an invasive and two native fire ants. Oikos, 90: 238-252. doi:10.1034/j.1600-0706.2000.900204.x

Morrison, L.W. (2002). Long-term impacts of an arthropod-community invasion by the imported fire ant, *Solenopsis invicta*. Ecology, 83:2337-2345.doi:10.1890/0012-9658(2002)083[2337:LTIOAA]2.0.CO;2

Obin, M.S. & Vander Meer, R.K. (1985). Gaster flagging by fire ants (*Solenopsis* spp.): functional significance of venom dispersal behavior. Journal of Chemical Ecology, 11: 1757-1768. doi:10.1007/BF01012125

Oster, G.F. & Wilson, E.O. (1978). Caste and ecology in the social insects. Princeton University Press, Princeton, New Jersey. 352 pp.

Parris, L.B., Lamont, M.M. & Carthy, R.R. (2002). Increased incidence of red imported fire ant (Hymenoptera: Formicidae) presence in loggerhead sea turtle (Testudines: Cheloniidae) nests and observations of hatchling mortality. Florida Entomologist, 85: 514-517. doi:10.1653/0015-4040(2002)085[0514:IIORIF]2.0.CO;2

Pascoe, A. (2002). Strategies for managing incursions of exotic animals to New Zealand. Micronesica Supplem., 6: 129-135.

Porter, S.D. (1992). Frequency and distribution of polygyne fire ants (Hymenoptera: Formicidae) in Florida. Florida. Entomologist, 75: 248-257.

Porter, S.D. & Savignano, D.A. (1990). Invasion of polygyne fire ants decimates native ants and disrupts arthropod community. Ecology, 71: 2095-2106. doi:10.2307/1938623

Porter, S.D., Van Eimeren, B. & Gilbert, L. (1988). Invasion of red imported fire ants (Hymenoptera: Formicidae): microgeography of competitive replacement. Annals of the

Entomological Society of America, 81: 913-918.

Reitz, S.R. & Trumble, J.T. (2002). Competitive displacement among insects and arachnids. Annual Review of Entomology, 47(1): 435-465. doi:10.1146/annurev.ento.47.091201.145227

Scheurer V.S. & Liebig G. (1998). *Tapinoma melanocephalum* Fabr. (Formicidae, Dolichoderinae) in gebäuden Beobachtungen zu ihrer Biologie und Bekämpfung. Anz. Schä dlingskde., Pflanzenschutz, Umweltschutz, 71: 147-148.

Shen, P., Zhao, X.L., Cheng, D.F., Zheng, Y.Q. & Lin, F.R. (2007). Impacts of the imported fire ant, *Solenopsisinvicta* invasion on the diversity of native ants. [in Chinese, English abstract]. Journal of the Southwestern China Normal University, 32: 93-97.

Smith E.H. & Whitman R.C. (1992). Field Guide to Structural Pests. National Pest Management Association, Dunn Loring, VA.

Smith, M.R. (1965). House-infesting ants of the eastern United States: their recognition, biology, and economic importance. Technical bulletin Nº1326, US Department of Agriculture.

Stewart, J. & Vinson, S.B. (1991). Red imported fire ant damage to commercial cucumber and sunflower plants. Southwestern Entomology, 16: 168-170.

Tomalski, M., Blum, M., Jones, T., Fales, H., Howard, D. & Passera, L. (1987). Chemistry and functions of exocrine secretions of the ants *Tapinoma melanocephalum* and *T. erraticum*. Journal of Chemical Ecology, 13: 253-263. doi:10.1007/BF01025886

Vinson, S.B. Impact of the invasion of the imported fire ant. Insect Science, 20: 439-455. doi:10.1111/j.1744-7917.2012.01572.x

Wang, L., Lu, Y.Y., Xu, Y.J. & Zeng, L. (2013). The current status of research on *Solenopsis invicta* Buren (Hymenoptera: Formicidae) in Mainland China. Asian Myrmecology, 5: 125-138.

Wilson, E. &Brown Jr., W. (1958). Recent changes in the introduced population of the fire ant *Solenopsis saevissima* (Fr. Smith). Evolution, 2: 211-218.

Wu, B.Q., Lu, Y.Y., Zeng, L. & Liang, G.W. (2008). Influences of *Solenopsis invicta* Buren invasion on the native ant communities in different habitats in Guangdong.[in Chinese, English abstract]. Chinese Journal of Applied Ecology, 19: 151-156.

Zheng, J., Mao, R. & Zhang, R. (2007). Comparisons of foraging activities and competitive interactions between the red imported fire ant (Hymenoptera: Formicidae) and two native ants under high soil-surface temperatures. Sociobiology, 50: 1165-1175.

Zheng, J.H. & Zhang, R.J. (2010).Interspecific competition between the red imported fire ant, *Solenopsis invicta* Buren and the ghost ant, *Tapinoma melanocephalum* (F.) for different food resources. [in Chinese, English abstract]. Journal of Environmental Entomology, 32: 312-317.

# 11

# Arboreal Ant Assemblages Respond Differently to Food Source and Vegetation Physiognomies: a Study in the Brazilian Atlantic Rain Forest

JJ Resende[1], PEC Peixoto[1], EN Silva[1], JHC Delabie[2,3] & GMM Santos[1]

1 - Universidade Estadual de Feira de Santana, Feira de Santana, Bahia, Brazil

2 - Universidade Estadual de Santa Cruz / CEPLAC/CEPEC, Ilhéus-Itabuna, Bahia, Brazil

3 - CEPLAC/Centro de Pesquisas do Cacau, Itabuna, Bahia Brazil

**Keywords**

Habitat preference, resource preference, matrix quality, Formicidae, Community Ecology

**Corresponding author**

Janete Jane Resende
Departamento de Ciências Biológicas
Universidade Estadual de Feira de Santana
Av. Transnordestina s/n Novo Horizonte
Feira de Santana, Bahia, Brazil
44036-900
E-Mail: antforjane@gmail.com

**Abstract**

This study aimed to analyze assemblages of arboreal ants in different vegetation physiognomies within the Tropical Moist Forest (Atlantic Rain Forest) domain. The study was carried out at the Michelin Ecological Reserve, State of Bahia, Northeast of Brazil. We used sardine (protein resource) and honey (carbohydrate resource) baits to collect ants foraging in three vegetation types: (1) preserved native forest, (2) forest in regeneration (capoeira) with many invasive plants and (3) a mixed agroystem of rubber and cocoa tree plantation. We recorded 69 ant species attracted to the baits, 21 of them exclusive to honey bait and 25 exclusive to the sardine baits. The vegetation physiognomies preserved forest and rubber/cacao agrosystem showed higher species richness in relation to the forest in regeneration (capoeira), suggesting that rubber tree plantations can be a good matrix for the maintenance of some ant species typical of the forest matrix. The type of resource used is important for the structuring of the arboreal ant assemblages. The ants that were attracted to protein resources showed a guild composition that is more differentiated between vegetation types that of ants attracted to glucose resources.

## Introduction

The richness and composition of ant assemblages have been related to the different structural aspects or the level of habitat preservation (Schoener 1971, Greenslade & Greenslade 1977, Levings 1983 & Andersen 1995). Ants are frequently chosen for studies that focus on the understanding of the effects of repetitive events of man-made habitat or ecosystem simplification on biodiversity (Matos *et al.* 1994, Majer 1996, Perfecto *et al.* 1997). Several studies have shown significant correlations between ant assemblages and habitat structural complexity, particularly in the tropics (Andersen & Majer, 2004, Delabie *et al.* 2006). For example, the richness of ants in the forest leaf litter has a strong correlation with plant diversity (Pereira *et al.* 2005) and in coffee agrosystems, the diversity of twig dwelling ants increases in habitats with more diverse shade tree cover (Armbrecht *et al.* 2004).

Human activities have often caused the simplification of natural environments, leading to local extinction of popula-

tions and species and, consequently, could negatively impact important ecological processes such as nutrient cycling, seed dispersal and pollination (Thomas 2000, De Marco & Coelho 2004). Simplified environments often harbor a lower richness and diversity of ants, with an ant fauna consisting of generalist species (Sobrinho & Schoereder 2006), unlike forested habitats that harbor ant assemblages with higher levels of diversity, consistent with the characteristics and the complexity of the vegetation (Majer 1996, Pereira *et al.* 2007).

Different sampling methods have been used in surveys of ants and there is no direct means of comparison between different collection procedures (Romero & Jaffe 1989). There are several methods used for sampling ants such as oil sardine, carbohydrates, meat and cassava flour baits, Winkler extractor, pitfall or manual collection, each one of them suited to select different classes of ants (soil dwelling, carnivorous, detritivorous, omnivorous) (Bestelmeyer *et al.* 2000; Freitas *et al.* 2004). Each of the collection procedures sampled a different set of ant species (Romero & Jaffe 1989) and the foraging activities may reflect or indicate the nutrients that are most limiting to in the respective nesting habitats, some ants prefer honey baits (carbohydrate) and other ants set prefer fish baits (nitrogen- protein) (Hashimoto *et al.* 2010). Ants provide an ideal system to test how macronutrient availability affects the costs and benefits of competitive dominance (Grover et al., 2007). Considerable evidence suggests that resource competition strongly influences population and community dynamics in ants (Hölldobler & Wilson 1990)

This study aimed to analyze arboreal ant assemblages in different vegetation physiognomies within the Atlantic Rain Forest domain. We tested two hypotheses. First, if the agroforestry system constitutes a good matrix for the maintenance of ant species, richness of ants would be expected to be similar to or even greater than that found in forest like physiognomies. Second, since trophic groups of ants that are glycoside and protein consumers are generalist and specialist, respectively, the ant assemblages of generalist ants are expected to be more similar while the second ones would be more dissimilar between habitats.

**Material and Methods**

*Study area*

The study was conducted at the Michelin Ecological Reserve (headquarters: 13° 50'S, 39° 10' W) Ituberá, state of Bahia, Brazil. The climate is of the type As according to the Köppen classification, tropical, rainy, hot, characterized by rainfall concentrated in summer and autumn, and average temperatures are never below 20° C. The landscape of the region is characterized by the dominance of cacao agroforestry, mixed rubber/cacao plantations, pastures and forest fragments with natural vegetation comprising primary forest (a small percentage) and different stages of forest regeneration (with a large proportion of secondary forest).

We collected ants in six different periods between October 2007 and September 2008. Ant assemblies attracted to carbohydrate and protein baits were collected in three vegetation physiognomies: (I) Conserved Atlantic Forest Fragment (PF), a typical Low Land Humid Tropical Forest measuring about 550 ha comprising blocks of native forest canopy of 15-25 m in height, with isolated trees reaching 30-40 m; (II) Forest Regeneration Fragment (Capoeira - SF), an early to intermediate stage of secondary succession tract of land, measuring about 10 ha of contiguous forest in regeneration, characterized by the presence of lianas, bromeliads, orchids, rocky ground, shrub vegetation and invasive plants and (III) agrosystem of mixed rubber tree/cacao plantation (AG), with over 20 years of age and 12 ha of area.

*Sampling methods*

The sampling consisted of 18 transects distributed among three vegetation types. In each vegetation type six

transects of 400 m were established. In each transect there were 20 sampling points spaced at 20 m intervals. Each sampling point consisted of a bait rich in proteins and lipids (sardine oil) and another bait rich in carbohydrates (honey) installed within the same tree at a height of 2 m and at least 20 cm away from one another.

After installation, the baits remained on the plants for about 30 min. They were then collected and the ants present were fixed in alcohol 70%. In total, 120 baits were placed in each vegetation type. The sorting, assembly and morphospeciation of ant specimens occurred in the Laboratory of Entomology, at Feira de Santana State University (UEFS). For the identification of ant species we used the classification of Bolton (2003), except for the genus *Nylanderia* (based on Lapolla *et al.* 2010), and the genera *Strumigenys* e *Basiceros* (based on Baroni-Urbani & De Andrade 2007).

Vouchers were deposited in the Entomological Collection Prof. Johann Becker, Museum of Zoology - UEFS (MZFS) at Feira de Santana and the collection of the Myrmecology Laboratory of CEPEC/CEPLAC (CPDC) at Ilhéus, state of Bahia.

*Data analyses*

The analyses of data took into consideration only presence or absence of the species as usual in ant community ecology studies (Longino 2000). The observed species richness was calculated using the rarefaction curve (Mao Tau) (Colwell *et al.* 2004). The total richness was estimated using the 1st order Jackknife estimator based on 50 randomizations (Heltshe & Forrester, 1983), performed with the program EstimateS, version 7.5.2 (Colwell, 2006).

The dissimilarity of ant assemblages between vegetation types was assessed by performing a Principal Coordinate Analysis (PCoA) using the program R (R Development Core Team 2010). This analysis was preceded of the calculation of the Jaccard similarity index for the pair-wise combination of data collected in all transects, considering as a transect the 20 samples collected by vegetation type and sampling date. Therefore, the analyses were based on 18 transects (three vegetation types times six dates).

**Results**

We recorded 69 species of ants attracted to the baits. Out of this total, 21 species were exclusive to honey baits and 25 to sardine baits (Table 1). With respect to vegetation types, 17 species were exclusive to Conserved Forest, 16 to the agrosystem and seven to the "Capoeira". Regarding the frequency of species, *Solenopsis* sp.2 was the most frequent on both sardine (31.6%) and honey (20%) baits in rubber/cacao tree agroforestry. The second most frequent species in this physiognomy was *Camponotus* sp.5 which was recorded on 11.6% of sardine baits.

In the Conserved Forest physiognomy, the most fre-

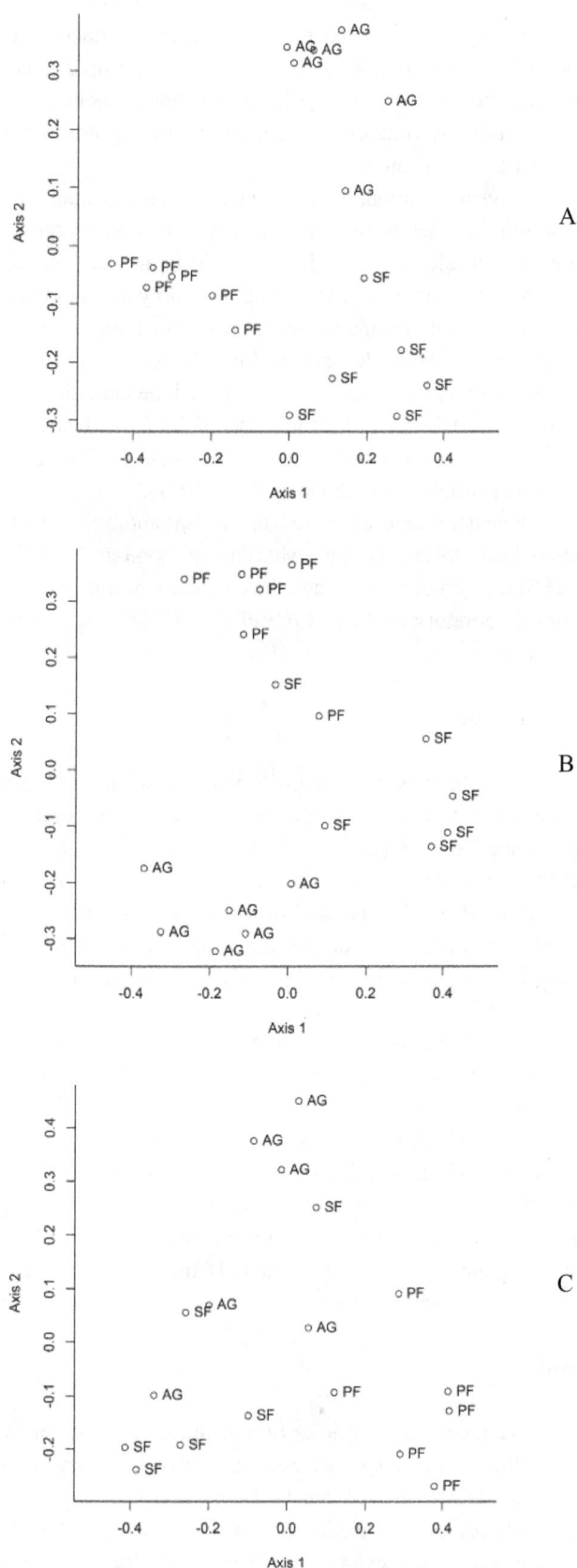

Fig. 1. Ordination by PCoA of species of ants collected in different phytophysiognomies, Reserva Ecológica da Michelin, Ituberá e Igrapiúna municipalities Bahia, Brazil. **PF** – Conserved forest; **SF** - Capoeira; **AG** – Agrosystem of mixed rubber tree/cacao plantation. (**A**) ants collected in honey and sardine baits; (**B**) ants collected in sardine baits and (**C**) ants collected in honey baits.

quent ants on sardine baits were *Crematogaster* sp.1 (9.1% of the baits) and *Ectatomma tuberculatum* (8.3%). The same happened in the "Capoeira" area with *Crematogaster* sp.1 present on 10% of the baits and *E. tuberculatum* in 9.1%. For honey baits, *Strumigenys* sp.9 (5.8%) was the most frequent species in conserved forest while *Crematogaster* sp.1 (6.6%) was the most abundant in "Capoeira".

Among the most speciose genera, *Pheidole* (14 species), *Pachycondyla* (8), *Camponotus* (7) and *Solenopsis* (5) were the most rich. The proportion of species belonging to these genera present in both types of baits remained very close, except for *Camponotus* which had two and seven species recorded on honey and sardine baits, respectively.

The two axis extracted from the PCoA analyses with data from the two bait types grouped explained 34.3% of the total variation in ant composition among samples (18.6% of variation explained by axis 1 and 15.6% of variation explained by axis 2). According to this analysis, the species collected in sardine differed from the species collected in honey baits. Considering only ants collected in sardine baits, there was also a very clear separation between vegetation types, indicating that the composition of ant species differs between vegetation types (17.6% of variation explained by axis 1 and 17.1% of variation explained by axis 2). On the other hand, the analysis including only ants collected in honey baits demonstrated a low distinction between vegetation types (18.4% of variation explained by axis 1 and 11.7% of variation explained by axis 2), indicating that the assemblage of species that visited the baits rich in carbohydrates were similar among the three habitats (Fig. 1).

The greatest observed and expected richness of ants were recorded in the preserved forest fragment and in the agrosystem of mixed rubber/cacao plantation (Fig. 2, Table 1).

The observed richness curves (Mao Tau) showed no stabilization in any of phytophysiognomies (Fig. 2). Never-

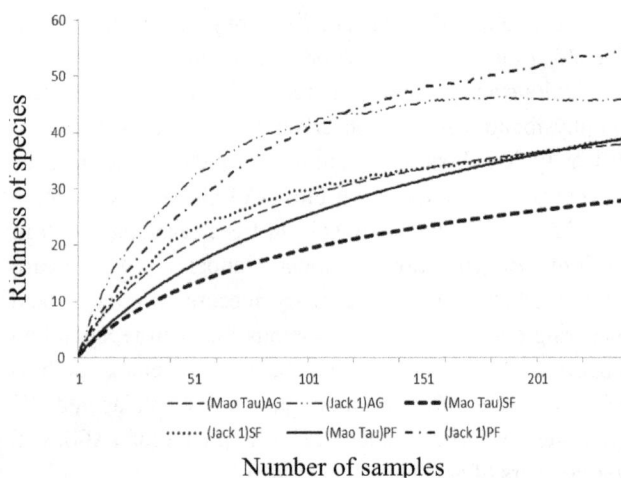

Fig. 2. Expected (Jack1) and observed (Mao Tau) richness curves for species of ants collected in a total of 240 samples. Reserva Ecológica da Michelin, Ituberá e Igrapiúna municipalities, Bahia, Brazil. **PF** – Conserved forest; **SF** - Capoeira; **AG** – Agrosystem of mixed rubber tree/cacao plantation.

theless, the expected richness according to the species accumulation curve (jack1) showed a sharp increase in the number of species as a function of sample size, followed by an asymptote in the agrosystem habitat. The shape of the curve suggests that in the agrosystem the sampled fauna is more homogeneous than in "Capoeira" and conserved forest. Therefore the majority of the species is sampled with a lower number of samples. Extrapolation of the curves also suggests that the number of total species in the agrosystem is lower than the two other habitats.

## Discussion

According to our results, the vegetation physiognomies of conserved forest and agrosystem presented species richness highest than "Capoeira", showing that agricultural habitats including the association between cacao and rubber trees forms a suitable matrix for the maintenance of ant species typical of forest environment. However, the juxtaposition of forest areas close to the agrosystem is an important point allowing to maintain the ant diversity (Delabie *et al*. 2007).

*Similarity between bait types and vegetation physiognomies*

The similarity between the ant assemblages sampled in sardine baits among the studied vegetation physiognomies, is probably a result of the shared occurrence of species of the genera *Azteca*, *Camponotus*, *Crematogaster*, *Pheidole* and *Solenopsis* which are considered dominant or subdominant if considering the structure of arboreal ant fauna (Wilson 1976; Majer *et al*. 1994; Brandão *et al*. 2009). Some species of these genera can significantly influence the structure of the arthropod community, exercising strong predation, especially on larvae of Lepidoptera and Coleoptera (Majer & Delabie 1993, Floren *et al*. 2002; Philpott & Armbrecht 2006).

This same ecological context may apply to explain the great similarity between the ant assemblages attracted to honey and sardine baits within the agrosystem, because almost the same species were present in both types of baits. Other species that were recorded in this physiognomy such as *E. tuberculatum*, *Pachycondyla venusta* and *Odontomachus haematodus* belong to genera typically considered as those of generalist predator species. Except for *P. venusta*, these ants may supplement their diet with nectar exudates from plants and honeydew producer insects (Hölldobler & Wilson 1990, Delabie 2001). An additional outcome supporting this explanation is the occurrence of the species *Tapinoma* sp.1 in the agrosystem of mixed rubber/cacao plantation. This species has as main feature the generalist behavior, being undemanding in terms of habitat quality and shifting easily from one food source to another.

The numerous species of ants visiting sardine baits in each vegetation physiognomy, such as those of the genera *Hypoponera*, *Megalomyrmex*, *Nylanderia*, *Wasmannia* in

Conserved Forest, *Linepithema* in Capoeira and *Dorymyrmex* in the agroystem, has an important ecological implication. In case where protein-based resources are scarce in these habitats, they will be almost exclusively used by specialist species that tend to defend this resource with aggressive behaviors, by exhibiting a rapid recruitment of workers and thus preventing access of other species. The species of the genus *Solenopsis*, for instance, have aggressive behavior and are common in disturbed habitats. On the other hand, carbohydrate-based resources are more common and visited by generalist species, which do not have any preference to the types of bait used, such as those of the genera *Camponotus*, *Pheidole* and *Solenopsis* (Hölldobler & Wilson 1990).

Although, in general, species richness correlates positively with habitat complexity, this correlation seems to depend on the habitat, because the agrosystem had a higher observed richness compared to "Capoeira", which is considered structurally more complex than the former. Lassau & Hochuli (2004) found similar results with a greater richness in less complex habitats, believing that the movement of ants may be more efficient in terms of energy, in less complex habitats. Gomes *et al* (2010) demonstrates that ant fauna is more influenced by vegetation integrity than by fragment size, distance to edge or forest cover surrounding fragments. Lopes *et al* (2012) shows that the species that compose the ant assemblies in different phytophysiognomies are a reflex of the environment, especially of the plant species, supporting the hypothesis that differences in the vegetational composition result in different position of the ant assembly.

The higher number of species in the agrosystem in comparison to "Capoeira" suggests the occurrence of occasional species with low occurrence, sometimes called "tourists". These species are treated by the estimators of richness as singletons or doubletons and can boost the expected richness estimation. In spite of this, a greater number of species considered dominant (genera *Camponotus*, *Crematogaster*, *Solenopsis* and *Pheidole*) was recorded for the agrosystem in comparison to "Capoeira".

It is important to remark that although the species accumulation curve based on Jack1 reached an asymptote in the agrosystem, the estimated richness in the Capoeira is in an ascending trend and could surpass the estimated richness of the agrosystem if the sampling effort were increased. This leaves the inference on the quality of the rubber/cacao agrosystem as a good matrix open, from the perspective of which habitat is richer.

However, looking at the list of species in Table 1, the following results are found. Out of the 69 ant species collected in all habitats, 13 (18.84%) are common to both conserved forest and agrosystem. Six species are protein consumers and seven carbohydrate consumers. Furthermore, the relative frequencies of most of these species are low and very similar in forest and agrosystem habitats. Thus, from the point of view of life history, we can infer that the agrosystem is a good quality matrix for at least some rare species that also occur in forest vegetation.

The choice of food resource (honey or sardine baits) by the ant assemblages, clearly differentiate the three vegetation physiognomies.

Both proteins and carbohydrates are of paramount importance for maintenance of ant populations. Hashimoto *et al.* (2010) demonstrated that different populations of ants forage more actively on these resources when they are scarce. On one hand, the proteic resources are fundamental for brood development and tissue synthesis (Hölldobler & Wilson 1990). On the other hand, carbohydrates are key resources for the maintenance of the ant activity (Davidson, 1998). Byk and Del-Claro (2011) demonstrated the benefits of the carbohydrate resources to ant populations. The glucidic foods influence ant species composition, abundance, number of individuals per colony, body weight, survivorship, growth, reproduction and interactions at the community level. Bihn et al. (2008) examined bait preferences of litter ants along a successional gradient of forests in the Atlantic Forest of Brazil and observed that ants preferred protein-based baits in secondary forests, and carbohydrate-based baits in old-growth forests. In addition, the ant preference for carbohydrate or protein is subject to change individually according a previous ingestion of one of this food source, extrafloral nectar or insect. This is the theory of ecological stoichiometry, which relates nutrient balance to ecological processes (Sterner & Elser 2002). Such nutritional complexity can mediate patterns of ecological interactions (Anderson *et al.* 2004). Therefore, if an ant individual is consuming extrafloral nectar, it is more propable it attacks a herbivore on the host plant aiming to balance the ingestion of protein after too much ingestion of carbohydrate (Grover et al., 2007).

Our data support our working hypothesis that the type of resource used by ant species is important in structuring the community. The distribution of ant species that are specialist and consumers of protein-based resources (sardine baits) differed more between the vegetation types than the distribution of ant species that are generalist and consumers of carbohydrate-based resources. The similarity of species collected in each vegetation physiognomy was higher among ant assemblages that use carbohydrate-based resources (Fig. 1c) than among ant assemblages that use protein-based resources, in an exclusive manner (Fig. 1b) or together with carbohydrate sources (Fig. 1c).

## Acknowledgements

To CNPq (Project 620021/2008-0 – Edital nº 16/2008 and PRONEX FAPESB-CNPq project PNX 011-2009); to Foundation for Research Support de in Bahia state (Fundação de Amparo à Pesquisa do Estado da Bahia - FAPESB (Project 5577/2009) and the Michelin Ecological Reserve (REM) supported this study. GMMS and JHCD thanks to CNPq for productivity Fellowship. To many colleagues that contributed to this study, specially Allan Rhalff Gomes Teixeira and Emerson Mota da Silva for helped with field sampling. We would like to thank two anonymous reviewers provided their insightful comments on the manuscript.

## References

Andersen, A.N. & Majer, J.D. (2004). Ants show the way Down Under: invertebrates as bioindicators in land management. Front. Ecol. Environ., 2: 291-298.

Andersen, A.N. (1995). A Classification of Australian Ant Communities, Based on Functional Groups Which Parallel Plant life-forms in relation to stress and disturbance. J. Biogeog., 22:15-29.

Anderson, T., Elser, J. J. & Hessen, D. O. (2004). Stoichiometry and population dynamics. Ecol. Lett., 7: 884–900. doi: 10.1111/j.1461-0248.2004.00646.x

Armbrecht, I., Perfecto, I.& Vandermeer, J. (2004). Enigmatic Biodiversity: Ant Diversity Responds to Diverse Resources. Science 304:284-286.

Baroni-Urbani, C. & De Andrade, M.L. (2007). The ant tribe Dacetine: limits and constituent genera, with descriptions of new species. Ann. Mus. Civ. Stor. Nat. Giacomo Doria, 99:1-191.

Bestelmeyer, B.T., Agosti, D., Leeanne, F., Alonso, T., Brandão, C.R.F., Brown, W.L., Delabie, J.H.C., & Silvestre, R. (2000). Field techniques for the study of ground-living ants: An Overview, description, and evaluation. *In*: D. Agosti, J.D. Majer, A. Tennant & T. de Schultz (eds), Ants: standart methods for measuring an monitoring biodiversity. Smithsonian Institution Press, Washington 122-144.

Bihn, J.H., Verhaagh, M., Brand, R. (2008). Ecological stoichiometry along a gradient of forest succession: Bait preferences of litter ants. Biotropica, 40: 597-599.

Bolton, B. (2003). Synopsis and classification of Formicidae. The American Entomological Institute. Gainesville 370p.

Brandão, C.R.F., Silva, R.R. & Delabie, J.H.C. (2009). Formigas (Hymenoptera), pp. 323–369. In: A.R.Panizzi & J.R.P. Parra (eds). Bioecologia e Nutrição de Insetos. Base para o Manejo Integrado de Pragas. Brasília, Embrapa Informação Tecnológica, 1263p.

Byk, J. & Del-Claro, K. (2011). Ant-plant interaction in the Neotropical savanna: direct beneficial effects of extrafloral nectar on ant colony fitness. Popul. Ecol., 53(2): 327-332.

Chazdon, R.L., Colwell, R.K., Denslow, J.S. & Guarigurata, M.R. (1998). Statistical methods for estimating species richness of woody regeneration in primary and secondary rain forests of northeastern Costa Rica. In Forest biodiversity research, monitoring and modelling. (F. Dallmeier & J. A.

Comiskey, eds). MAB - Man and Biosphere Series, UNES-CO, Paris 285-309.

Colwell, R. K. (2006). EstimateS: Statistical estimation of species richness and shared species from samples, version 8.0. Disponível em: http://viceroy.eeb.uconn.edu/EstimateS

Colwell, R. K., Mao, C.X. & Chang, J. (2004). Interpolating, extrapolating, and comparing incidence-based species accumulation curves. Ecology 85:2717-2727.

Davidson D.W. (1998) Resource discovery versus resource domination in ants: a functional mechanism for breaking the trade-off. Ecol. Entomol., 23:484–490

De Marco P. & Coelho, F.M. (2004). Services performed by the ecosystem: Forest remnants influence agricultural cultures' pollination and production. Biodiv. Conserv., 13: 1245-1255.

Delabie, J.H.C. (1993). Formigas exóticas na Bahia. Bahia, Análise Dados. 3:19-22.

Delabie, J.H.C. (2001). Trophobiosis between Formicidae and Hemiptera (Sternorrhyncha and Auchenorrhyncha): an overview. Neotrop. Entomol., 30(4): 501-516.

Delabie, J.H.C., Jahyny, B., Nascimento, I.C., Mariano, C., Lacau, S., Campiolo, S., Philpott, S.M. & Leponce, M. (2007). Contribution of cocoa plantations to the conservation of native ants (Insecta: Hymenoptera: Formicidae) with a special emphasis on the Atlantic Forest fauna of southern Bahia, Brazil. Biodiv. Conserv., 16: 2359-2384.

Delabie, J.H.C., Fisher, B.L., Majer, J.D. & Wright, I.W. (2000). Sampling Effort and Choice of Methods. In: D. Agosti; Majer, J.D.; Tennant, A.; Schultz, T.R. (Eds). Ants: standard methods for measuring and monitoring biodiversity. Smithsonian Institution Press, Washington, DC., USA 145-154.

Delabie, J.H.C., Paim, V.R.L., Nascimento, I.C., Campiolo, S. & Mariano, C.S.F. (2006). As Formigas como Indicadores Biológicos do Impacto Humano em Manguezais da Costa Sudeste da Bahia. Neotrop. Entomol., 35(5):602-615.

Dunn, R.R., Agosti, D., Andersen, A.N., Arnan, X., Bruhl, C.A., Cerdá, X., Ellison, A.M., Fisher, B.L., Fitzpatrick, M.C., Gibb, H., Gotelli, N.J., Gove, A.D., Guenard, B., Janda, M., Kaspari, M., Laurent, E.J., Lessard, J.P., Longino, J.T., Majer, J.D., Menke, S.B., McGlynn, T.P., Parr, C.L., Philpott, S.M., Pfeiffer, M., Retana, J., Suarez, A.V., Vasconcelos, H.L., Weiser, M.D. & Sanders, N.J. (2009). Climatic drivers of hemispheric asymmetry in global patterns of ant species richness. Ecol. Lett., 12: 324-33.

Floren, A., Biun, A.& Linsenmair, K.E. (2002). Arboreal ants as key predators in tropical lowland forest trees. Oecologia, 131: 137-144

Freitas, A.V.L, Francini, R.B. & Brown, K.S. (2004). Insetos como indicadores ambientais. Pp. 125-151. In: Cullen Jr., L.;

Rudran, R. & Valladares-Padua, C. (Org.). Métodos de estudos em biologia da conservação e manejo da vida silvestre. Editora UFPR, Curitiba. 665p.

Gomes, J. P., Iannuzzi, L. & Leal, I.R. (2010). Resposta da comunidade de formigas aos atributos dos fragmentos e da vegetação em uma paisagem da Floresta Atlântica nordestina. Neotrop. Entomol., 39(6):898-905

Greenslade, P.J.M. & Greenslade, P. (1977). Some effects of vegetation cover and disturbance on a tropical ant fauna. Insectes Soc., 24: 163-182.

Grover, C.D., Kay, A.D., Monson, J.A., Marsh, T.C., Holway, D.A. (2007). Linking nutrition and behavioural dominance: carbohydrate scarcity limits aggression and activity in Argentine ants. Proc. R. Soc. B, 274: 2951–2957. doi:10.1098/rspb.2007.1065

Hashimoto Y., Morimoto Y., Widodo, E.S., Maryati Mohamed & Fellowes, J.R. (2010). Vertical habitat use and foraging activities of arboreal and ground ants (Hymenoptera: Formicidae) in a Bornean tropical rainforest. Sociobiology, 56(2): 1-14.

Heltshe, J. & Forrester, N.E. (1983). Estimating species richness using the jackknife procedure. Biometrics, 39: 1-11

Hölldobler, B. & Wilson, E.O. (1990). The ants. Harvard University Press, Cambridge, 732p.

Lapolla, J., Brady, S.G. & Shattuck, S.O. (2010). Phylogeny and taxonomy of the Prenolepis genus-group of ants (Hymenoptera: Formicidae). System. Entomol., 35: 118-131.

Lassau, S.A. & Hochuli, D.F. (2004). Effects of habitat complexity on ant assemblages. Ecography, 27: 157-164.

Levings, S.C. (1983). Seasonal, annual, and among-site variation in the ground ant community of a deciduous tropical forest: some causes of patchy species distributions. Ecol. Monogr., 53: 435-455.

Longino, J.T. 2000. Que fazer com os dados. p.186 - 203. In: Agosti, D.; Majer, J. D.; Alonso, L. E. & Schultz, T. R. (Eds.) Formigas: Métodos padrão para medição e monitoramento da biodiversidade. Smithsonian Institution Press, Washington, DC 280.pp

Lopes, J.F.S., Hallack, N.M. dos R., Sales, T.A. de, Brugger, M.S., Ribeiro, L.F., Hastenreiter, I.N. & Camargo, R. da S. (2012). Comparison of the Ant Assemblages in Three Phytophysionomies: Rocky Field, Secondary Forest, and Riparian Forest—A Case Study in the State Park of Ibitipoca, Brazil, Psyche, vol. 2012, Article ID 928371, 7 pages, 2012. doi:10.1155/2012/928371

Majer, J. D. & Delabie, J.H.C. (1993). An evaluation of Brazilian cocoa farm ants as potential biological control agents, J. Plant Prot. in the Tropics, 10(1): 43-49.

Majer, J. D., Delabie, J.H.C. & Smith, M.R.B. (1994). Arboreal ant community patterns in brazilian cocoa farms, Biotropica, 26(1):73-83.

Majer, J.D. (1996). Ant recolonization of rehabilitated bauxite mines at Trombetas, Pará, Brazil. J. Appl. Ecol., 12: 257-273

Matos, J. Z., Yamanaka, C.N., Castellani, T.T. & Lopes, B.C. (1994). Comparação da fauna de formigas de solo em áreas de plantio de *Pinus elliottii*, com diferentes graus de complexidade estrutural (Florianópolis, SC). Biotemas, 7(1-2): 57-64.

Pereira, M.P.S., Queiroz, J.M., Souza, G.O. & Mayhé-Nunes, A.J. (2007). Influência da heterogeneidade da serapilheira sobre as formigas que nidificam em galhos mortos em floresta nativa e plantio de eucalipto. Neotrop. Biol. Conserv. 2(3): 161-164.

Pereira, M.P.S., Queiroz, J.M., Valcarcel, R. & Nunes, A.J.M. (2005). Fauna de formigas no biomonitoramento de ambientes de área de empréstimo em reabilitação na ilha da madeira, RJ. In: Simpósio Nacional e Congresso 3 Latino-Americano de Recuperação de Áreas Degradadas. Curitiba. Anais Sobrade, 6: 5-12.

Perfecto, I., Vandermeer, J., Hanson, P.& Cartin, V. (1997). Arthropod biodiversity loss and the transformation of a tropical agro-ecosystem. Biodiv. Conserv., 6: 935-945.

Philpott, S. M., Perfecto, I. & Vandermeer, J. (2006). Effects of management intensity and season on arboreal ant diversity and abundance in coffee agroecosystems. Biodiv. Conserv. 15: 139–155.

Resende, J.J., Santos, G.M.M., Nascimento, I.C., Delabie, J.H.C. & Silva, E.M. (2011). Communities of Ants (Hymenoptera – Formicidae) in Different Atlantic Rain Forest Phytophysionomies. Sociobiology, 58: 779-799.

Romero, H. & Jaffé, K. (1989). A comparison of methods for sampling ants (Hymenoptera, Formicidae) in savannas. Biotropica, 21: 348-325.

Schoener, T.W. (1971). Theory of feeding strategies, Ann. Rev. Ecol. Syst., 2: 369-404.

Sobrinho T.G. & Schoereder, J.H. (2006). Edge and shape effects on ant (Hymenoptera:Formicidae) species richness and composition in forest fragments. Biodiv. Conserv. 16: 1459-1470.

Sterner, R.W. & Elser, J.J. (2002). Ecological stoichiometry. Princeton, NJ: Princeton University Press. 584pp.

Thomas, C. D. (2000). Dispersal and extinction in fragmented landscapes. Proc. R. Soc. Lond. B Biol. Sci. 267: 139-145.

Wilson, E. O. (1976). Which are the most prevalent ant genera? Studia Ent., 9: 1-4.

**Table 1.** Number of records and relative frequency (%) of ants collected in honey and sardine baits in three vegetation physiognomies in the Reserva Ecológica da Michelin. Ituberá e Igrapiúna municipalities, Bahia, Brazil **PF** – Preserved forest; **SF** - Secondary Forest (Capoeira); **AG** – Agrosystem of mixed rubber tree/cacao plantation. Food Preferences (C= Primarily Carbohydrate consumer, P= Primarily Protein consumer, C/P = Both; *sensu* Brandão et al. 2009).

| Sampled Species | Food Preferences | Honey bait | | | Sardine bait | | |
|---|---|---|---|---|---|---|---|
| | | PF | SF | AG | PF | SF | AG |
| AMBLYOPONINAE | | | | | | | |
| *Prionopelta* sp.1 | P | - | 1 (0.8) | - | - | - | - |
| ECTATOMMINAE | | | | | | | |
| *Ectatomma* sp.1 | P | - | 2 (1.6) | - | - | - | - |
| *Ectatomma brunneum* F. Smith, 1858 | P | - | 3 (2.5) | - | - | - | - |
| *Ectatomma tuberculatum* F. Smith, 1858 | P | - | 4 (3.3) | 8 (6.6) | 10 (8.3) | 11 (9.1) | 7 (5.8) |
| *Gnamptogenys* sp.3 | P | - | - | - | 3 (2.5) | - | - |
| DOLICHODERINAE | | | | | | | |
| *Azteca* sp.1 | C/P | 2 (1.6) | - | 3 (2.5) | 7 (5.8) | 1 (0.8) | 6 (5) |
| *Azteca* sp.28 | C/P | 4 (3.3) | - | - | 3 (2.5) | 2 (1.6) | 3 (2.5) |
| *Azteca* sp.3 | C/P | - | - | - | - | - | 2 (1.6) |
| *Azteca* sp.4 | C/P | - | - | 3 (2.5) | - | - | - |
| *Dorymyrmex* sp.1 | C | - | - | - | - | - | 4 (3.3) |
| *Linepithema* sp.1 | C | - | 3 (2.5) | 4 (3.3) | - | 2 (1.6) | - |
| *Linepithema* sp.3 | C | - | - | - | - | 1 (0.8) | - |
| *Tapinoma* sp.1 | | 1 (0.8) | 2 (1.6) | 4 (3.3) | - | - | 1 (0.8) |
| FORMICINAE | | | | | | | |
| *Camponotus* sp.2 | C | - | - | - | - | 5 (4.1) | - |
| *Camponotus* sp.3 | C | 1 (0.8) | - | - | - | 4 (3.3) | - |
| *Camponotus* sp.4 | C | - | - | - | 3 (2.5) | - | - |
| *Camponotus* sp.5 | C | - | - | 3 (2.5) | - | - | 14 (11.6) |
| *Camponotus* sp.6 | C | - | - | - | - | - | 5 (4.1) |
| *Camponotus* sp.7 | C | - | - | - | - | - | 3 (2.5) |
| *Camponotus* sp.8 | C | - | - | - | - | - | 1 (0.8) |
| *Brachymyrmex* sp.1 | C/P | 3 (2.5) | - | 3 (2.5) | - | - | - |
| *Brachymyrmex* sp.4 | C/P | - | 3 (2.5) | - | 1 (0.8) | 1 (0.8) | - |
| *Nylanderia* sp.2 | C/P | - | - | - | 1 (0.8) | - | - |
| *Nylanderia* sp.3 | C/P | - | - | 1 (0.8) | - | - | - |
| *Nylanderia* sp.4 | C/P | - | - | 1 (0.8) | 3 (2.5) | - | - |
| MYRMICINAE | | | | | | | |
| *Cephalotes atratus* (Linneus, 1758) | C | - | 1 (0.8) | - | - | - | - |
| *Cephalotes* sp.1 | C | - | - | - | - | 3 (2.5) | - |
| *Cephalotes* sp.2 | C | 1 (0.8) | - | - | - | - | - |
| *Crematogaster* sp.1 | C | - | 8(6.6) | 3 (2.5) | 11 (9.1) | 12 (10) | 8 (6.6) |
| *Crematogaster* sp.2 | C | 1 (0.8) | - | 2 (1.6) | 1 (0.8) | 1 (0.8) | - |
| *Crematogaster* sp.3 | C | - | - | 3 (2.5) | - | 1 (0.8) | 12 (10) |
| *Crematogaster* sp.5 | C | - | - | - | 2 (1.6) | 1 (0.8) | - |
| *Crematogaster* sp.6 | C | - | - | 10 (8.3) | - | - | - |
| *Pheidole* sp.1 | C/P | 2 (1.6) | 1 (0.8) | - | - | - | - |
| *Pheidole* sp.2 | C/P | - | - | - | 1 (0.8) | - | - |

Table 1 (continued)

| | | | | | | | |
|---|---|---|---|---|---|---|---|
| *Pheidole* sp.5 | C/P | - | - | - | - | - | 3 (2.5) |
| *Pheidole* sp.10 | C/P | - | 1 (0.8) | 1 (0.8) | - | 3 (2.5) | 1 (0.8) |
| *Pheidole* sp.12 | C/P | - | - | - | 3 (2.5) | - | - |
| *Pheidole* sp.14 | C/P | - | 6 (5) | - | 4 (3.3) | - | - |
| *Pheidole* sp.15 | C/P | 1 (0.8) | 1 (0.8) | - | - | - | - |
| *Pheidole* sp.16 | C/P | - | - | - | 5 (4.1) | - | 3 (2.5) |
| *Pheidole* sp.17 | C/P | 1 (0.8) | - | 2 (1.6) | - | - | - |
| *Pheidole* sp.18 | C/P | 1 (0.8) | - | 1 (0.8) | - | - | - |
| *Pheidole* sp.20 | C/P | 1 (0.8) | - | - | - | - | 3 (2.5) |
| *Pheidole* sp.23 | C/P | 1 (0.8) | - | - | - | - | 1 (0.8) |
| *Pheidole* sp.24 | C/P | 2 (1.6) | 1 (0.8) | - | 1 (0.8) | - | - |
| *Pheidole* sp.26 | C/P | 3 (2.5) | - | 2 (1.6) | - | - | - |
| *Carebara pilosa* Fernández, 2004 | P | 2 (1.6) | - | - | - | - | - |
| *Megalomyrmex* sp.1 | P | - | - | - | 2 (1.6) | - | - |
| *Solenopsis* sp.1 | P | - | - | - | - | - | 1 (0.8) |
| *Solenopsis* sp.2 | P | - | 5 (4.1) | 24 (20) | 2 (1.6) | 4 (3.3) | 38 (31.6) |
| *Solenopsis* sp.3 | P | - | - | 2 (1.6) | - | 5 (4.1) | - |
| *Solenopsis* sp.4 | P | 2 (1.6) | 2 (1.6) | 4 (3.3) | 7 (5.8) | 1 (0.8) | - |
| *Solenopsis* sp.5 | P | - | - | 2 (1.6) | - | - | 5 (4.1) |
| *Strumigenys* sp.8 | P | - | - | - | 1 (0.8) | - | - |
| *Strumigenys* sp.9 | P | 7 (5.8) | | | | | |
| *Wasmannia auropunctata* (Roger, 1863) | C/P | - | - | - | 1 (0.8) | - | - |
| *Basiceros (Octostruma)* sp.2 | P | - | - | 1 (0.8) | - | - | - |
| PONERINAE | | | | | | | |
| *Hypoponera* sp.1 | P | - | - | - | 1 (0.8) | - | - |
| *Hypoponera* sp.2 | P | - | - | - | 2 (1.6) | - | |
| *Pachycondyla apicalis* (Latreille, 1802) | P | 2 (1.6) | - | - | - | - | - |
| *Pachycondyla constricta* (Mayr, 1884) | P | - | - | - | - | - | 1 (1.6) |
| *Pachycondyla complexo villosa* (Fabricius, 1804) | P | - | - | 5 (4.1) | - | - | - |
| *Pachycondyla harpax* (Fabricius, 1804) | P | - | - | - | 1 (0.8) | - | - |
| *Pachycondyla venusta* (Forel, 1912) | P | - | 2 (1.6) | 2 (1.6) | - | 3 (2.5) | 1 (0.8) |
| *Pachycondyla villosa* (Fabricius, 1804) | P | - | - | - | - | - | 2 (1.6) |
| *Pachycondyla* sp.1 | P | 1 (0.8) | - | - | - | - | - |

# 12

# Richness of Termites and Ants in the State of Rio Grande do Sul, Southern Brazil

E Diehl[1], E Diehl-Fleig[1], EZ de Albuquerque[2], LK Junqueira[3]

1 - Universidade Federal do Rio Grande do Norte, Natal, RN, Brazil.
2 - Museu de Zoologia, Universidade de São Paulo, São Paulo, SP, Brazil.
3 - Pontifícia Universidade Católica de Campinas, Campinas, SP, Brazil.

**Keywords**
Social insects, distribution, altitude.

**Corresponding author**
Eduardo Diehl-Fleig
Instituto do Cérebro, Universidade
Federal do Rio Grande do Norte
Av. Nascimento de Castro 2155
Natal, RN, Brazil
59056-450
E-mail: edf.formiga@gmail.com

**Abstract**
Previous studies on the effects of environmental factors, such as altitude, latitude, temperature, deforestation, forest fragmentation, fire, and flood on the community structure of termites and ants were conducted in various regions of Brazil; few of them were carried out in the southernmost Brazilian state of Rio Grande do Sul. Here we describe termites and ants diversity at different sites along the four geomorphologic units of this state. We recorded 16 taxa of termites, of which three are new state records, increasing to 19 the number of termite species known to occur in the state. Accordingly, we also found 73 species and 115 morphospecies of ants, of which only one was a new record, raising to 265 taxa the number of ant species known to occur in the state. As expected, we found a higher species richness of ants than termites. The low richness of both groups relative to other Brazilian regions could be a consequence of the subtropical to temperate climate in the state, since most portions of the state are below 30° latitude, the study areas be above 500 m altitude, and other environmental characteristics of each site. We suggest a positive relationship between species richness of termites and altitude, while ant richness indicated an inverse relationship. However, our data are not conclusive, due to the low number of replications in each altitude, particularly for termites. This study is unique in presenting an updated checklist of termites and ants in the state of Rio Grande do Sul.

## Introduction

Studies on biodiversity are urging in face of diversity loss (Wilson, 1997; McGeoch & Chown, 1998). Termites (Mill, 1982; Constantino, 1992; Eggleton et al., 1995; Black & Okwakol, 1997; Lavelle et al., 1997; Jones, 2000; Gathorne-Hardy et al., 2001; Jones et al., 2003) and ants (Majer, 1983; Hölldobler & Wilson, 1990; Andersen, 1997; Silva & Brandão, 1999; Ward, 2000; Ribas et al., 2012) have been considered biodiversity indicators due to their biological and ecological characteristics. These groups have also been considered as surrogate groups in evaluations of conservation status, degradation, or recovery of terrestrial ecosystems (New, 1996; Majer, 1996; Andersen, 1997; Majer & Nichols, 1998; Lobry de Bruyn, 1999; Jones & Eggleton, 2000; Bandeira &

Vasconcello, 2002; Diehl et al., 2004).

Like ants, termites are entirely eusocial and have profound ecological significance in the tropics (Engel et al., 2009). Studies have shown that ants and termites play a crucial role to create soil structure, influence aeration, water infiltration and nutrient cycling acting as ecosystem engineers (Lobry de Bruyn & Conacher, 1990; Lavelle et al., 1997; Mora et al., 2005; Jouqueta et al., 2011; Del Toro et al., 2012). In some cases, termites and ants are also responsible for increased crop yield under dry conditions through soil water infiltration due to their tunnels and improved soil nitrogen (Evans et al., 2011).

Environmental disturbances, such as deforestation, forest fragmentation, fire, and floods may affect the community structure of termites (De Souza & Brown, 1994; Eggleton et al., 1994, 1995; Davies, 2002; Bandeira et al., 2003; Inoue et al., 2006) and ants (Hölldobler & Wilson, 1990; Majer & Nichols, 1998; Diehl et al., 2004; Schmidt & Diehl, 2008). Previous studies have shown that local species richness of ants and termites is influenced by environmental factors, such as temperature, rainfall, and vegetation (Benson & Harada, 1988; Silva & Brandão, 1999; Gathorne-Hardy et al., 2001; Oliveira & Del-Claro, 2005; Del-Claro, 2008; Jones & Eggleton, 2011). There has also been some evidence that increasing altitude and decrease of temperature decrease species richness (Kusnezov, 1957; Collins, 1980; Ward, 2000; Gathorne-Hardy et al., 2001) and that the phylogenetic structure might be involved in this process, specially in the temperate forests (Machac et al., 2011).

The Neotropics rank in the third position in termite richness among the biogeographic regions, with the Oriental and Ethiopian ones presenting the highest number of species recorded so far (Krishna et al., 2013). About 72% of the neotropical species occur in South America (Constantino, 2014) and the sites with good termite surveys are between the parallels 0° S and 20° S, which include the biomes Amazon and Atlantic Forests, Cerrado and Caatinga (Constantino & Acioli, 2006). A review of the termite fauna in across Brazil can be found in Constantino & Acioli (2006). Recent studies recorded 33 species for five fragments of semideciduous Atlantic Forest in northern Brazil, varying locally from 11 to 27 species (Souza et al., 2012). With different levels of perturbation, 26 species were recorded in a region of semi-arid Caatinga (Vasconcellos et al., 2010). In southern Brazil, termite fauna is poorly known. Studies mention only 16 termite species to Rio Grande do Sul (Araújo, 1977; Constantino, 1998; Fontes, 1998; Castro & Diehl, 2003; Diehl et al., 2005b; Florencio & Diehl, 2006). This low number can be due to the climate (Eggleton et al., 1994; Constantino & Acioli, 2006), or lack of taxonomists and local studies (Diehl-Fleig et al., 1995; Constantino & Acioli, 2006).

Despite its importance, little is known about the edaphic insects, specially ants and termites, in southern Brazil. Thus, here we present the richness and a checklist of termites and ants in the four geomorphic units of Rio Grande do Sul, reporting new records for state.

The checklist also includes species from other inventories conducted in the state of Rio Grande do Sul by researchers apart from our group leaded by E. Diehl. Most aimed at collecting species of *Acromyrmex* and *Atta* or urban ant fauna and few have focused on ant species inventories in forests and other relevant ecosystems (please see supplementary material for further details at):

http://periodicos.uefs.br/ojs/index.php/sociobiology/rt/suppFiles/313/408

DOI: 10.13102/sociobiology/.v61i2.145-154.s451

## Material and Methods

### Study areas

Geomorphological units of Rio Grande do Sul

The southernmost Brazilian state of Rio Grande do Sul has four geomorphologic units: Southern Plateau, Central Depression, Sul-Riograndense Shield, and Coastal Plain (Fig 1). Two main vegetation types originally occurred in the state: forests and grasslands occupying 34% and 46% of the state area, respectively. The remaining was occupied by coastal vegetation, wetlands, and other vegetation types. The vegetation reflects soil type and origin and forests usually occur where the soil is deep and fertile. Grasslands develop where the soil is siliceous or shallow. While the southern region of the state is dominated by grasslands, the northern grasslands are interspersed with Araucaria forest (Rambo, 1994).

The weather in southern Brazil is subtropical humid, according to the Köppen classification. While the average annual temperature in the Coastal Plain is around 18°C, temperature approaches 20°C in the Central Depression, which also has mild winters. Conversely, the annual average temperature in the Southern Plateau does not exceed 15°C, due to its high altitude, winters are harsh, with often negative temperatures, severe frosts, and even deep snow. The Campanha, in the Sul-Riograndense Shield, is an open landscape and receives all continental winds in the winter, but due to their large grassland areas it receives much sunlight resulting in summers with high temperatures (Rambo, 1994).

Sampling sites (Fig 1) are described below and are mostly under different levels of disturbance:

1) remnants of seasonal semideciduous forest and suburban areas around São Leopoldo (29° 45' - 29° 47' S, 51° 05' - 51° 10' W; 15 m a.s.l.) in the Central Depression (Mayhé-Nunes & Diehl-Fleig, 1994; Diehl-Fleig & Fleig, 1997; Haubert et al., 1998; Flores et al., 2002; Diehl et al. 2006; Florencio & Diehl, 2006; Marchioretto & Diehl 2006);

2) periurban and urban areas in Canela (29° 21' S, 50° 49' W; 819 m a.s.l.) and Gramado (29° 22' S, 50° 52' W; 825 m a.s.l.) in the Southern Plateau (Diehl-Fleig, 1997);

3) dunes and restingas in Torres (29° 20' - 29° 23' S, 49° 43' - 49° 46' W; 0 - 16 m a.s.l.) in the northern Coastal Plain (Diehl-Fleig et al., 2000; Hameister et al., 2003; Albuquerque et al. 2005; Diehl, unpub. ms.);

4) grasslands in the Caçapava do Sul (30° 47' S, 52° 24' W; 220 m a.s.l.) in soils with different copper levels, including a copper mine and a native savanna area in the Camaquã River Basin in the Sul-Riograndense Shield (Diehl et al., 2004);

5) restingas in Morro da Grota (30° 21' S, 50° 01' W; 263 m a.s.l.) near Lagoa dos Patos and a typical forest of granite

soil, sandy and rocky areas in Pedreira beach (30° 21' S, 50° 02' W; at the sea level), on the banks of the Guaíba Lake, both at the Itapuã State Park in Viamão, which is within both Sul-Riograndense Shield and Coastal Plain (Sacchett & Diehl, 2004; Diehl et al., 2005a; Diehl, unpub. ms.);

6) Eucalyptus plantation in restingas in Capivari do Sul (30° 11' S, 50° 21' W; 12 m a.s.l.) and Tramandaí (29° 59' S, 50° 13' W; 8 m a.s.l.) in the northern Coastal Plain (Fonseca & Diehl, 2004);

7) wetland used as pastures for cattle or irrigated rice in Santo Antônio da Patrulha (29° 54' S, 50° 33' W; 25 m a.s.l.) in the Coastal Plain (Diehl et al., 2005b; Moraes & Diehl, 2009);

8) an area of mixed ombrophilous forest in the São Francisco de Paula National Forest (29° 23' - 29° 27' S, 50° 23' - 50° 25' W; 930 m a.s.l.) and in the Pró-Mata Research and Nature Conservation Center (29° 28' S, 50° 13' W; 900 a.s.l.) in the Southern Plateau (Diehl et al., 2005c; Pinheiro et al., 2010);

9) vineyards (29° 56' S, 51° 33' W; 645 m a.s.l.) in Bento Gonçalves, within the northeastern Central Depression (Sacchett et al., 2009);

10) yerba mate plantations in Mato Leitão (29° 31' S, 52° 07' W; 81 m a.s.l.), Ilópolis (28° 55' S, 52° 07' W; 683 m a.s.l.), and Putinga (29° 00' S, 52° 09' W; 435 m a.s..l) in the northeastern Central Depression (Junqueira et al., 2001; Steffens, 2006);

11) primary forest and reforestation area in Alto Ferrabraz, Sapiranga (29° 34' S, 50° 56' W; 500 m a.s.l.). This is a transition zone between the Southern Plateau and the northeastern Central Depression (Haubert, 2006);

12) secondary forest, with Acacia plantation and an area with reforestation in Rolante (29° 36' 32.2" S, 50° 31' 39.1" W; 370 m a.s.l.) in the Central Depression (Schmidt & Diehl, 2008);

13) grasslands in Campos de Cima da Serra (29° 10' S, 50° 10' W; 1,200 m a.s.l) in the Southern Plateau (Albuquerque & Diehl, 2009);

At each termite sampling site, we established a transect (100 m x 3 m), further divided into 20 sections of 15 m². We then sampled ten non-contiguous sections on the right and left sides of the transect alternately, with a sampling effort of 60 minutes per section. In each section, we search for termites in leaf litter and humus at the base of trees and roots, under both rocks and fallen trees, inside logs, hollow twigs, decomposing fallen branches, epigean and arboreal nests. We also set up one Termitrap® bait per section (Almeida & Alves, 1995) at 15 cm deep, 60 days before samplings. Furthermore, we also searched for termites in ten blocks (340 cm³) of soil horizon A per section.

Termites were stored in individual amber glass with 80% ethanol. Termites were identified to the genus level

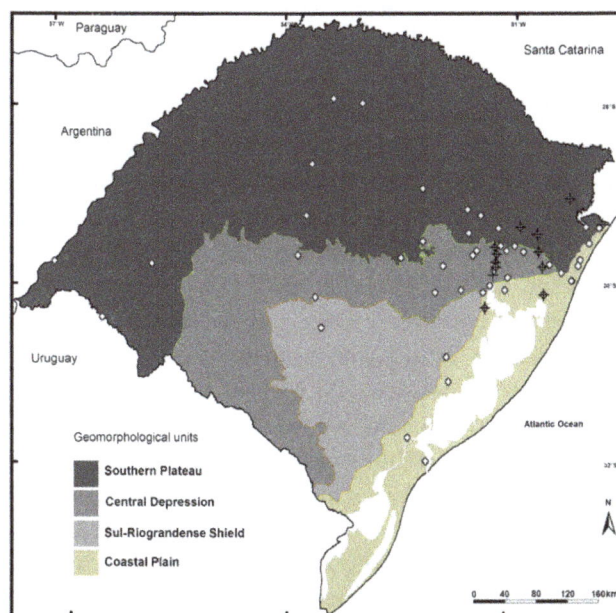

Fig 1. Distribution of combined termite and ant sampling (crossed circles) and other ant inventoried sites (open circles) in the four geomorphologic units of Rio Grande do Sul, southern Brazil by Diehl group (geomorphologic map adapted from SEPLAG, 2008).

(Fontes, 1995; Constantino, 1999; 2002) and eventually to the species level or morphospecies. We also consulted specialist taxonomists to confirm the identification of termites.

We used five techniques to sample ants: 1) manual capture of all ants on the ground, under stones, in hollow trees, and branches; 2) baits with sardines in vegetable oil on a piece of filter paper (6 cm²); 3) leaf litter samples, firstly sieved in the field and further using Winkler extractors, in which they remained for 72 hours; 4) pitfall traps with and without attractive, which consisted of 200-300 mL plastic cups buried up to the top containing 70% ethanol, remaining on the ground for 24 to 48 hours. Attractive traps were composed by sardines in vegetable oil pierced in a galvanized wire rod; 5) underground trap with attractive, which was a small plastic pot (3.3 cm wide by 5 cm high) with small holes on the sides. Within each trap a piece of 1 cm³ sardines in vegetable oil was added. Traps were buried 20 cm deep by 48 hours. In grasslands we used direct sampling, sardine baits and pitfall traps. In dunes, we only used direct sampling and sardine baits. At each site, we established two to three 100 m transects, along which we sampled every 10 m. Sampling effort was at least 2 hours in each site.

We classified ants into subfamilies following Bolton (2014) and by genus following Bolton (1994) and Fernández (2003). The subfamilies Ponerinae and Ecitoninae were treated according to the most recent reviews - see supplementary material for details. We used several identification keys and also consulted specialists to identify species. We assigned specimens to morphospecies when specific identification was not possible. At least three specimens of each species/ morphospecies were mounted with entomological pins and the remaining preserved in 70% ethanol.

Specimens of termites and ants are housed at the Collection of Social Insects (Isoptera and Formicidae) of the Universidade do Vale do Rio dos Sinos (UNISINOS), São Leopoldo, Rio Grande do Sul, Brazil.

The checklist presented results of ca. 30 years of collections leaded by E. Diehl added to most ant inventories conducted by few other researchers in the state of Rio Grande do Sul. We reviewed papers published from 1972 to 2014 included in the databases ISI Web of Science, Scielo and Google Scholar that contained the keyword combination: ant OR Formicidae AND Rio Grande do Sul. Only papers referring nominal species and site location (municipality or more specific location) were considered, except those with doubtful identification (approximate identification, to be confirmed, compared to, of group, subgenus or species complex). Three studies on Atta and Acromyrmex distribution were excluded because site location was not provided. Species previously registered to the state was provided by the catalog of Kempf (1972) with additions of Brandão (1991).

**Results**

We recorded 16 isopterans from two families (Table 1). We only found one species of the Kalotermitidae genus *Rugitermes*. We also recorded a new genus of Termitidae, besides seven morphospecies and seven species from four subfamilies (Apicotermitinae, Nasutitermitinae, Syntermitinae, and Termitinae).

Termite species richness was very low. We recorded 10 species in the Southern Plateau (181-1,200 m a.s.l.) and 15 in the Central Depression (20-30 m), whereas in the Sul-Riograndense Shield (23-29 m) and northern Coastal Plain (5-15 m) we found the lowest richness: four and six species, respectively.

We found 26 ant genera from seven subfamilies (Dolichoderinae, Dorylinae, Ectatomminae, Formicinae, Heteroponerinae, Myrmicinae, and Ponerinae) that were stored in 70% ethanol. Furthermore, we recorded 51 genera from nine subfamilies (Amblyoponinae, Dolichoderinae, Dorylinae, Ectatomminae, Formicinae, Heteroponerinae, Myrmicinae, Ponerinae, and Pseudomyrmecinae), which were stored in entomological drawers. In total, we recorded 73 species and 115 morphospecies in Rio Grande do Sul. Compiling only nominal ant species registered by our group and other researchers, 127 species distributed among 55 genera from 10 subfamilies occur in the state (an extended list is available as supplementary material).

A single species, *Monomorium floricola*, is reported as new record for the state of Rio Grande do Sul. Thus, ant species richness in Rio Grande do Sul raises to 265, considering published data on nominal species from the catalogs (Kempf, 1972; Brandão, 1991), from our research group and the other groups. This number would likely increase, if we

**Table 1.** Termites recorded in the state of Rio Grande do Sul, Brazil.

| Family/ Subfamily | Taxon |
| --- | --- |
| **Kalotermitidae** | *Rugitermes* sp. (1, 3, 4)* |
| **Termitidae** | |
| Apicotermitinae | Apicotermitinae 1 (1, 2, 3, 4) |
| | Apicotermitinae 2 (1, 2, 3, 4) |
| | *Grigiotermes bequaerti* (Snyder & Emerson, 1949) (3) |
| | *Grigiotermes* sp. (3) |
| | *Ruptitermes* sp. (3) |
| | *Tetimatermes* sp. (3) |
| | New Genus 1 (3) |
| Nasutitermitinae | *Araujotermes caissara* Fontes, 1982 (1) |
| | *Cortaritermes fulviceps* (Silvestri, 1901) (1, 2, 3, 4) |
| | *Nasutitermes aquilinus* (Holmgren, 1910) (3, 4) |
| | *Nasutitermes jaraguae* (Holmgren, 1910) (3, 4) |
| Syntermitinae | *Cornitermes cumulans* (Kollar, 1832) (2, 3, 4) |
| Termitinae | *Dihoplotermes inusitatus* Araújo, 1961 (3, 4) |
| | *Neocapritermes* sp. (1, 3, 4) |
| | *Termes* sp. (3, 4) |

*Geomorphologic units: 1, Coastal Plain; 2, Sul-Riograndense Shield; 3, Central Depression; 4, Southern Plateau

consider the morphospecies not yet identified to the species level and other published studies with nominal species and site location indicated (see studies available as supplementary material).

Ant richness and composition varied along the geomorphologic units. We found 88 species/morphospecies in the Southern Plateau (900-1,200 m a.s.l.), 40 in the transition between the Southern Plateau and the Central Depression (500 m), 108 in the Central Depression (10-683 m), 51 in the Sul-Riograndense Shield (220 m), 81 in the transition between the Sul-Riograndense Shield and the Coastal Plain (16-263 m), and 97 in the Coastal Plain (0-80 m). Of the 188 species/morphospecies, most (40.43%) were restricted to a single geomorphologic unit, while only ca. 2.5% were common to all units (*Pheidole* sp.3, *Pheidole* sp.6, *Pheidole* sp.15, *Wasmannia* sp. and *Wasmannia* sp.1).

Excluding three sites where ant fauna was inventoried exclusively in agricultural or urban areas, there is some indication that species richness and composition could vary in response to an altitudinal gradient. The highest ant richness, 56, 81, and 73 species were found in sites at the sea level, 10 m, and 26 m, respectively. Conversely, we found 51, 35, 39, 55, and 33 species in sites at 220 m, 370 m, 500 m, 930 m, and 1,200 m, respectively. Of the 158 ant species/morphospecies in these eight areas, none were present in all sites, only ca. 3% were common to seven areas. The great majority (35.44 %) of ants were restricted to a single site.

# Discussion

According to Krishna et al. (2013) there are 2,937 termite species worldwide, of which 569 are found in the Neotropics. The On-line Termite Database maintained by Constantino (2014) refers 2,882 termite species in the world, of which 562 species occur in the Neotropics (database was last updated in September 2012). Estimates of termite richness for Brazil are not yet accurate, ranging from 250 to 364 species (Cancello & Schlemmermeyer, 1999; Constantino, 1999; 2014; Fontes & Araújo, 1999). However, the estimated species richness should increase with termite sampling in other areas of the country. Our results provide records of a new genus likely belonging to Apicotermitinae, another unidentified species of this subfamily (Apicotermitinae 1), and *Dihoplotermes inusitatus*. Thus, the number of known termite species in Rio Grande do Sul increases to 19.

Drywood termites of the family Kalotermitidae have major economic importance and a wide geographic distribution. These termites are poorly known because they inhabit places difficult to access (Jones & Eggleton, 2011). The largest and most diverse isopteran family is Termitidae (Eggleton & Tayasu, 2001). We found species from four out of the eight termite subfamilies previously known to occur in Rio Grande do Sul (Apicotermitinae, Nasutitermitinae, Syntermitinae, and Termitinae). A negative correlation between altitude and species richness has been reported for various organisms (Rahbek, 2005; McCain, 2009; McCain & Grytnes, 2010). This correlation seems to be dependent on altitude in tropical and subtropical areas. Altitude is positively correlated with richness in latitudes below 30° and below 500 m, but negative in latitudes above 30° and at altitudes above 500 m (Kusnezov, 1957; Ward, 2000).

We found indication that termite species richness relates positively to altitude. Previous studies (Inoue et al., 2006) found the species richness of Nasutitermitinae increased with altitude, while species richness and abundance of Macrotermitinae decreased with altitude. The other groups were apparently not influenced by altitude. Recently, Palin et al. (2010) evaluated the influence of the variation along an Amazon–Andes altitudinal gradient in Peru on richness, abundance, and diversity of functional groups of termites. They registered 49 species and verified that, in general, the diversity declined with increased elevation. The functional groups responded differently to the upper distribution limit: for the soil-feeding it was between 925 and 1,500 m a.s.l., while the wood-feeding termites was between 1,550 and 1,850 m a.s.l. And this differential response led the authors to suggest that the energy requirements for each group are a key factor in shaping their occurrence associated with the altitude and temperature.

Termite species richness seems to be influenced by local environmental conditions, such as altitude, temperature, rainfall, and vegetation (Collins, 1980; Gathorne-Hardy et al., 2001; Davies, 2002; Inoue et al., 2006; Jones & Eggleton, 2011). The species richness of termites we found is as low as reported in subtropical and temperate areas. For example, previous studies found two to eight morphospecies in the Atlantic Forest of southeastern Brazil, whereas an average of 30 morphospecies were found in tropical forests near Bahia, northeastern Brazil (Cancello et al., 2002).

The far southernmost part of South America has often been considered as a separate biogeographic sub-region. It is depauperate in termite species due to its high latitude (Eggleton, 2000), and countries like Chile and Uruguay have even fewer species registered: six and nine, respectively, while 37 are reported for Paraguay (Constantino, 2014). Argentina, on the other hand, has 57 species reported (Constantino, 2014), even though this number could increase to 95 (Torales et al., 2008). Argentina local termite species varies from 12 in the Chaco province (Godoy et al., 2012) to 26 species in northeast Argentina (Laffont et al., 2004).

We present the first two records of the rare, monotypic genus *Dihoplotermes* to Rio Grande do Sul. A colony of *D. inusitatus* was found inhabiting a nest *Cornitermes bequaerti* in gallery forests in Mato Grosso (Mathews, 1977), whereas other studies (Constantino, 1999) found *D. inusitatus* in the Cerrado and disturbed habitats in the Southeast. We found the first record of *D. inusitatus* within a nest of *C. cumulans* in grasslands surrounding the São Francisco de Paula National Forest. However, we could not find this species in further surveys. This is a rare species that could have went locally extinct after a *Pinus* plantation covered the area. The other record of *D. inusitatus* was also in a nest of *C. cumulans* in the suburb of São Leopoldo, more than 500 km away from the first site. With the increase of urbanization in the region, this rare species is subject to rapid local extinction.

The family Formicidae comprises more than 15,700 species and subspecies (AntWeb, 2013); recent estimates suggest that this number should exceed 23,000 species (AntWeb, 2013). However, previous studies mention 208 (Kempf, 1972) or 224 species Brandão (1991) to Rio Grande do Sul. Since its last catalog update (Brandão, 1991), 23 years ago, our study is the first to present an updated checklist of the ant species in Rio Grande do Sul. If we take into account the determination of species level of the all morphospecies in our collection at Unisinos and in other studies, the species list would likely increase. Only few list are available for other states, such as Santa Catarina (Ulysséa et al., 2011), where ant richness (366 species and 17 subspecies) is greater than the 265 species now updated to Rio Grande do Sul. Apart from different collection efforts, historical aspects linked to the myrmecology development are part of this numerical difference reported.

Ant inventories in Rio Grande do Sul until the early 2000s focused primarily in the genera *Atta* and *Acromyrmex* (e.g., Diehl-Fleig, 1997). This attines have long been known for their pest status and major impact on agriculture, fairly common and economically relevant throughout the state. It is clear that ant fauna inventories in RS are still very scarce and the geomorphic units are undersampled, particularly the Pampa biome. This grassland ecosystem is unique in fauna and vegetation among Brazilian biomes and less than 42% remains preserved (Roesch et al., 2009). So far, ant fauna has been surveyed mostly in the Atlantic Forest biome in the state (Diehl et al., 2005c; Albuquerque & Diehl, 2009; Pinheiro et al., 2010) while a single study included the Pampa (Marques & Schoereder, 2014).

The diversity of habitats, vegetation, altitude, latitude, and climate between sampling sites should be the main factors responsible for the differences found in termite and ant species richness along the study areas and also other tropical areas in Brazil (Oliveira & Del-Claro, 2005). Currently, 106 ant genera are known to occur in Brazil (Antwiki, 2013), while the termite diversity is much lower, only 74 genera (Constantino, 2014). As expected, we found a higher richness of ant genera than termite along the studied areas. Species diversity decreases with increasing distance from the equator (Kusnezov, 1957; Eggleton et al., 1994; Ward, 2000). Therefore, the lowest species richness of termites and ants found by us was expected. Additionally, the low richness may be due to the subtropical climate, the low latitude ($< 30°$), and high altitudes ($> 500$ m), as might be case for our findings that ant richness was lower at higher altitudes. Moreover, sampled sites varied in the degree of habitat complexity, from single crop to native forest, which also play a key role in shaping species richness and composition with different outcomes (e.g., Lassau & Hochuli, 2004; Silva et al., 2007; Pacheco & Vasconcelos, 2012). This study is unique in compiling the richness of both termites and ants so far known in the state of Rio Grande do Sul and we hope it provides a landmark in exposing the necessity for future studies and establishing efforts in unraveling our eusocial insect diversity.

## Acknowledgements

We would like to express our gratitude to three anonymous reviewers, to Dr. Jacques H. Delabie, Dr. Antônio J. Mayhé-Nunes, Dr. Rodrigo dos S.M. Feitosa, Dr. Carlos E. Sanhudo and Thiago Ranzani who helped us to confirm either/or identify ant species, in addition to providing several dichotomous keys. We also thank Dr. Reginaldo Constantino and Dr. Luiz R. Fontes for identification of termites. Daniela F. Florencio, Laura V.A. Menzel and Carlos E. Sanhudo helped in field and laboratory work. FAPESP provided financial support to E.Z. de Albuquerque (Proc. N. 2010/02560-5). E. Diehl was supported by CNPq. Formigas do Brasil research group provided much of the ant bibliographic review. We thank Unisinos for the laboratory facilities..

## References

Albuquerque, E.Z.; Diehl-Fleig, Ed. & Diehl, E. (2005). Density and distribution of nests of *Mycetophylax simplex* (Emery) (Hymenoptera: Formicidae) in areas with mobile dunes on the northern coast of Rio Grande do Sul, Brazil. Revista Brasileira de Entomologia, 49(1): 123-126. doi: 10.1590/S0085-56262005000100013

Albuquerque, E.Z. de, Diehl, E. (2009). Análise faunística das formigas epígeas (Hymenoptera, Formicidae) em campo nativo no Planalto das Araucárias, Rio Grande do Sul. Revista Brasileira de Entomologia, 53: 123-126. doi: 10.1590/S0085-56262009000300014

Almeida, J.E.M. & Alves, S.B. (1995). Seleção de armadilhas para captura de *Heterotermes tenuis* (Hagen). Anais da Sociedade Entomologica do Brasil, 24: 619-624.

Andersen, A.N. (1997). Using ants as bioindicators: multiscale issues in ant community ecology. Conservation Ecology 1: 8. Retrieved from: http://www.consecol.org/vol1/iss1/art8

AntWeb. (2013). http://www.antweb.org. (accessed date: 8 December, 2013).

Antwiki (2013). http://www.antwiki.org/wiki/Distribution_and_Diversity. (accessed date: 27 November, 2013).

Araújo, R.L. (1977). Catálogo dos Isoptera do Novo Mundo. Rio de Janeiro: Academia Brasileira de Ciências, 92 p.

Bandeira, A.G. & Vasconcellos, A. (2002). A quantitative survey of termites in a gradient of disturbed highland forest in Northeastern Brazil (Isoptera). Sociobiology, 39: 429-439.

Bandeira, A.G., Vasconcellos, A., Silva, M.P. & Constantino, R. (2003). Effects of habitat disturbance on the termite fauna in a highland humid forest in the Caatinga domain, Brazil. Sociobiology, 42: 117-127.

Benson, W.W. & Harada, A.Y. (1988). Local diversity of tropical and temperate ant faunas (Hymenoptera: Formicidae). Acta Amazonica, 18(3-4): 275-289.

Black, H.I.J. & Okwakol, M.J.N. (1997). Agricultural intensification, soil biodiversity and agroecosystem function in the tropics: the role of termites. Applied Soil Ecology, 6: 37-53. doi: 10.1016/S0929-1393(96)00153-9

Bolton, B. (1994). Identification guide to the ant genera of the world. Cambridge: Harvard University Press, 222 p

Bolton, B. (2003). Synopsis and classification of Formicidae. Florida: Memoirs of the American Entomological Institute, 370 p

Brandão, C.R.F. (1991). Adendos ao catálogo abreviado das formigas da região Neotropical (Hymenoptera: Formicidae). Revista Brasileira de Entomologia. 35: 319-412.

Cancello, E.M. & Schlemmermeyer, T. (1999). Reino Animalia:

Ordem Isoptera. In C.R.F. Brandão & E.M Cancello (Eds.), Biodiversidade do Estado de São Paulo, Brasil: síntese do conhecimento ao final do século XX, invertebrados terrestres (pp. 82-91). São Paulo: FAPESP.

Cancello, E.M.; Oliveira, L.C.M.; Reis, Y.T. & Vasconcellos, A. (2002). Termite diversity along the Brazilian Atlantic Forest. In Proceedings of the XIV International Congress of IUSSI (pp. 164). Sapporo: Hokkaido Univ.

Castro, Z. & Diehl, E. (2003). Gêneros de térmitas em ninhos epígeos no campus da Unisinos, São Leopoldo, RS. Acta Biologica Leopoldensia, 25: 93-102.

Collins, N.M. (1980). The distribution of soil macrofauna on the west ridge of Gunung (Mount) Mulu, Sarawak. Oecologia, 44: 263–275. doi: 10.1007/BF00572689

Constantino, R. (1992). Abundance and diversity of termites (Insecta: Isoptera) in two sites of primary rain forest in Brazilian Amazonia. Biotropica, 24: 420-430.

Constantino, R. (1998). Catalog of the living termites of the new world (Insecta: Isoptera). Arquivos de Zoologia, 35: 135–260. doi: 10.11606/az.v35i2.12014

Constantino, R. (1999). Chave ilustrada para identificação dos gêneros de cupins (Insecta: Isoptera) que ocorrem no Brasil. Papéis Avulsos de Zoologia, 40: 387-448.

Constantino, R. (2002). An illustrated key to Neotropical termite genera (Insecta: Isoptera) based primarily on soldiers. Zootaxa, 67: 1-40.

Constantino, R. (2014). On-line termite database. http://164.41.140.9/catal/. (accessed date: 10 February, 2014).

Constantino, R. & Acioli, A.N.S. (2006). Termite diversity in Brazil (Insecta: Isoptera). In F.M.S. Moreira, J.O. Siqueira & L. Brussaard (Eds.), Soil biodiversity in Amazonian and other Brazilian ecosystems. (pp. 117-128). London: CAB International. doi: 10.1079/9781845930325.0117

Davies, R.G. (2002). Feeding group responses of a Neotropical termite assemblage to rain forest fragmentation. Oecologia, 133: 233-242. doi: 10.1007/s00442-002-1011-8

De Souza, O.F.F. & Brown, V.K. (1994). Effects of habitat fragmentation on Amazonian termite communities. Journal of Tropical Ecology, 10: 197-206. doi: 10.1017/S0266467400007847

Del-Claro, K. (2008). Biodiversidade Interativa: a ecologia comportamental e de interações como base para o entendimento das redes tróficas que mantém a viabilidade das comunidades naturais. In J. Seixas & J. Cerasoli (Eds.), UFU, ano 30 – Tropeçando Universos (artes, humanidades, ciências) (pp. 599-614). Uberlândia: EDUFU.

Del Toro, I., Ribbons, R.R. & Pelini, S.L. (2012). The little things that run the world revisited: a review of ant-mediated ecosystemservices and disservices (Hymenoptera: Formicidae).

Myrmecological News, 17: 133-146

Diehl-Fleig, E., Silva, M.E. da & Castilhos-Fortes, R. (1995). O problema dos cupins no Rio Grande do Sul. In E. Berti Filho & L.R. Fontes (Eds.), Alguns aspectos atuais da biologia e ecologia dos cupins (pp. 53-56). Piracicaba: FEALQ.

Diehl, E. (1997). Ocorrência de Acromyrmex em áreas com distintos níveis de perturbação antrópica no Rio Grande do Sul. Acta Biologica Leopoldensia, 19: 165-171.

Diehl, E. & Diehl-Fleig, Ed. (1997). Primeiro registro de Zacryptocerus depressus Klug e de Z. incertus Emery (Hymenoptera: Formicidae) no Rio Grande do Sul. Acta Biologica Leopoldensia, 19: 225-228.

Diehl-Fleig, E., Sanhudo, C.E.D. & Diehl-Fleig, Ed. (2000). Mirmecofauna de solo nas dunas da Praia Grande e no Morro da Guarita no município de Torres, RS, Brasil. Acta Biologica Leopoldensia, 22: 37-43.

Diehl, E., Sanhudo, C.E.D. & Diehl-Fleig, Ed. (2004). Ground-dwelling ant fauna of sites with soil high level of copper. Brazilian Journal of Biology, 64: 33-39. doi: 10.1590/S1519-69842004000100005

Diehl, E., Sacchett, F., Albuquerque, E.Z.. (2005a). Riqueza de formigas de solo na Praia da Pedreira, Parque Estadual de Itapuã, Viamão, RS, Brasil. Revista Brasileira de Entomologia, 49: 552-556. doi: 10.1590/S0085-56262005000400016

Diehl, E., Junqueira, L.K. & Berti-Filho, E. (2005b). Ant and termite mound coinhabitants in the wetlands of Santo Antonio da Patrulha, Rio Grande do Sul, Brazil. Brazilian Journal of Biology, 65: 431-437. doi: 10.1590/S1519-69842005000300008

Diehl, E., Florencio, D.F., Schmidt, F.A. & Menzel, L.V.A. (2005c). Riqueza e composição das comunidades de formigas e térmitas na Floresta Nacional de São Francisco de Paula (FLONA-SFP), RS. Acta Biologica Leopoldensia 27: 99-106.

Diehl, E.; Göttert, C.L. & Flores, D.G. (2006). Comunidades de formigas em três espécies utilizadas na arborização urbana em São Leopoldo, Rio Grande do Sul, Brasil. Bioikos, 20: 25-32.

Eggleton, P., Williams, P.M. & Gaston, K.J. (1994). Explaining global termite diversity: productivity or history? Biodiversity and Conservation 3: 318-330. doi: 10.1007/BF00056505

Eggleton, P., Bignell, D.E., Sands, W.A., Waite, B., Wood, T.G. & Lawton, J.H. (1995). The species richness of termites (Isoptera) under differing levels of forest disturbance in the Mbalmayo Forest Reserve, southern Cameroon. Journal of Tropical Ecology, 11: 85-98. doi: 10.1017/S0266467400008439

Eggleton, P. (2000). Global patterns of termite diversity. In T. Abe, D.E. Bignell & M. Higashi (Eds.), (pp. 25-51). Termites: evolution, sociality, symbioses, ecology. Dordrecht: Kluwer Academic Publishers

Eggleton, P. & Tayasu, I. (2001). Feeding groups, lifetypes and the global ecology of termites. Ecological Research, 16:

941-960. doi: 10.1046/j.1440-1703.2001.00444.x

Engel, M.S., Grimaldi, D.A. & Krishna, K. (2009). Termites (Isoptera): their phylogeny, classification, and rise to ecological dominance. American Museum Novitates, 3650: 1-27.

Evans, T.A., Dawes, T.Z., Ward, P.R. & Lo, N. (2011). Ants and termites increase crop yield in a dry climate. Natural Communities, 2: 262 doi: 10.1038/ncomms1257

Fernández, F. (2003). Introducción a las hormigas de la región Neotropical. Bogotá: Instituto de Investigación de Recursos Biológicos Alexander Von Humboldt, 424 p

Florêncio, D.F. & Diehl, E. (2006). Termitofauna (Insecta, Isoptera) em Remanescentes de Floresta Estacional Semidecidual em São Leopoldo, Rio Grande do Sul. Brasil. Revista Brasileira de Entomologia, 50: 505-511. doi: 10.1590/S0085-56262006000400011

Flores, D.G., Goettert, C.L., Diehl, E. (2002). Comunidade de formigas em *Inga marginata* (Bignoniaceae) em uma área suburbana. Acta Biológica Leopoldensia, 24: 147-155.

Fonseca, R.C. & Diehl, E. (2004). Riqueza de formigas (Hymenoptera, Formicidae) epigéicas em povoamentos de *Eucalyptus* spp. (Myrtaceae) de diferentes idades no Rio Grande do Sul, Brasil. Revista Brasileira de Entomologia, 48: 95-100. doi: 10.1590/S0085-56262004000100016

Fontes, L.R. (1995). Sistemática geral de cupins. In E. Berti Filho & L.R. Fontes (Eds.), Alguns aspectos atuais da biologia e controle de cupins (pp. 11-17). Piracicaba: FEALQ.

Fontes, L.R. (1998). Novos aditamentos ao "Catálogo dos Isoptera do Novo Mundo", e uma filogenia para os gêneros neotropicais de Nasutitermitinae. In L.R. Fontes & E. Berti Filho (Eds.), Cupins: o desafio do conhecimento (pp. 309-412). Piracicaba: FEALQ.

Fontes, L.R. & Araújo, R.L. de. (1999). Os cupins. In F.A.M. Mariconi (Ed.), Insetos e outros invasores de residências (pp. 35-90). Piracicaba: FEALQ.

Gathorne-Hardy, F., Syaurani & Eggleton, P. (2001). The effects of altitude and rainfall on the composition of the termites (Isoptera) of the Leuser Ecosystem (Sumatra, Indonesia). Journal of Tropical Ecology, 17: 379-393. doi: 10.1017/S0266467401001262

Godoy, M.C.; Laffont, E.R.; Coronel, J.M.; Etcheverry, C. (2012). Termite (Insecta, Isoptera) assemblage of a gallery forest relic from the Chaco province (Argentina): taxonomic and functional groups. Arxius de Miscellània Zoològica 10: 55–67.

Gonçalves, C.R. (1961). O gênero *Acromyrmex* no Brasil. Studia Entomologica, 4(1-4): 113-180.

Hameister, T.M., Diehl-Fleig, Ed. & Diehl, E. (2003). Comunidades de formigas (Hymenoptera: Formicidae) epígeas no Morro de Itapeva, município de Torres, RS. Acta Biologica Leopoldensia, 25: 187-195.

Haubert F., Diehl, E. & Mayhé-Nunes, A. (1998). Mirmecofauna de solo do município de São Leopoldo, RS: Levantamento preliminar. Acta Biologica Leopoldensia, 20: 103-108.

Haubert, F. 2006. Riqueza e composição da mirmecofauna de solo no Morro Alto Ferrabraz, município de Sapiranga, RS. São Leopoldo, 2006. 100f. [Dissertação de Mestrado - PPG em Biologia: Diversidade e Manejo de Vida Silvestre/UNISINOS].

Hölldobler, B. & Wilson, E.O. (1990). The Ants. Cambridge: Harvard University Press, 732 p

Inoue, T., Takematsu, Y., Yamada, A., Hongoh, Y., Johjima, T., Moriya, S., Sornnuwat, Y., Vongkaluang, C., Ohkuma, M. & Kudo, T. (2006). Diversity and abundance of termites along an altitudinal gradient in Khao Kitchagoot National Park, Thailand. Journal of Tropical Ecology, 22: 609-612. doi: 10.1017/S0266467406003403

Jones, D.T. (2000). Termite assemblages in two distinct montane forest types at 1000 m elevation in the Maliau Basin, Sabah. Journal of Tropical Ecology, 16: 271-286. doi: 10.1017/S0266467400001401

Jones, D.T. & Eggleton, P. (2000). Sampling termite assemblages in tropical forests: testing a rapid biodiversity assessment protocol. Journal of Applied Ecology, 37: 191-203. doi: 10.1046/j.1365-2664.2000.00464.x

Jones, D.T. & Eggleton, P. (2011) Global biogeography of termites: a compilation of sources. In D.E. Bignell, Y. Roisin & N. Lo (Eds.) Biology of termites: a modern synthesis (pp. 477-517). Dordrecht: Springer. doi: 10.1007/978-90-481-3977-4_17.

Jones, D.T., Susilo, F.X., Bignell, D.E., Hardiwinotos, S., Gillinson, A.N. & Eggleton, P. (2003). Termite assemblages collapse along a land-use intensification gradient in lowland central Sumatra, Indonesia. Journal of Applied Ecology, 40: 380-391. doi: 10.1046/j.1365-2664.2003.00794.x

Jouqueta, P., Traoréc, S., Choosaid, C., Hartmanna, C. & Bignelle, D. (2011). Influence of termites on ecosystem functioning. Ecosystem services provided by termites. European Journal of Soil Biology, 47: 215-222. doi: 10.1016/j.ejsobi.2011.05.005

Junqueira, L.K., Diehl, E. & Diehl-Fleig, Ed. (2001). Formigas (Hymenoptera: Formicidae) visitantes de *Ilex paraguariensis* (Aquifoliaceae). Neotropical Entomology, 30: 161-164. doi: 10.1590/S1519-566X2001000100024

Kempf, W.W. (1972). Catálogo abreviado das formigas da região Neotropical (Hymenoptera: Formicidae). Studia Entomologica, 15(1-4): 3-344.

Krishna, K., Grimaldi, D.A., Krishna, V., Engel, M.S. (2013). Treatise on the Isoptera of the World: 1. Introduction. American Museum of Natural History Museum Bulletin, 377: 1-200. doi: 10.1206/377.1

Kusnezov, N. (1957). Numbers of species of ants in faunae of different latitudes. Evolution, 11: 298-299.

Laffont, E.R., Torales, G.J., Arbino, M.O., Godoy, M.C., Porcel, E.A. & Coronel, J.M. (1998). Termites associadas a *Eucalyptus grandis* W.Hill Ex Maiden en el noroeste de la província de Corrientes (Argentina). Revista de Agricultura, 73: 201-214.

Lassau, S.A. & Hochuli, D.F. (2004). Effects of habitat complexity on ant assemblages. Ecography, 27: 157-164. doi: 10.1111/j.0906-7590.2004.03675.x

Lavelle, P., Bignell, D., Lepage, M., Wolters, V., Roger, P., Ineson, P., Heal, O.W. & Dhillion, S. (1997). Soil function in a changing world: the role of invertebrate ecosystem engineers. European Journal of Soil Biology, 33: 159-193.

Lobry de Bruyn, L.A. & Conacher, A.J. (1990). The role of termites and ants in soil modification: a review. Australian Journal of Soil Research, 28: 55-93. doi: 10.1071/SR9900055

Lobry de Bruyn, L.A. (1999). Ants as bio-indicators of soil function in rural environments. Agriculture, Ecosystems and Environment, 74(1-3): 425-441. doi: 10.1016/S0167-8809-(99)00047-X

McCain, C.M. (2009). Global analysis of bird elevational diversity. Global Ecology and Biogeography, 18: 346-360. doi: 10.1111/j.1466-8238.2008.00443.x

McCain, C.M. & Grytnes, J.A. (2010). Elevational gradients in species richness. In Encyclopedia of Life Sciences (ELS). Chichester: John Wiley & Sons, Ltd. doi: 10.1002/9780470015902.a0022548

Machac, A., Janda, M., Dunn, R.R. & Sanders, N.J. (2011). Elevational gradients in phylogenetic structure of ant communities reveal the interplay of biotic and abiotic constraints on diversity. Ecography, 34: 364-371. doi: 10.1111/j.1600-0587.2010.06629.x

Majer, J.D. (1983). Ants: bio-indicators of mine site rehabilitation, land-use, and land conservation. Environment Management, 7: 375-383. doi: 10.1007/BF01866920

Majer J.D. (1996). Ant recolonization of rehabilited bauxite mines at Trombetas, Pará, Brazil. Journal of Tropical Ecology, 12: 257-273. doi 10.1017/S02667400009445.

Majer, J.D. & Nichols, O.G. (1998). Long-term re-colonization patterns of ants in Western Australian rehabilitated bauxite mines with reference to their use as indicators of restoration success. Journal of Applied Ecology, 35: 161-182. doi: 10.1046/j.1365-2664.1998.00286.x

Marchioretto, A. & Diehl, E. (2006). Distribuição espaciotemporal de uma comunidade de formigas em um remanescente de floresta inundável às margens de um meandro antigo do rio dos Sinos, São Leopoldo, RS. Acta Biologica Leopoldensia, 28: 25-32.

Marques, T. & Schroereder, J.H. (2014). Ant diversity partitioning across spatial scales: Ecological processes and implications for conserving Tropical Dry Forests. Austral Ecology, 39: 72-82. doi: 10.1111/aec.12046

Mathews, A.G.A. (1977). Studies on termites from the Mato Grosso State, Brazil. Rio e Janeiro: Academia Brasileira de Ciências, 267 p.

Mayhé-Nunes, A.J. & Diehl, E. (1994). Distribuição de *Acromyrmex* (Hymenoptera: Formicidae) no Rio Grande do Sul. Acta Biologica Leopoldensia, 16: 115-118.

McGeoch, M.A. & Chown, S.L. (1998). Scaling up the value of bioindicators. Trends in Ecology and Evolution, 13: 46-47.

Mill, A.E. (1982). Populações de térmitas (Insecta: Isoptera) em quatro habitats no baixo rio Negro. Acta Amazonica, 12: 53-60.

Mora, P., Miambi, E., Jiménez, J.J., Decaëns, T. & Rouland, C. (2005). Functional complement of biogenic structures produced by earthworms, termites and ants in the neotropical savannas. Soil Biology and Biochemistry, 37: 1043-1048. doi: 10.1016/j.soilbio.2004.10.019

Moraes, A.B. & Diehl, E. (2009). Comunidades de formigas em dois ciclos de cultivo de arroz irrigado na planície costeira do Rio Grande do Sul. Bioikos, 23: 29-37.

New, T.R. (1996). Taxonomic focus and quality control in insect surveys for biodiversity conservation. Australian Journal of Entomology, 35: 97-106. doi: 10.1111/j.1440-6055.1996.tb01369.x

Oliveira, P.S. & Del-Claro, K. (2005). Multitrophic interactions in the Brazilian savanna: ant hemipteran systems, associated insect herbivores, and host plant. In D. Burslem, M. Pinard & S. Hartley (Eds.), Biotic interaction in the Tropics: their role in the maintenance of species diversity (pp. 414-438). London: Cambridge University Press.

Pacheco, R. & Vasconcelos, H.L. (2012). Habitat diversity enhances ant diversity in a naturally heterogeneous Brazilian landscape. Biodiversity and Conservation, 21: 797-809. doi: 10.1007/s10531-011-0221-y

Palin, O.F., Eggleton, P., Malhi, Y., Girardin, C.A.J, Rozas-Dávila, A. & Parr, C.L. (2011). Termite diversity along an Amazon-Andes elevation gradient, Peru. Biotropica, 43: 100-107. doi: 10.1111/j.1744-7429.2010.00650.x

Pinheiro, E.R., Duarte, L.S., Diehl, E. & Hartz, S.M. (2010). Edge effects on epigeic ant assemblages in a grassland-forest mosaic in southern Brazil. Acta Oecologica, 36: 365-371. doi: 10.1016/j.actao.2010.03.004

Rahbek, C. (2005). The role of spatial scale and the perception of large-scale species-richness patterns. Ecology Letters, 8: 224-239. doi: 10.1111/j.1461-0248.2004.00701.x

Rambo, B. (1994). A fisionomia do Rio Grande do Sul. São Leopoldo: Editora Unisinos, 473 p

Roesch, L.F.W, Vieira, F.C.B, Pereira, V.A., Schünemann,

A.L., Teixeira, I.F., Senna, A.J.T. & Stefenon, V.M. 2009. The Brazilian Pampa: a fragile biome. Diversity, 1: 182-198.

Ribas, C.R., Campos, R.B.F., Schmidt, F.A. & Solar, R.R.C. (2012). Ants as indicators in Brazil: a review with suggestions to improve the use of ants in environmental monitoring programs. Psyche, 2012: Article ID 636749. doi: 10.1155/2012/636749.

Sacchett, F. & Diehl, E. (2004). Comunidades de formigas de solo no Morro da Grota, Parque Estadual de Itapuã, RS. Acta Biologica Leopoldensia, 26: 79-92.

Sacchett, F., Botton, M. & Diehl, E. (2009). Ant species associated with the dispersal of *Eurhizococcus brasiliensis* (Hempel in Wille) (Hemiptera: Margarodidae) in vineyards of the Serra Gaúcha, Rio Grande do Sul, Brazil. Sociobiology, 54: 943-954.

Schmidt, F.A. & Diehl, E. (2008). What is the effect of soil use on ant communities? Neotropical Entomology, 37: 381-388. doi: 10.1590/S1519-566X2008000400005

SEPLAG. (2008). Atlas socioeconômico do Rio Grande do Sul. http://www.scp.rs.gov.br/atlas/default.asp. (accessed date: 30 January, 2014).

Silva, E.G. & Bandeira, A.G. (1999). Abundância e distribuição vertical de cupins (Insecta, Isoptera) em solo de Mata Atlântica, João Pessoa, Paraíba. Revista Nordestina de Biologia, 13(1/2): 13-36.

Silva, R.R. Da & Brandão, C.R.F. (1999). Formigas (Hymenoptera: Formicidae) como indicadoras da qualidade ambiental e da biodiversidade de outros invertebrados terrestres. Biotemas, 12: 55-73.

Silva, R.R., Feitosa, R.S.M. & Eberhardt, F. (2007). Reduced ant diversity along a habitat regeneration gradient in the southern Brazilian Atlantic Forest. Forest Ecology and Management, 240(1-3): 61-69. doi: 10.1016/j.foreco.2006.12.002

Souza, H.B.A., Alves, W.F. & Vasconcellos, A. (2012). Termite assemblages in five semideciduous Atlantic Forest fragments in the northern coastland limit of the biome. Revista Brasileira de Entomologa, 56: 67-72. doi: 10.1590/S0085-56262012005000013

Torales, G.J., Coronel, J.M., Godoy, M.C., Laffont, E.R & Romero, V.L. (2008). Additions to the taxonomy and distribution of Isoptera from Argentina. Sociobiology, 51: 31-48.

Ulysséa, M.A., Cereto, C.E., Rosumek, F.B., Silva, R.R. & Lopes, B.C. (2011). Updated list of ant species (Hymenoptera, Formicidae) recorded in Santa Catarina State, southern Brazil, with a discussion of research advances and priorities. Revista Brasileira de Entomologia, 55: 603-611. doi: 10.1590/S0085-56262011000400018

Vasconcellos, A., Bandeira, A.G., Moura, F.M.S., Araújo, V.F.P., Gusmão, M.A. & Constantino, R. (2010). Termite assemblages in three habitats under different disturbance regimes in the semi-arid Caatinga of NE Brazil. Journal of Arid Environments, 74: 298-302. doi: 10.1016/j.jaridenv.2009.07.007

Steffen, L.E. (2006). Riqueza e composição das comunidades de formigas em quatro formas de cultivo de erva-mate (*Ilex paraguariensis* St. Hil. 1822) na encosta superior do Nordeste do Rio Grande do Sul. São Leopoldo, 2006. 51f. [Dissertação de Mestrado - PPG em Biologia: Diversidade e Manejo de Vida Silvestre /UNISINOS].

Ward, P.S. (2000). Broad-scale patterns of diversity in leaf litter ant communities. In D. Agosti, J.D. Majer, L.E. Alonso & T.R. Schulz (Eds.), Standard methods for measuring and monitoring biodiversity (pp. 99-121). Washington: Smithsonian Inst. Press.

Wilson, E.O. (1997). A situação atual da diversidade biológica. In E.O. Wilson & F.M. Peter (Eds.), Biodiversidade (pp. 3-24). Rio de Janeiro: Nova Fronteira.

# 13

# Co-existence of ants and termites in *Cecropia pachystachya* Trécul (Urticaceae)

AC Neves[1], CT Bernardo[2], FM Santos[3]

1 - Universidade Federal de Minas Gerais, Belo Horizonte, MG, Brazil.

2 - Universidade de Brasília, Brasília, DF, Brazil.

3 - Universidade Estadual de Campinas, Campinas, SP, Brazil.

**Keywords**
Myrmecophytism, *Azteca*, Pantanal, *Nasutitermes ephratae*

**Corresponding author**
Ana Carolina de Oliveira Neves
Departamento de Biologia Geral
Instituto de Ciências Biológicas
Universidade Federal de Minas Gerais
Caixa Postal 486
31270-901, Belo Horizonte, MG, Brazil
E-mail: ananeves@gmail.com

**ABSTRACT**

Individuals of *Cecropia pachystachya* Trécul (Urticaceae) host *Azteca* (Hymenoptera: Formicidae) colonies in their hollow internodes and feed them with glycogen bodies produced in modified petiole bases (trichilia). In turn, ants keep trees free from herbivores and lianas. Here, we report for the first time the association of nests of *Nasutitermes ephratae* Rambur (Isoptera: Termitidae) with these trees, in South-Pantanal (Brazil). We aimed to describe the *Cecropia*-ant-termite relationship and to investigate how their coexistence is made possible. We hypothesize that: 1) The frequency of termite nests in *C. pachystachya* is lower than in neighbor trees; 2) Termite nests occur in trees with lower density of foraging ants; 3) The time that ants take to find and remove live termite baits in *C. pachystachya* trees is lower in leaves (close to trichilia) than in trunks; 4) Termite nests are fixed preferentially in the smallest and less branched trees; and 5) Termite nests are fixed preferentially distant from the canopies. Unexpectedly, termitaria occurred in *C. pachystachya* at the same frequency as in other tree species; there was no relationship between ant patrol activity and the occurrence of termite nests in *C. pachystachya;* and they occurred mainly in the tallest and more branched trees. However, termite nests generally were fixed in the trunk, fork or basal branches, where there is better physical support and ant patrol is more modest. The segregation of termite and ant life-areas may represent a escape strategy of termites in relation to ants inhabiting *C. pachystachya*, specially during nest establishment. The isolation of termites in fibrous nests and galleries may complete their defense strategy.

## Introduction

The neotropical genus *Cecropia* (Urticaceae) includes 60-70 species, 80% of them are trees inhabited by obligatory simbiotic ants of at least four subfamilies: Dolichoderinae, Formicinae, Myrmicinae e Ponerinae (Davidson & Fisher 1991, Davidson & McKey 1993, Folgarait *et al.* 1994). The relationship between *Cecropia spp.* and *Azteca spp.* (Hymenoptera: Formicidae) – one of the most conspicuous and well studied mirmecofilic interactions – is considered a symbiotic relationship because ants live only in the large hollow internodes of some species, feeding on Müllerian bodies produced in modified petiole bases (trichilia) (Wheeler 1942, Janzen 1969, Dejean *et al.* 2009). These food bodies are rich in carbohydrates, lipids, proteins and primarily glycogen, which remarkably is the principal storage carbohydrate found in animals and is

extremely rare in plants (Rickson 1971, Rico-Gray & Oliveira 2007). *Azteca* ants supplement their diet with invasive insects, which they aggressively attack (Dejean *et al.* 2009). There is strong evidence that ants provide nutrients to *Cecropia* individuals (Putz & Holbrook 1988, Sagers *et al.* 2000) and keep them free from herbivores and vines, thus acting as allelopathic agents and increasing tree competitive ability (Janzen 1966, Downhower 1975, Schupp 1986, Vasconcelos & Casimiro 1997).

Ants involved in obligate mutualisms reduce their foraging area to plant's surface and develop specialized behaviors to protect food resources, becoming extremely aggressive (Carroll & Janzen 1973). They are able to detect physical disturbances and chemical signals, such as volatile substances from leaves, responding with rapid recruitment of numerous soldiers (Agrawal *et al.* 1998, 1999, Dejean *et al.* 2009). Some studies show that herbivores prefer *Cecropia* trees unoccupied by *Azteca* individuals, including trees smaller than 2 m tall, which are usually not inhabited by ants or that present a smaller ant patrol activity (Downhower 1975, Schupp 1986, Vasconcelos & Casimiro 1997). However, some insects escape from predation even in trees occupied by ants. Ant genus such as *Cephalotes, Crematogaster* and *Pseudomyrmex,* were found foraging on *Cecropia pachystachya* Trécul (Urticaceae) branches (Vieira *et al.* 2010) and *Camponotus, Solenopsis* and *Procryptocerus* were found in *Cecropia insignis* (Liebm.) inhabited by *Azteca* ants (Longino 1991). Herbivore larvae such as *Ophtalmoborus* (Coleoptera: Curculionidae) are usually present in the spikes of *Cecropia* pistillate inflorescences (Berg *et al.* 2005). Schupp (1986) showed that the leaf damage made by chewer beetles was lower in ant-occupied *Cecropia obtusifolia* Bertol. than in unoccupied individuals. On the other hand, according to the same author, gall flies (Diptera) and phloem-feeding homoptera (Hemiptera) activity was not affected by presence of ants in these plants. He attributed ant ineffectiveness to gall flies small size and rapid oviposition, and to the motionless feeding pattern of homoptera that must be contacted by patrolling ants, while beetles are detectable from a distance due to leaf vibration caused by chewing and body movements. Also, when contacted, homoptera 'explode' off the leaf and alight undetected elsewere on the plant.

Ants are the main predators of termites (Hölldobler & Wilson 1990). There are at least six ant genera specialized in feeding on such insects, and arboreal termitaria are negatively influenced by predator ants (Wilson 1971, Gonçalves *et al.* 2005). Termites are often attacked during mating flights or after the accidental breakage of nests and galleries (Weber 1964, Shepp 1970, Carroll & Janzen 1973). Some ant colonies, such as *Solenopsis, Carebara, Centromyrmex* and *Hypoponera,* invade termite nests and attack their eggs, nymphs and adults (Lemaire *et al.* 1986, Delabie 1995, Dejean & Feneron 1999). However, early in the 20th century about 200 ant species were described living in pacific association with termites (Wheeler 1936), including the mutualism between *Amitermes laurensis* Mjoeberg (Hymenoptera: Termitidae) and two ant species of the genus *Camponotus,* which inhabit termite mounds and protect them from attack by the "meat ant" *Iridomyrmex sanguineus* Forel (Hymenoptera: Formicidae) (Higashi & Ito 1989). Little is known about the opposite situation, i. e., the occupation of ant colonies by termites (Quinet *et al.* 2005).

*Nasutitermes ephratae* Rambur (Hymenoptera: Termitidae) and other arboreal termite species are often found in Southern-Pantanal. The presence of this termite species in *Cecropia pachystachya* is remarkable because such trees host entire colonies of aggressive ants, mainly of the genus *Azteca,* which do not establish peaceful association with termites. Thus, the objective of this study was to describe the *Cecropia*-ant-termite relationship and to investigate how their coexistence is possible.

We hypothesize that: 1) The frequency of termite nests in *C. pachystachya* is lower than in neighbor trees; 2) Termite nests occur in trees with lower density of foraging ants; 3) The time that ants take to find and remove live termite baits in *C. pachystachya* trees is lower in leaves than in trunks; 4) Termite nests are fixed preferentially in the smallest and more branched trees; and 5) Termite nests are fixed preferentially distant from the canopies (where trichilia are concentrated).

## Material and Methods

### Study area

The Pantanal is a seasonal floodplain in tropical South America, located in the upper Paraguay River basin. It occupies an area of 147.572 km², between 80 and 150 m above sea level (Alho & Gonçalves, 2005). There is a rainy season from November to March corresponding to 72% of the total annual rainfall (1,182.5 mm) and a dry season from April to October. The average annual temperature is 25.5 ° C, but the absolute maximum exceeds 40° C and the minimum is close to 0° C. The average relative humidity is 82% (Soriano *et al.* 1997). Annual floods occur during summer (February to May), although there are also multi-annual floods, which produce long periods of pronounced dry and wet seasons. Flooding results mainly from drainage difficulties caused by the low declivity of the terrain, which varies from 3 to 5 cm/km (east-west), and from 1 to 30 cm/km (north-south) (Alho & Gonçalves, 2005).

Sampling was done along the highway MS-184 (Park Road), between the coordinates 19°38'56.4"S, 057°01'37.6"W and 19°22'20.28"S, 57°02'32.40"W, in the municipality of Corumbá, Mato Grosso do Sul state, Brazil. The tree comunity along the road was dominated by *C. pachystachya*, *Vitex cymosa* (Verbenaceae), *Copernicia alba* (Arecaceae), *Enterolobium contortisiliquum* (Fabaceae) and *Tabebuia* spp (Bignoniaceae).

### Study species

*Cecropia pachystachya* occurs throughout the Pantanal, where it is common in riparian flooded areas and non flooded forested patches ("capões" and "cordilheiras") (Pott & Pott 1994). Their large-hollow internodes provide nesting space for ants, and the thin spots in their upper wall (prostomata) allow ants to circulate from inside to outside the tree (Berg *et al.* 2005). They also present trichilia formed by patches of dense indumentum abaxially at the base of the petiole of adult leaves, which produce food corpuscles called Müllerian bodies (Berg *et al.* 2005).

In the study area, *C. pachystachya* are inhabited by *Azteca ovaticeps* Forel, *A.isthmica* Wheeler and *A. alfari* Emery (Vieira *et al.* 2010). *Azteca* are territorial ants that feed on Müllerian bodies and form dense colonies in *Cecropia*. During foraging activity, workers move randomly through the tree until they find any invader. They recruit soldiers by alarm pheromones and aggressively attack the invader (even when it is dead), killing and throwing it from the tree, and eventually feeding on it (Carroll & Janzen 1973, Quinet *et al.* 2005, Dejean *et al.* 2009).

*Nasutitermes ephratae* is a Neotropical termite that feeds on plant-debris and inhabits lowland areas (Thorne 1980, Vasconcellos & Moura 2009). It builds arboreal, spherical or ellipsoidal, carton nests in trunks or branches, with internal or external galleries spreading from nests toward food sources (Thorne 1980, Thorne & Haverty 2000). There are records of aggressive encounters between *Nasutitermes* individuals and *Azteca* ants (Braekman *et al.* 1983; Noirot & Darlington 2000, Quinet *et al.* 2005). Although these termites feed mainly in litter wood, they are usually found, attacked and removed by ants that inhabit *Cecropia* trees when used as baits (e. g. Oliveira *et al.* 1987, Dejean *et al.* 2009). *Nasutitermes* soldiers defend themselves throwing a viscous secretion produced by frontal glands (see Eisner *et al.* 1976).

### Sampling methods

To compare the frequency of termitaria between *C. pachystachya* and its neighbor trees, we covered a path of ~ 30 km of a dirt road counting trees from several species, with and without nests. We sampled only trees whose trunks and crowns were fully visible (not hidden by other trees or vines), a total of 140 individuals of *C. pachystachya* and 78 from other tree species.

To verify whether ant patrol activity differed between *Cecropia* trees with or without termitaria, we counted the number of ants that crossed an area of 5 x 2 cm$^2$, defined by a carton frame disposed in the tip of a branch and in a leaf (near petiole insertion). The count was done after a soft tap in the branch. The procedure lasted for two minutes and was done between 8 am and 11 am, in 17 trees without termite nests and in eight trees with termite nests.

To test how long ants took to find and remove live termite baits in different parts of the trees, we used white school glue to adhere *N. ephratae* soldiers on leaves (in the abbaptial surface), petioles (next to trichilia) and trunks. We repeated this procedure in 20 *C. pachystachya* individuals and observed for up to seven minutes.

To characterize the architecture of *C. pachystachya* individuals with and without termitaria, we estimated the height of trees (H), we counted the number of branches per tree (N) and calculated the ratio branches: tree height (N/H) as an indicator of canopy density, in 90 trees, 22 of which had termitaria. Histograms of H and N/H of trees were made, highlighting the ocurrence of termite nests in the population of *C. pachystachya*.

In order to have an indication of termitarium distance from the canopy, we estimated the height (h) of 27 nests and then calculated the ratio nest height: tree height (h/H) and

separated the results into four categories. Thus, the closer the ratio h/H was to zero, the nearest to ground the nest was. We also determined the tree structure in which 18 nests were fixed (trunk, fork, lower or upper branches).

### Data analysis

We compared the frequency of termite nests between *C. pachystachya* and its neighbour tree species with the chi-square test. Ant patrol activity in leaves and branches were compared between *C. pachystachya* individuals with and without termitaria using the Student's t test.

### Results and Discussion

In the study area, termitaria were composed by a carton nest with rigid galleries made of dirty and fibrous material, spreading toward branches (Fig 1).

While ants were present in all the *Cecropia pachystachya* trees with which we had direct contact, termite nests were present in 42.9% of the *C. pachystachya* individuals and in 36.8% of the other tree species, with no statistical difference between them ($\chi^2 = 0.105$, p = 0.74, n=218).

Ants crossed the defined 10 cm$^2$ area in *C. pachystachya* individuals with and without termite nests, respective-

Fig 1: *Cecropia pachystachya* Trécul (Urticaceae) in South-Pantanal, Brazil, with a nest of *Nasutitermes ephratae* Rambur (Termitideae). Note the galleries spreading in trunk and branches.

ly, 70.75±27.12 and 56.47±18.16 times in branches, and 3.63±2.56 and 10.94±3.80 in leaves (average ± standard error of mean). Ant patrol activity was similar between trees with and without termite nests in branches (t = -0.44, p=0.60, df =23, n=25) and in leaves (t =1.25, p =0.22, df =23, n=25).

Thirty five per cent of termites used as baits were found and attacked by ants. In some cases, especially in trunks, they were found but were not attacked. Attacks occurred in 65% of 20 trials in petioles, 25% in leaves and 15% in trunks. The time elapsed for attacks were 112.54±32.53 seconds (n=13) in petioles, 151.40±56.94 s (n=5) in leaves and 67.77±28.26 s (n=3) in trunks. It was not possible to compare statistically the time of ant atack due to the small size of samples. We also observed that the galleries spreading from nests were not examined or attacked by ants.

In the sampled population (n=90), *C. pachystachya* individuals varied from 3-12 m tall, but termitaria (n=22) ocurred only in trees to 7-11 m tall, with a modal class of frequency in 8 m that coincided with the population mode (Fig 2).

The number of branches per height varied from 0.20 – 6.86. Termitaria ocurred in trees with one to 6.86 branches per meter. Again, the modal classes of trees with termitarium coincided with the modal class of branching in the population, i. e. 1.51 – 2.50 branches per meter (Fig 3).

Almost fifteen percent (14.8%) of termite nests were fixed in the lower one-quarter of trees, 63% in the second, 18,5% in the third and only 2.7% in the fourth one-quarter (from bottom to up). Most of them were located at the base of the lower branches (60%), followed by the fork and upper branches (15% each), and finally the trunk (10%).

Opposite to our hypothesis' predictions, *Nasutitermes ephrateae* termitaria occur in *Cecropia pachystachya* individuals in the same frequency than in non-mirmecophyte trees, as well as in *C. pachystachya* individuals with intense or modest ant patrol activity. Although it is an unexpected result, considering the mirmecophilic life story of *Cecropia*, we sugest that physical mechanisms mediate the relationship between *N. ephrateae* and ants of the genus *Azteca*, namely the occupation of distinct parts of the tree by termites and ants, and the physical isolation of termitaria (nests and galleries) by fibrous structures.

Although the distribution of *C. pachystachya* individuals with termitarium have the same modal class than the population as a whole (regarding tree height and number of branches per height), termite nests did not occur randomly. Termite nests were found only in a restricted range of tree height within the population studied. By their turn, ants are expected to occur in *C. pachystachya* trees taller than 2 m after Downhower (1975) and Schupp (1986).

One might expect that termites fixed their nests preferentially on small trees, which are not inhabited by ants or

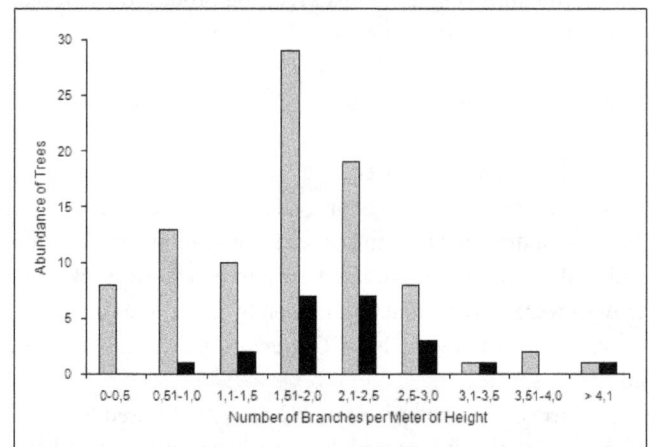

Fig 3: Abundance of individuals of *Cecropia pachystachya* Trécul (Urticaceae) in Pantanal Sul, Brazil, regarding their ratio branches: height of the trees. Black bars: trees with termitarium of *Nasutitermes ephratae* Rambur (Termitideae); gray bars: trees without termitarium.

present a smaller ant patrol activity, being prefered by herbivores (Downhower 1975, Schupp 1986). However, besides the scape of predation, physical support must be a determinant of termitaria establishment, as it is suggested by the absence of termite nests in *C. pachystachya* individuals smaller than 7 m and its scarcity in trees with less than 1.51 branches per meter of height. In fact, Cunha (2000) found that *Constrictotermes cyphergaster* Silvestri (Isoptera: Termitidae) usually fixed their nests in trees with intermediate trunk circumference.

Although termite nests predominated in trees with 1.51 to 2.5 branches per height (the modal class of branch density in the population), they were disproportionally underrepresented in the smaller classes. Termitaria predominance in more branched trees once more indicates that its occurrence is not random. Densely branched individuals may provide more opportunity for nesting and better support conditions, and although they have more dense canopies, it does not necessarily imply in larger density of trichilia and therefore greater

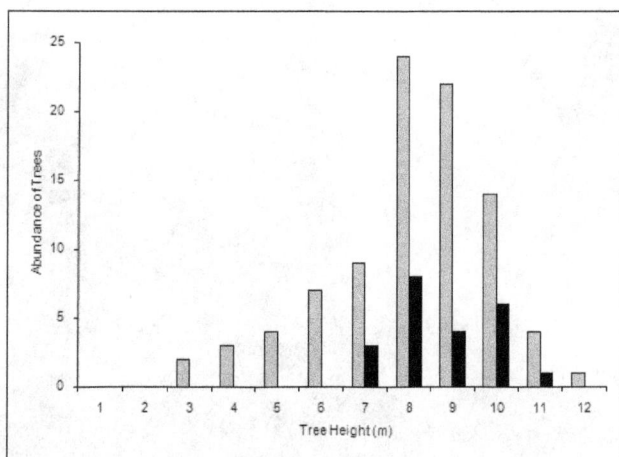

Fig 2. Abundance of individuals of *Cecropia pachystachya* Trécul (Urticaceae) in Pantanal Sul, Brazil, regarding their height. Black bars: trees with termitarium of *Nasutitermes ephratae* Rambur (Termitideae); gray bars: trees without termitarium.

patrol by ants, since the lower branches often had no leaves or ants. In other species of mirmecophyte *Cecropia* it was observed that ant patrol differed between leaves in different portions of the canopy, as well as herbivory. In *Cecropia peltata* Linnaeus, 53.9 % of ants were located in leaves in the upper canopy and only 11.2 % in the leaves of basal branches, resulting in proportional rates of herbivory (Downhower 1975). In *C. pachystachya* termite nests were more frequent in the lower half of the trees and in lower branches, i. e. distant from young-active trichilia in the top canopy, where ant patrol may be more intense. This suggests that there is a displacement of the location of termites and ants, avoiding overlap. This finding is corroborated by the fact that termites are rarely attacked by ants in trunks, but they are attacked and removed when experimentally placed in leaves or petioles, which are close to trichilia.

One hypothesis to explain the coexistence of *Azteca* sp. and *N. efrateae* in *C. pachystachya* is that they do not engage in agonistic encounters. However termites were readily attacked and removed from leaves and petioles, and sometimes from trunks when experimentally introduced. Ant's attacks to termite would probably be more frequent, especially in  branches, if they did not construct galleries with fibrous and dust material.

We suggest that, after a period of large vulnerability during the establishment of termitaria, nest and galleries built with rigid material contribute to the coexistence of these groups of insects, allowing termites to move protected inside them, even in the canopy. This strategy is consistent with the one adopted by other termite species that coexist with ants without predation. In cases that ants colonize termitaria, these insects often dwell distinct portions. For example, *Crematogaster rochai* Forel (Hymenoptera: Formicidae) inhabit termite nests of *N. ephrateae* and *Nasutitermes corniger* Motschulsky (Hymenoptera: Termitidae) and maintains a physical separation with its hosts by plugging the cells that they inhabit with fibrous material (Quinet *et al.* 2005).

The observed trend of tree occupation is also in agreement with the distribution of vines in *Cecropia*, that, as well as herbivores, are cut and removed by ants (Janzen 1969, Downhower 1975, Vasconcelos & Casimir 1997). Vines are often found on the stem and the lower parts of the lower branches, but not higher up (Berg *et al.* 2005). This was attributed to the pattern of branches' growth and loss – lower branches depart from the trunk in angles of 45 degrees and lower leaves are continuously lost –, thus affecting the growth of vines (Berg *et al.* 2005). We suppose that in mirmecophyte *Cecropia* it may also be attributed to the pattern of tree occupation by ants and their vine cutting activity.

In conclusion, this study shows that *N. ephrateae* nests occur at the same frequency in *Cecropia* individuals than in non mirmecophyte trees, as well in *Cecropia* individuals with itense or modest ant foraging activity. The determinants of nest occurrence are probably the mechanical support offered by trees and nest distance from trichillia, i. e. from the area

more intensively patrolled by ants, that may influence specially the colonization time. After that, isolation of termites inside fibrous nests and galleries may contribute to their coexistence with ants. So, our results indicate that the coexistence between termites and ants in *C. pachystachya* may be possible due to spatial segregation of their colonies, avoiding agonistic interactions.

To the best of our knowledge, this is the first report of the occurrence of termites on *Cecropia* trees, although this is a common association in the study area. Further research on the interaction between these insects would help to understand how their coexistence is made possible, as well as the influence of the flood pulse in the Pantanal in the evolution of this ecological relationship.

## Acknowledgments

We thank two anonymous referees for valuable comments in the manuscript; R. Constantino and I. Leal for termite and ant identification; G. A. F. Medina for language editing and comments on the manuscript; and students of the undergraduate course of Biological Sciences in the of Universidade Federal de Mato Grosso do Sul (UFMS) for helping with sampling. This study was conducted during the Ecology Field course fully supported by the Graduate Program in Ecology and Conservation (UFMS). A. C. O. Neves was granted Conselho Nacional de Desenvolvimento Científico e Tecnológico (CNPq) and Coordenação de Aperfeiçoamento de Pessoal de Nível Superior / Fundação de Amparo à Pesquisa do Estado de Minas Gerais (Capes/Fapemig) for scholarships.

## References

Agrawal, A.A. (1998) Leaf damage and associated cues induce aggressive ant recruitment in a neotropical ant-plant. Ecology, 79: 2100-2112.

Agrawal, A.A. & Dubin-Thaler, B.J. (1999) Induced Responses to Herbivory in the Neotropical Ant-Plant Association Between *Azteca* Ants and *Cecropia* Trees: Response of Ants to Potential Inducing Cues. Behavioral Ecology and Sociobiology, 45: 47-54

Alho, C.J.R. & Gonçalves, H.C. (2005) Biodiversidade do Pantanal: Ecologia e Conservação. Campo Grande: Editora Uniderp, 135 p.

Berg, C.C, Rosselli, P.F. & Davidson, D.W. (2005) Flora Neotropica: *Cecropia*, vol 94. New York: New York Botanical Garden Press, 236 p.

Braekman, J.C., Daloze, D., Dupont, A., Pasteels, J.M., Lefeuve, P., Borderau, C., Declercq, J.P. & Van Meerssche, M. (1983) Chemical composition of the frontal gland secretion from soldiers of *Nasutitermes lujae* (Termitidae: Nasutermitinae). Tetrahedron, 39: 4237-4241.

Carroll, C.R. & Janzen, D.H. (1973) Ecology of foraging by ants. Annual Review of Ecology and Systematics, 4: 231-257.

Cunha, H.F. (2000) Estudo de colônias de *Constrictotermes cyphergaster* (Isoptera, Termitidae: Nasutitermitinae) no Parque Estadual da Serra de Caldas Novas, GO. Dissertação de mestrado, Univ. Federal de Goiás, Instituto de Ciências Biológicas/DBG, Goiânia, 51p.

Davidson, D.W. & Fisher, B.L. (1991) Symbiosis of ants with *Cecropia* as a function of light regime. In: Huxley, C. & Cutler, D. K. (eds.). *Ant-Plant Interactions* (pp: 289-309). New York: Oxford University Press.

Davidson, D.W. & McKey, D. (1993) The evolutionary ecology of symbiotic ant-plant relationships. Journal of Hymenoptera Research, 2: 13-83.

Dejean, A. & Fénéron, R. (1999) Predatory behaviour in the ponerine ant, *Centromyrmex bequaerti*: a case of termitolesty. Behavioural Processes, 47: 125-133.

Dejean, A., Grangier, J., Leroy, C. & Orivel, J. (2009) Predation and aggressiveness in host plant protection: a generalization using ants of the genus *Azteca*. Naturwissenschaften, 96: 57-63.

Delabie, J.H.C. (1995) Inquilinismo simultâneo de duas espécies de *Centromyrmex* (Hymenoptera: Formicinae: Ponerinae) em cupinzeiros de *Syntermes* sp. (Isoptera: Termitidae: Nasutiterminae). Revista Brasileira de Entomologia, 39: 605-609.

Downhower, J.F. (1975) The distribution of ants on *Cecropia* leaves. Biotropica, 7: 59-62.

Eisner, T., Kriston, I. & Aneshansley, D.J. (1976) Defensive Behavior of a Termite (*Nasutitermes exitiosus*). Behavioral Ecology and Sociobiology, 1: 83-125.

Folgarait, P.J., Johnson, H.L. & Davidson, D.W. (1994) Responses of *Cecropia* to experimental removal of mullerian bodies. Functional Ecology, 8: 22-28.

Gonçalves, T.T., Reis, R., DeSouza, O. & Ribeiro, S.P. (2005) Predation and interference competition between ants (Hymenoptera: Formicidae) and arboreal termites (Isoptera: Termitidae). Sociobiology, 46: 409-419.

Higashi, S. & Ito, F. (1989) Defense of termitaria by termitophilous ants. Oecologia, 80:145-147.

Hölldobler, B. & Wilson, E.O. (1990) The ants. Berlin: Harvard University Press, 732 p.

Janzen, D.H. (1966) Coevolution of mutualism between ants and acacias in Central America. Evolution, 20: 249-275.

Janzen, D.H. (1969) Allelopathy by myrmecophytes: the ant *Azteca* as an allelopathic agent of *Cecropia*. Ecology, 50: 147-153.

Janzen, D.H. (1973) Dissolution of mutualism between *Cecropia* and its *Azteca* ants. Biotropica, 5: 15-28.

Lemaire, M., Lange, C., Lefevre, J. & Clement, J.L. (1986) Strategie de camouflage du prédateur *Hypoponera eduardi* dans les

sociétés de *Reticulitermes* européens. Actes Coll. Insectes Sociaux., 2: 97–101.

Longino. J.T. (1991) *Azteca* ants in *Cecropia* trees: taxonomy, colony structure and behaviour. In: Cutler, D. F. & C. R. Huxley (eds.) *Ant-Plant Interactions* (pp: 198-212). New York: Oxford University Press.

Noirot, C. & Darlington, J.P.E.C. (2000) Termites nests: architecture, regulation and defence. In: Abe, T., *et al.* (eds.) *Termites: Evolution, Sociality, Symbioses, Ecology* (pp: 121-39). Dordrecht: Kluwer Academic Publishers.

Oliveira, P.S., Oliveira-Filho, A.T. & Cintra, R. (1987) Ant foraging on ant-inhabited *Triplaris* (Polygonaceae) in western Brazil: a field experiment using live termite-baits. Journal of Tropical Ecology, 3: 193-200.

Pott, A. & Pott, V.J. (1994) Plantas do Pantanal. Brasília: Embrapa-SPI, 320 p.

Putz, F.E. & Holbrook, N.M. (1988) Further observations on the dissolution of mutualism between *Cecropia* and its ants: The Malaysian case. Oikos, 53: 121-125.

Quinet, Y., Tekule, N. & Biseau, J.C. (2005) Behavioural interactions between *Crematogasterbrevispinosa rochai* Forel (Hymenoptera: Formicidae) and two *Nasutitermes* species (Isoptera: Termitidae). Journal of Insect Behavior, 18: 1-17. doi: 10.1007/s10905-005-9343-y.

Rico-Gray, V. & Oliveira, P.S. (2007) The ecology and evolution of ant-plant Interactions. Chicago: University of Chicago Press, 320 p.

Rickson, F.R. (1971) Glycogen plastids in Müllerian body cells of *Cecropia peltata* – a higher green plant. Science, 173 (3994): 344-347. doi: 10.1126/science.173.3994.344

Sagers, C.L., Ginger, S.M. & Evans, R.D. (2000) Carbon and nitrogen isotopes trace nutrient exchange in an ant-plant mutualism. Oecologia, 123: 582–586.

Sheppe, W. (1970) Invertebrate predation on termites of the African savanna. Insectes Sociaux, 17: 205-18.

Schupp, E.W. (1986) *Azteca* protection of *Cecropia*: ant occupation benefits juvenile trees. Oecologia, 70: 319-385.

Soriano, B.M.A., Oliveira, H., Catto, J.B., Comastri-Filho, J.A., Galdino, S. & Salis, S. M. (1997) Plano de Utilização da Fazenda Nhumirim (Documento 21). Corumbá: Embrapa-CPAP, 72 p. http://www.cpap.embrapa.br/publicacoes/online/DOC21.pdf (accessed date: 12 November, 2013).

Thorne, B.L. (1980) Diferences in nest architecture between the neotropical arboreal termites *Nasutitermes corniger* and *N. ephrateae* (Isoptera, termitideae). Psyche, 87: 235-244.

Thorne, B.L. & Haverty, M.I. (2000) Nest growth and survivorship in three species of neotropical *Nasutitermes* (Isoptera: Termitidae). Environmental Entomology, 29: 256-264.

Vasconcelos, H.L. & Casimiro, A.B. (1997) Influence of *Azteca alfari* ants on the exploitation of *Cecropia* trees by a leaf-cutting ant. Biotropica, 29: 84-92.

Vasconcellos, A. & Moura, F.M.S. (2010) Wood litter consumption by three species of *Nasutitermes* termites in an area of the Atlantic Coastal Forest in northeastern Brazil. Journal of Insect Science, 10: 72. Retrived from: http://insectscience.org/10.72/

Vieira, A.S., Faccenda, O. Antonialli-Junior, W.F. & Fernandes, W.D. (2010) Nest structure and occurrence of three species of *Azteca* (Hymenoptera, Formicidae) in *Cecropia pachystachya* (Urticaceae) in non-floodable and floodable pantanal areas. Revista Brasileira de Entomologia, 54: 441- 445. doi: 10.1590/S0085-56262013000100013.

Weber, N.A. (1964) Termite prey of some African ants. Entomological News, 75: 197-204.

Wheeler, W.M. (1936) Ecological relations of ponerinae and other ants to termites. Proceedings of the American Academy of Arts and Science, 71: 159-243.

Wheeler, W.M. (1942) Studies of neotropical ant-plants interactions ant their ants. Bulletin of the Museum of Comparative Zoology Harvard, 90: 1-262.

Wilson, E. (1971) The Insect Societies. Cambridge: Belknap Press of Harvard University Press, 560 p.

# 14

# Activity Patterns of the Red Harvester Ant in a Mexican Tropical Desert

L. Ríos-Casanova[1], G. Castaño[2], V. Farías-González[1], P. Dávila[1], H. Godínez-Alvarez[1]

1 - UBIPRO, FES-Iztacala, Universidad Nacional Autónoma de México, Tlalnepantla, Estado de México, México.

2 - Facultad de Ciencias, Campus Juriquilla, Universidad Nacional Autónoma de México, Querétaro, México.

**Keywords**

*Pogonomyrmex barbatus*, deserts, soil surface temperature, Tehuacan Valley

**Corresponding author**

Leticia Ríos-Casanova
UBIPRO, FES-Iztacala
Universidad Nacional Autónoma de México
Av. de los Barrios 1, Los Reyes Iztacala
Tlalnepantla, Estado de México, México
54090
E-mail: leticiarc@campus.iztacala.unam.mx

**Abstract**

Red harvester ant (*Pogonomyrmex barbatus*) inhabits deserts of USA and Mexico. Its activity patterns are well known in temperate deserts, but they have not been studied in tropical ones. We studied these patterns in the Tehucan Valley, a tropical desert in central Mexico. It had bimodal activity patterns in spring, summer, and fall while unimodal patterns in winter. These patterns differ from those reported for this species in temperate deserts where activity stopped in winter. Our results suggest that *P. barbatus* extends its activity periods and remains active all year round in the Tehuacan Valley.

## Introduction

Red harvester ant (*Pogonomyrmex barbatus* Smith) is one of the most common and abundant species in deserts of USA and Mexico (Johnson, 2000). This ant harvests high proportions of seeds from annual and perennial plants, affecting their abundance and distribution. Activity patterns determine periods in which ants forage seeds, therefore it is essential to document them and how they change along year in sites located at different latitudes.

Activity patterns of several *Pogonomyrmex* species, including *P. barbatus*, have been studied in North American temperate deserts. Diurnal ants are active in the morning and late afternoon with a period of decline around midday when temperatures reach maximum values, showing a bimodal pattern of activity most of the year (Whitford, 1978; MacKay & MacKay, 1989; Pol & Lopez de Casenave, 2004). Ants diminish activity during winter changing to inactivity or to a unimodal pattern by being active during the hottest hours of the day (Whitford & Ettershank 1975; Hölldobler & Wilson,

1990; Crist & MacMahon, 1991). Although these activity patterns are known in temperate deserts, they have not been studied in tropical ones. *Pogonomyrmex barbatus* is distributed in North American temperate and tropical deserts; however, no reports exist about its activity in the tropics. We study activity patterns of this ant species in the Tehuacan Valley, a tropical desert in central Mexico. This study represents the southernmost site where activity has been studied until now (Figure 1).

Because seasonal and daily temperature variation is lower in tropical than in temperate regions (MacKay & MacKay, 1989), we expect *P. barbatus* will be active all year round at the Tehuacan Valley compared to northern locations. This paper reports the number of ants returning to the nest and soil surface temperature for all seasons of the year. Although activity patterns depend on environmental factors such as seed availability, daily and seasonal temperatures, we only measured soil surface temperature because it is considered the most important factor regulating ant activity in deserts (Whitford, 1999).

## Material and methods

This study was conducted at San Rafael Coxcatlán (18° 12' - 18° 14' N, 97° 07'- 97° 09' W; 1000m a. s. l.; Fig 1) in the Tehuacán Valley, central Mexico. The mean annual temperature is 25°C and the mean annual rainfall is 394.5 mm (Valiente, 1991). The main vegetation type is tropical decidu-ous forest dominated by *Fouquieria formosa*, *Bursera aptera*, and *Ceiba parvifolia* (Ríos-Casanova et al., 2006).

We selected 8 nests separated by at least 10 m each, to count the number of ants returning to nest during 5-min pe-riods. Counts were conducted from 0700–1900 h during one day in fall (September 2010), winter (December 2010), spring (March 2011), and summer (July 2011) for a total of 384 ob-servation periods. Nocturnal activity was not recorded. Soil surface temperature was recorded every hour (0700–1900 h) by placing one thermometer at 30 cm from nest entrance and burying the mercury containing bulb at 1 cm depth.

## Results

Ants showed a bimodal activity pattern during fall, spring, and summer (Fig. 2). Foraging activity started when surface temperature rose to 21°C. The first activity peak (60-71 ants/5 min) occurred between 11-13 h, when soil surface temperature was 36-42°C. The second peak (10-40 ants/5 min) occurred in the afternoon (16-17 h), when temperature was 39-47°C. Foragers stopped activity between peaks in spring when maximum temperature was ca. 50°C while they were al-ways active between peaks in summer and fall. The maximum temperature in these seasons was 45-47°C.

Ants only showed one activity peak in winter (20 ants/5 min), which occurred during midday when temperature was 40°C, which was the highest temperature recorded in the season (Fig. 2).

## Discussion

Our results showed that *P. barbatus* has bimodal activ-ity patterns in spring, summer, and fall, and unimodal activity patterns in winter. Activity peaks in bimodal patterns occurred when soil surface temperature was 36-47°C. Activity stopped when temperature rose to 50°C. Activity peak in unimodal patterns occurred when temperature was 40°C.

The activity pattern found in our study differs from that reported for *P. barbatus* in a temperate desert (García-Pérez et al., 1994), where bimodal activity patterns occurred in spring and summer, and unimodal patterns ocurred in fall. This ant species was not active in winter (Garcia-Perez et al., 1994). Our findings therefore suggest that *P. barbatus* inhabit-ing a tropical desert extends its activity periods and remains active all year round, according to our predictions.

Despite differences in activity patterns, it seems that soil surface temperature is one of the main environmental fac-

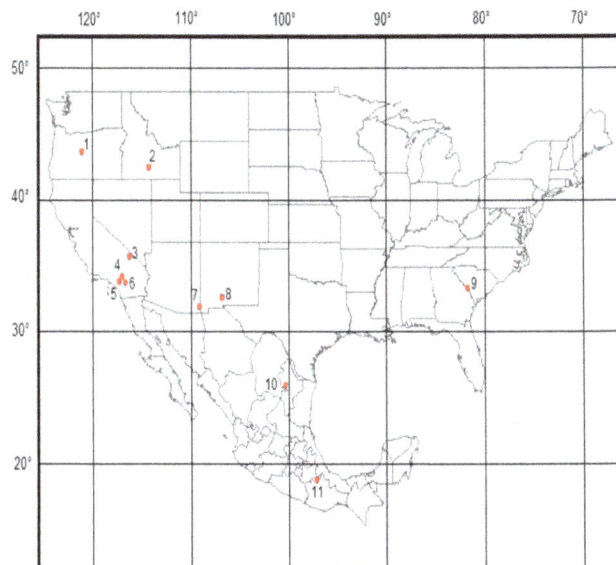

**Fig 1**. Localities where *Pogonomyrmex* activity has been studied. 1 = *P. owyheei*, 2 = *P. occidentalis* (Bernstein, 1979), 3 = *P. califor-nicus*, *P. rugosus*, (Bernstein, 1974), 4 = *P. montanus*, 5 = *P. rugo-sus*, 6 = *P. subnitidus* (MacKay & MacKay, 1989), 7 = *P. barbatus* (Gordon, 1983), 8 = *P. rugosus* (Whitford & Ettershank, 1975, 9 = *P. badius* (Golley & Gentry, 1964), 10 = *P. barbatus* (García-Pérez, et al., 1994), 11 = *P. barbatus* (this study).

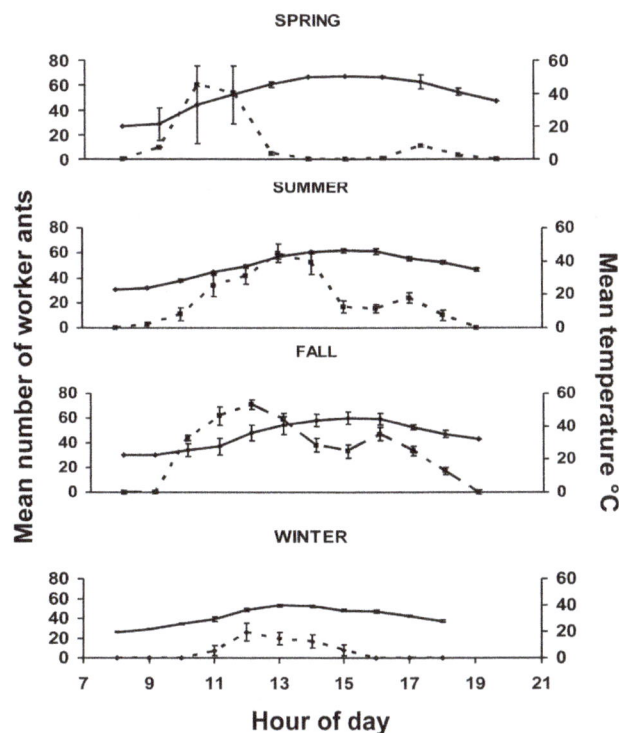

**Fig 2**. Mean soil surface temperature ± 1 standard error (solid line) and average number of active ants (*P. barbatus*) outside the nest ± 1 standard error (dashed line) hourly and four seasons of the year in San Rafael Coxcatlán, Puebla, Mexico.

tors regulating activity of this ant species in temperate and tropical deserts. Johnson (2000) in analyzing physiological thermal tolerance of *P. barbatus* from the Chihuahuan Desert found that ant foragers stopped their activity at 47°C because they were unable to survive at higher temperatures. Our data on soil surface temperature indicate that ant activity stopped at 50°C. Other *Pogonomyrmex* species such as *P. montanus*, *P. subnitidus*, *P. apache*, and *P. rugosus* from temperate North American deserts also reduce or stopped their activity when surface temperature was > 50°C (MacKay & MacKay, 1989; Hölldobler & Wilson, 1990; Whitford, 1999). These findings suggest that activity patterns of *Pogonomyrmex* species are regulated by soil surface temperature. This may be interpreted as an adaptation of these ants to high desert temperature, independently of latitude, which may be a phylogenetically conserved trait (MacMahon et al., 2000). Our results however should be interpreted cautiously because they assume that soil surface temperature is the only factor regulating ant activity. Other factors such as seed availability, presence of other harvester ants, predation, and plant relationships which also regulate activity were not measured in this study (Wilby & Shachak, 2000; Pol & López de Casenave, 2004). These factors should be considered in future studies to determine their relative importance on ant foraging activity.

In short, *P. barbatus* extends its activity period and remains active all year round in the Tehuacan Valley. Evaluation of activity patterns throughout its distribution is essential to understand seed foraging impact on plant abundance and distribution.

## Acknowledgments

We thank authorities of San Rafael Coxcatlán, Puebla for their permission to work in their lands. This work was supported by grants from Facultad de Estudios Superiores Iztacala, Universidad Nacional Autónoma de México, Proyecto PAPCA No. 2010-2011 to L.R-C. and V.F-G. We thank two anonymous referees for their comments and criticisms on the manuscript.

## References

Bernstein, R. (1974). Seasonal food abundance and foraging activity in some desert ants. American Naturalist, 108:490-498.

Bernstein, R. (1979). Schedules of foraging activity in species of ants. Journal of Animal Ecology, 48:921-930.

Crist, T.O. & MacMahon, J.A. (1991). Foraging patterns of *Pogonomyrmex occidentalis* (Hymenoptera: Formicidae) in a shrubsteppe ecosystem: the roles of temperature, trunk trails, and seed resources. Environmental Entomology, 20: 265-275.

García-Pérez, J.A., Rebeles-Manrique, A. & Peña-Sánchez, R. (1994). Seasonal changes in trails and the influence of

temperature in foraging activity in a nest of the ant *Pogonomyrmex barbatus*. Southwestern Entomology, 2: 181-187.

Golley F. B. & Gentry, J.B. (1964). Bioenergetics of the Southern harvester ant, *Pogonomyrmex badius*. Ecology 45: 217–225.

Gordon, D. (1983). The relationship of recruitment rate to activity rhythms in the harvester ant *Pogonomyrmex barbatus* (F. Smith) (Hymenoptera: Formicidae). Journal of Kansas Entomological Society, 56: 277-285.

Hölldobler, B. & Wilson, E.O. (1990). The Ants. The Belknap Press of Harvard University Press, Cambridge. 732 p.

Johnson, R.A. (2000). Seed-harvester ants (Hymenoptera: Formicidae) of North America: An overview of ecology and biogeography. Sociobiology, 36: 89-122.

MacKay, W.P. & MacKay, E.E. (1989). Diurnal foraging patterns of *Pogonomyrmex* harvester ants (Hymenoptera: Formicidae). Southwestern Naturalist, 34: 213-218.

MacMahon, J.A., Mull, J.F. & Crist, T.O. (2000). Harvester ants (*Pogonomyrmex* spp.): their community and ecosystem influences. Annual Review of Ecology and Systematics, 31: 265-291. doi: 10.1146/annurev.ecolsys.31.1.265

Pol, R. & López de Casenave, J. (2004). Activity patterns of harvester ants *Pogonomyrmex pronotalis* and *Pogonomyrmex rastratus* in the Central Monte Desert, Argentina. Journal of Insect Behavior, 17: 647-661. doi: 0892-7553/04/0900-0647/0

Ríos-Casanova, L., Valiente-Banuet, A. & Rico-Gray, V. (2006). Ant diversity and its relationship with vegetation and soil factors in an alluvial fan of the Tehuacán Valley, Mexico. Acta Oecologica, 29: 316-323. doi:10.1016/j.actao.2005.12.001

Valiente, B.L. (1991). Patrones de precipitación en el Valle semiárido de Tehuacán, Puebla, México. Tesis de Licenciatura. Facultad de Ciencias, UNAM, México. 61 pp.

Whitford, W.G. (1978). Structure and seasonal activity of Chihuahua Desert ant communities. Insectes Sociaux, 25:79-88.

Whitford, W.G. (1999). Seasonal and diurnal activity patterns in ant communities in a vegetation transition region of southern New Mexico (Hymenoptera: Formicidae). Sociobiology, 3: 477-492.

Whitford, W.G. & Ettershank, G. (1975). Factors affecting foraging activity in Chihuahuan desert harvester ants. Environmental Entomology, 4: 689-696.

Wilby, A. & Shachak, M. (2000). Harvester ant response to spatial and temporal heterogeneity in seed availability: patterns in the process of granivory. Oecologia, 125: 495-503. doi:10.1007/s004420000478

# 15

# Mutualistic relationships between the shield ant, *Meranoplus bicolor* (Guérin–Méneville) (Hymenoptera: Formicidae) and honeydew–producing hemipterans in guava plantation

I Burikam, D Kantha

*Kasetsart University, Nakhon Pathom, Thailand*

**Keywords**
cotton aphid, striped mealybug, coccinellids, syrphid fly, *Psidium guajava*

**Corresponding author**
Intawat Burikam
Department of Entomology
Faculty of Agriculture at Kamphaeng Saen, Kasetsart University
Nakhon Pathom 73140, Thailand
E–Mail: intawat.b@ku.ac.th

**Abstract**

Mutualistic relationships between the shield ant, *Meranoplus bicolor* (Guérin–Méneville), and two species of hemipteran, *Aphis gossypii* Glover and *Ferrisia virgata* (Cockerell), were investigated in an unsprayed guava plot at Kamphaeng Saen, Nakhon Pathom, Thailand. The reciprocal benefits were observed in both field and laboratory studies. *M. bicolor* activity coincided with peak seasonal activity of both hemipterans during June–August. We indicated two sets of support evidence in *M. bicolor* honeydew preference: (i) statistically higher value of adjusted honeydew weight collected by ant workers from *A. gossypii* compared with that from *F. virgata* (*p*–value = .005), and (ii) the higher value of the strength of effect ($\eta^2$ = .62) in the total variance of multi-species association. The physical property on honeydew viscosity was discussed concerning ant preference. We used two–group, ant–tended and ant–excluded, between–subjects multivariate analysis of variance (MANOVA) in order to show hemipteran benefits. Both hemipteran populations increased in the ant–tended treatment, together with lesser amounts of two species of coccinellids, *Menochilus sexmaculatus* (Fabricius) and *Coccinella transversalis* Fabricius, and one species of syrphid fly, *Pseudodorus clavatus* (Fabricius), compared with the ant exclusion treatment (*p*–value <.001). The facultative mutualistic relationships of *M. bicolor* and the two hemipteran species were mentioned.

## Introduction

The shield ant, *Meranoplus bicolor* (Guérin–Méneville) (Hymenoptera: Formicidae), is a common ground nesting species of the subfamily Myrmicinae, and is widely distributed throughout the entire Oriental Region (Schödlh, 1998). The workers not only forage on dead arthropods as scavengers, but also collect honeydew as carbohydrate source from hemipterans, e.g. the cotton aphid, *Aphis gossypii* Glover (Hemiptera: Aphidae) and the striped mealybug, *Ferrisia virgata* (Cockerell) (Hemiptera: Pseudococcidae), in agricultural ecosystem. However, the trophobiotic relationships or mutualism between *M. bicolor* and honeydew–producing hemipterans are unknown.

Mutualism between ants and honeydew–producing hemipterans has been identified as a continuum of relationships ranging from mutualistic to antagonistic (Stadler & Dixon, 2005; Billick et al., 2007), and hemipterans tending by ants are mostly facultative or opportunistic (Delabie, 2001). Generally, ants benefit from associations with hemipterans by obtaining carbohydrate–rich food source in the form of "honeydew" secreted from hemipterans (e.g.: Nixon, 1951; Way, 1963; Hölldobler & Wilson, 1990). Specifically, the benefits to ants have been focused on the foraging behavior of worker ants (Stadler & Dixon, 2005; Grover et al., 2007; Kay et al., 2010). Some have concentrated on fitness benefits in terms of ant colony growth (Grover et al., 2007; Helms & Vinson, 2008; Wilder et al., 2011). In return the benefits, ants may reduce hemipteran contamination of their waste products, removing dead individuals, protecting natural enemies, and transport hemipterans to new feeding sites, resulting in the abundance of hemipteran populations (e.g.: Way, 1963; Nielsen et al., 2010; Stadler & Dixon, 2005). Ants exploit hemipterans not only for their honeydew, but also as a protein source when foraging on them as a common prey (Buckley, 1987; Hölldobler & Wilson, 1990; Delabie, 2001). However, this type of antagonistic relationships will not be treated here; we are looking at a concrete evidence of mutually benefits among both partners.

In this study we verified, in both field and laboratory experiments, the reciprocal benefits of *M. bicolor* and two species of honeydew–producing hemipterans, *A. gossypii* and *F. virgata*. We concentrated for over three–month period in the guava plantation of Horticulture Department, Kasetsart University, Nakhon Pathom, Thailand, observing the mutualism of ant–hemipterans including the abundance of natural enemies, mainly predators. We tested three hypotheses: (i) ants receiving benefits in terms of honeydew from mutualistic associations in guava agroecosystem; (ii) ants protecting hemipterans from natural enemies therefore the densities of natural enemies decrease in the presence of ants; and (iii) in consequence of the two hypotheses mentioned earlier, resulting in the increments of hemipteran densities in ant–hemipteran associations compared with the ant–exclusion arrangement.

**Materials & Methods**

*Study species*

The study was conducted during April–December 2012 in the unsprayed varietal collection plots (varieties: Phant Si Thong, Kim Ju, and Vhan Pi Roon), consisting of 336 guava trees, *Psidium guajava,* of Horticulture Department, Faculty of Agriculture at Kamphaeng Saen, Kasetsart University, Thailand (14.0358 °N, 99.9826 °E). The predominant ground–nesting ant species in the study area was the native *M. bicolor*, with only a few colonies of the invasive ant species the tropical fire–ant, *Solenopsis geminata* (Fabricius) near the perimeter of the plantation. The honeydew–producing hemipterans were *A. gossypii* and *F. virgata*. The natural enemies, mainly predators, were two species of coccinellid beetles, *Menochilus sexmaculatus* (Fabricius) and *Coccinella transversalis* Fabricius (Coleoptera: Coccinellidae), and one species of syrphid fly, *Pseudodorus clavatus* (Fabricius) (Diptera: Syrphidae).

*Ant benefit and honeydew preference*

The direct benefit of *M. bicolor* was obtained by weighing a certain number of foraging ants, and then calculating the difference of weight gain between foragers descending and ascending the guava branches. We measured weight gains of *M. bicolor* after visiting hemipteran colonies as honeydew receiving. We randomly chose foraging ants from the field to weigh for honeydew loading; 50 ant foragers ascending the guava branch before reaching hemipteran colonies, and the other 50 individuals descending the branch with full load of honeydew. Honeydew loads were measured from the weight differences of ants filled with honeydew and ants ascending the guava branch. Individual worker of *M. bicolor* was cap-

tured in an empty hard gelatin capsule (size 0; outer diameter 7.65 mm, height 21.7 mm, and volume 0.68 ml of Torpac Inc., NJ), and shortly after, the capsule containing the arrested ant was weighed on a digital balance. The actual ant weight was obtained from the subtraction of the capsule weight. We weighed, from field collected, two sets (n = 200) foraging ants visiting *A. gossypii*, and one set (n = 100) of ants visiting *F. virgata*. The honeydew loads were confirmed with the laboratory set up by feeding of *M. bicolor* workers with honeydew. A set of field collected workers (n = 100) leaving their nests for foraging were randomly chosen, holding in captivity for 24 hours without food, and subsequently captured inside the gelatin capsule for weighing. After weighing, half of the 24–hour arrested *M. bicolor* was offered with guava leaves occupied by honeydew exudates of *A. gossypii*, and the other half of ants with honeydew from *F. virgata*. The ants were allowed to feed on honeydew until they either refused to feed or left the guava leave. All *M. bicolor* workers were weighed for the second time in order to obtain honeydew loads before releasing back to their former habitats.

*Hemipteran benefits*

We randomly selected 30 guava trees, age 6 years old, approximately 1.65 m in height and 2.5–2.75 m in diameter from the pesticide–free guava plot as our study units. One of two similar branches was randomly chosen from each selected tree to perform ant–exclusion treatment, using sticky barrier around the base of the branch covering 20 cm in length. The target branch was first wrapped around with plastic wrap, and then applied with generic horticultural glue (colorless and odorless). The objective of the gluey barrier is to prevent ants and other crawling insects from reaching hemipteran colonies at the guava shoots, allowing only the entering of air–borne insects, including winged aphids, mealybug crawlers, ladybugs, and syrphid flies. The barriers were examined periodically, and reapplied the glue as required, in order to maintain the effectiveness as ant barriers throughout the experimental period. The other branch was left unmanipulated as the ant–presence treatment. There was the total of 60 experimental units. This ant–exclusion/presence experiment was started in May, beginning with equal numbers of both *A. gossypii* and *F. virgata* between the two treatments on the same guava tree. Insect observations were made during peak seasonal activities of both hemipterans and their natural enemies in June–August 2012.

On each chosen guava tree, we randomly selected one terminal shoot from the total of 3–5 shoots of each experimental unit, in order to make observations. All terminal shoot belonging to each experimental unit had an equal chance to be picked on each data collection day. The number of hemipterans: *A. gossypii*, *F. virgata*; and larvae of predators: *M. sexmaculatus*, *C. transversalis*, and *P. clavatus*, occupying the branch terminal side of 30 cm in length of both presence and absence of *M. bicolor* were counted at various intervals throughout the duration of the experiment from April–December 2012. We counted the insects at interval of 3–5 days, with the total of 7 times per month during June–August, co-incided with the peak activities of both hemipterans and their predators, and every 15 days in other months. However, the observation data or multivariate responses of the five dependent variables were derived from the average of 7 times x 3 months = 21 field observations during peak activities of the insects in June–August 2012. We recorded the number of *M. bicolor* moving up or down (bidirectional) passed a fixed point on the treatment branch with no gluey barrier for 3–min period, to ascertain ant activity throughout the overall experimental period from April–December 2012. All observations in the field were done during 08:30–11:30 hours.

*Statistical analyses*

To answer the question on the difference of honeydew weights or ant's honeydew preference between *M. bicolor* collecting *A. gossypii* honeydew compared with those of *F. virgata*, we used analysis of covariance (ANCOVA) of IBM SPSS Statistics (Verma, 2013; Meyer et al., 2013). Body weight of M. bicolor workers with empty stomach (24–hour unfed workers) or weight before receiving honeydew was treated as covariate, and the criterion variable or dependent variable was ant weight after eating honeydew from each hemipteran species. The analysis of covariance approach was used in order to adjust the initial variations of *M. bicolor* worker size.

The honeydew–producing hemipteran benefits were demonstrated by interference of ants, predominantly *M. bicolor*, with sticky barrier applying around the base of the main branch in order to exclude the ant. The abundance of hemipterans and natural enemies were compared between presence and absence of *M. bicolor*. We anticipated more hemipterans and less natural enemies in the ant–attended guava branches.

Most studies of ant–hemipteran interactions included either ant or hemipteran removals from the study plants, and made comparisons with the unmanipulated partners. The conclusions, in general, relied on statistical analysis by the uses of univariate analysis of variance (ANOVA), which concentrated on one dependent variable, with attempts to make findings from multiple analyses of ANOVA (e.g.: Flatt & Weisser, 2000; Billick et al., 2007; Daane et al., 2007; Mgocheki & Addison, 2009; Styrsky & Eubanks, 2010). Herein we used multivariate analysis of variance (MANOVA) of IBM SPSS Statistics (Meyer et al., 2013; Rencher & Christensen, 2012), in order to draw one solid conclusion of ant–hemipteran mutualism based on the comparison of five dependent variables from two groups, presence and absence of *M. bicolor* on guava branches. These five dependent variables or multivariate responses were number of insects: i.e. nymphs and adults of *A. gossypii*; nymphs and adults of *F. virgata*; larvae of

*M. sexmaculatus*; larvae of *C. transversalis*; and larvae of *P. clavatus*. All insect counts were transformed into log (y + 1) format; where y = number of insect, in order to agree with statistical assumptions.

Several outputs were requested from the MANOVA analysis of IBM SPSS. Box's Test of Equality of Covariance Matrices expected to see if the dependent variable covariance matrices are equal across the levels of the presence–absence of *M. bicolor*. Bartlett's Test of Sphericity was demanded to ascertain sufficient correlation between dependent measures in order to proceed with the analysis. The core MANOVA output was inquired for the multivariate null hypothesis evaluation of no differences between presence and absence of *M. bicolor* on the composite dependent (number of insects) variate. When the multivariate test is statistically significant, we can proceed with some assessments of each dependent variable. We performed the Tests of Between–Subjects Effects to evaluate the statistical significance of each dependent variable separately. Bonferroni–corrected alpha level was applied to avoid alpha inflation in order to evaluate these presence and absence of *M. bicolor* effects. We divided .05 by the number of ANOVAs and obtained .05/5 or a Bonferroni-corrected alpha level of .01.

**Results and Discussion**

*M. bicolor* generally foraged on honeydew of hemipterans as carbohydrate source throughout the year in guava plantation at Kamphaeng Saen. Monthly averages (± SE) of *M. bicolor* activity from April–December 2012 are presented in Fig 1. *M. bicolor* activity coincided with population fluctuations of both hemipterans (*A. gossypii*, and *F. virgata*), with peaks seasonal activities in June–August (Fig 1). There were very high correlation coefficients (r's) between ant activity and either *A. gossypii* or *F. virgata* density at r =.97 (*p*–value < .001; n = 9) and r = .93 (*p*–value < .001; n = 9), respectively.

*M. bicolor* dominated the other ground–nesting ant species, *Solenopsis geminata*, in the studied guava plot, although *S. geminata* has been considered as one of the most invasive ant species worldwide (Wetterer, 2011), but not in this guava ecosystem with history of pesticide applications. There were no *S. geminata* workers observed on the experimental guava trees. The tolerance to pesticides of *M. bicolor* was probably due to the protection of long fine hair covering the entire body (Schödlh, 1998), together with the defensive behavior of *Meranoplus* by curling up the body and feigned dead when disturbed (Hölldobler, 1988).

*Ant benefit and honeydew preference*

In the studied guava plantation, foragers of *M. bicolor* leaving their nests weighed approximately 2.48 mg (SE = .08; n = 150). After visiting hemipteran colonies, *M. bicolor* with honeydew loaded, descending the branch back to their nests weighed on average 8.69 mg (SE = 0.1; n = 150). The honeydew loading is about 6.21 mg (8.69 – 2.48) or roughly estimate around 2.5–fold (6.21 ÷ 2.48) of the mean forager weight departure from their nests. The weighting capacity of *M. bicolor* workers was reconfirmed in a confined study of laboratory feeding of ant workers to different kinds of honeydew from both hemipteran species. After 24 hours in captivity, *M. bicolor* workers weighed 3.61 mg (SE = 0.12; n = 100) on average. We selected larger workers with more tolerance and easier for seizing in order to withstand the 24–hour starvation before obtaining honeydew. These workers were fully fed with honeydew from different hemipteran species, and later weighed approximately 10.70 mg (SE = 0.15; n = 100). The overall expected value of honeydew loading is 7.09 mg (10.70 – 3.61), with an estimate of 2.96–fold (7.09 ÷ 3.61) of the average worker weight after 24 hour in caging. The former 2.5–fold honeydew loading from field foragers was slightly lesser; this was probably due to the offering of honeydew by trophallaxis among workers before returning to their nests (Pfeiffer & Linsenmair, 2007).

One–way between–subjects ANCOVA assessing the difference of honeydew loadings from two hemipteran species of *M. bicolor* workers showed that the covariate effect or weight of 24–hour captured *M. bicolor* before honeydew feeding was statistically significant, $F_{(1, 97)} = 786.297$, $p$–value < .001. Moreover, a statistically significant effect of honeydew source, from either *A. gossypii* or *F. virgata* honeydew, was obtained, $F_{(1, 97)} = 8.387$, $p$–value = .005.

Mean weight of ants before eating honeydew of *F. virgata* group was higher than that of *A. gossypii* group, leading to higher full up honeydew loading from *F. virgata* compared with that from *A. gossypii* (Fig 2). However, the use of AN-COVA approach removed the covariate effect and unveiled the reversal outcome. Mean weight of *M. bicolor* plus honeydew from *A. gossypii* was significantly higher when corrected for weight prior to receiving honeydew (adjusted mean = 10.85; SE = .072; 95% CI = 10.707–10.992) than mean weight of

worker ant with honeydew loaded from *F. virgata* (adjusted mean = 10.55; SE = .072; 95% CI = 10.408–10.693) (Fig 2). This could indicate that *M. bicolor* workers prefer honeydew from *A. gossypii* to that from *F. virgata*.

Ants are expected to concentrate their honeydew collection activities on hemipteran species offering higher reward in terms of both quantitative and qualitative effects. Hemipteran species that produce larger amount of honeydew, or having honeydew with the presence of preferred sugars or amino acids should be more attractive to certain ant species (Cushman, 1991; Völkl et al., 1999; Yao, 2014). Ant preference for particular sugars in hemipteran honeydew can be species specific (Blüthgen & Fiedler, 2004). Several ant species react strongly to honeydew that holds large amounts of melezitose (Völkl et al., 1999), while others prefer sucrose to melezitose (Blüthgen & Fiedler, 2004). On the other hand, *A. gossypii* honeydew consisted of mainly sucrose, fructose, and erlose (Lawo et al., 2009), with no appearance of melezitose.

Honeydew composition of *F. virgata* is unknown; however, some studies of mealybugs' honeydew show composition of fructose, glucose, sucrose, and small amounts of melezitose and raffinose, together with a variety of amino acids (Gray, 1952; Salama & Rizk, 1969).

Another difference in honeydew quality beside the composition of sugars and amino acids is a physical property specifically honeydew viscosity. In our study, honeydew excreted by *F. virgata* was more viscous than that by *A. gossypii,* which their honeydew seemed to be watery liquid. A study in Argentine ant showed that workers fed eightfold longer on gel sucrose composition, and removed fivefold less sucrose than workers feeding on liquid sucrose (Silverman & Roulston, 2001). The later would agree with lesser amounts of honeydew loading of M. bicolor from *F. virgata* than that from *A. gossypii* in this study.

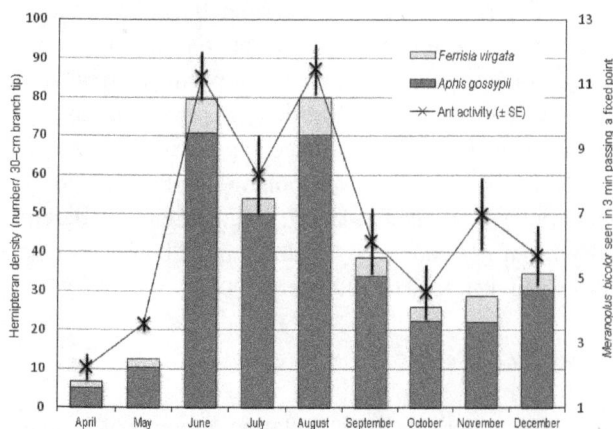

Fig 1. Monthly average of *Meranoplus bicolor* activity (± SE, vertical line), and hemipteran densities (*Ferrisia virgata* and *Aphis gossypii*) from 30 ant–tended guava branches at Kamphaeng Saen, Nakhon Pathom, Thailand in year 2012.

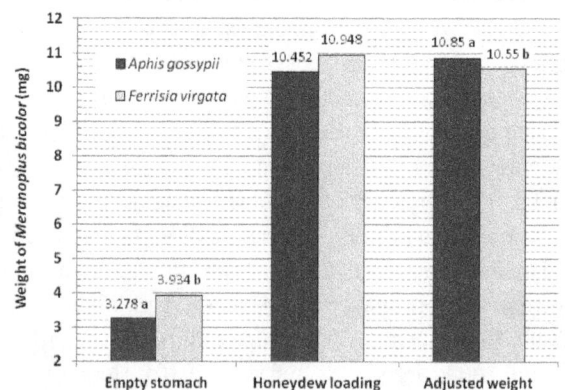

Fig 2. Mean weight of *Meranoplus bicolor* workers (mg) with empty stomach (24–hour without food), mean weight with honeydew loading, and adjusted mean weight from different hemipteran species, *Aphis gossypii* and *Ferrisia virgata*. Value on top of column chart indicates data label; different letters followed mean values represent statistically significant differences (*p*–values ≤ .005).

*Hemipteran benefits*

A two–group between–subjects MANOVA was done on logarithmic transformed data [log (y + 1); y = observation data] of five dependent variables: no. of *A. gossypii*; no. of *F. virgata*; no. of *M. sexmaculatus* larvae; no. of *C. transversalis* larvae; and no. of *P. clavatus* larvae. The independent variable or treatment was the presence-absence of ants, particularly *M. bicolor*, in guava plantation. There were two treatments, i.e. ant–tended and ant–excluded. In general, the ant–excluded treatment with sticky barrier was quite effective against *M. bicolor*, the slow–moving ant species. Even though some ants could accidentally reach the colonies of hemipterans on the exclusion treatment from adjacent branches due to the contact with nearby branches via wind blowing, however, these ants could not return back to their nests or could not be able to recruit additional ant foragers.

The sample consisted of 60 guava branches divided into equal amounts of presence and absence of *M. bicolor*. The output of Box's Test of Equality of Covariance Matrices was statistically significant (Box's $M$ = 69.998; $p$-value < .001), showing that the dependent variable covariance matrices were not equal across the levels of the presence–absence of *M. bicolor*. Therefore, Pillai's trace was used to evaluate all multivariate effects (Meyer et al., 2013). Bartlett's Test of Sphericity was statistically significant (approximate chi-square = 99.838; $p$-value < .001), indicating sufficient correlation between the dependent variables to proceed with the MANOVA. Using Pillai's trace as the criteria, the combined dependent variable was significantly affected by the presence-absence of *M. bicolor*, Pillai's trace = .807, $F_{(5, 54)}$ = 45.244, $p$-value < .001. There were reliable multivariate differences between ant–tended and ant–excluded treatments on the combined dependent variate. The partial eta squared = .807 (partial $\eta^2$), equivalent to the full eta squared ($\eta^2$) in this two–group design (Levine & Hullett, 2002), indicating that we had a very high proportion of the total variance (.807, or about 81%) explained by the activity of *M. bicolor*.

Each dependent measure or each observed insect density was assessed individually in order to determine the strength of the statistically significant multivariate effect. The result of the tests of the univariate effects is shown in Table 1. We had statistically significance univariate effects on all dependent variables (Table 1; $p$-values < .001). Of all insect species under investigation, *A. gossypii* provided the highest effect size ($\eta2$ = .62), while *M. sexmaculatus* had highest effect size in terms of natural enemies ($\eta2$ = .55) (Table 1).

The descriptive information for the univariate analysis is presented in Fig 3; providing each dependent measure's observed means, and total averages obtaining from 30 guava trees in the study. The presence of *M. bicolor*–tended hemipterans had a considerable impact on insect populations not only hemipteran themselves, but also their natural enemies. On ant–tended treatment, we detected higher densities of both

hemipteran species, together with lesser amounts of all natural enemies compared with the ant–excluded treatment (Fig 3). There was more abundant in density of roughly 7.6–fold [{antilog (1.824) – 1} ÷ {antilog (1.002) – 1} or 68.50 ÷ 9.05 = 7.57] of *A. gosypii* than *F. virgata* from 30 guava trees in the study (Fig 3).

In general, we would say that *M. bicolor* preferred to associated with *A. gossypii* more than *F. virgata,* in this meaning the preference of honeydew collecting, as indicated by higher value of the strength of effect or effect size (Levine & Hullett, 2002; Meyers et al., 2013), i.e. $\eta^2$ = .62 and $\eta^2$ = .52, respectively (Table 1). This could be the second evidence in supporting the previous study of honeydew preference of 24–hour captured *M. bicolor*. Among the three natural enemies or predators, *M. sexmaculatus* had more strength of effect ($\eta^2$ = .55), i.e. would be more effective predator, than the other two competitors ($\eta^2$s = .39, and .32) in this *M. bicolor*–hemipteran association (Table 1). Even though the surphid fly, *P. clavatus,* was more abundant than the other two coccinellid predators, but its appearance in the guava plot was restrict to June till August. In addition, there were no *P. clavatus* larvae found preying on the striped mealybug, *F. virgata,* in our study.

The mutualistic relationships or trophobiotic interactions between either *A. gossypii* or *F. virgata* with ants have been classified as facultative and very common phenomenon by Delabie (2001). There are two main reasons from this study in supporting the above mentioned: firstly, both hemipteran species are polyphagous and cosmopolitan species (Blackman & Eastop, 2000; da Silva-Torres et al., 2013), any mutualistic relationship with ants should be opportunistic or facultative rather than obligatory; and secondly, *M. bicolor* is the most common species of the genus *Meranoplus* in the Oriental Region (Schödlh, 1998), and is widely distributed as ground nesting species in disturbed habitats of agricultural

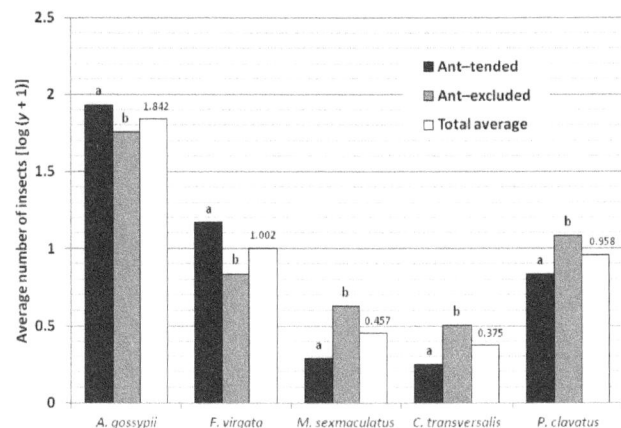

Fig 3. Effect of *Meranoplus bicolor*-exclusion on hemipterans (*Aphis gossypii* and *Ferrisia virgata*) and their natural enemies (*Menochilus sexmaculatus, Coccinella transversalis* and *Pseudodorus clavatus*). Different letters on top of dark columns represent statistically significant differences ($p$–values < .001); numbers on top of clear columns indicate data labels of total average or grand mean from 30 guava trees.

**Table 1.** Tests of univariate effects of *Meranoplus bicolor* activity on five dependent variables[a]: *Aphis gossypii*; *Ferrisia virgata*; *Menochilus sexmaculatus*; *Coccinella transversalis*; and *Pseudodorus clavatus*. Data are visual counts of insects occupying 30–cm length of branch terminal (n = 60)

| Source | Dependent var. | Type III SS | df | MS | F | p–value | Eta squared ($\eta^2$) |
|---|---|---|---|---|---|---|---|
| Ant | *A. gossypii* | .466 | 1 | .466 | 96.188 | .000[b] | .624 |
| | *F. virgata* | 1.681 | 1 | 1.681 | 61.859 | .000 | .516 |
| | *M. sexmaculatus* | 1.744 | 1 | 1.744 | 72.151 | .000 | .554 |
| | *C. transversalis* | .977 | 1 | .977 | 27.876 | .000 | .325 |
| | *P. clavatus* | .925 | 1 | .925 | 37.525 | .000 | .393 |
| Error | *A. gossypii* | .281 | 58 | .005 | | | |
| | *F. virgata* | 1.576 | 58 | .027 | | | |
| | *M. sexmaculatus* | 1.402 | 58 | .024 | | | |
| | *C. transversalis* | 2.034 | 58 | .035 | | | |
| | *P. clavatus* | 1.430 | 58 | .025 | | | |
| Corrected total | *A. gossypii* | .746 | 59 | | | | |
| | *F. virgata* | 3.257 | 59 | | | | |
| | *M. sexmaculatus* | 3.146 | 59 | | | | |
| | *C. transversalis* | 3.011 | 59 | | | | |
| | *P. clavatus* | 2.355 | 59 | | | | |

[a] The multivariate test of combined dependent measure was statistically significant; Pillai's trace = .807, $F_{(5,\,54)}$ = 45.244, p–value < .001.
[b] .000 indicates < .001.

ecosystem, therefore the acquiring for food in the vicinity should be by selection of most abundant resources.

This study showed that ant attending had a considerable effect on hemipterous pest densities in guava plantation. There were more individuals of hemipterans in the ant tending guava branches, together with lesser amounts of natural enemies mainly predators because of ant guarding activities. In return the benefit, ants received carbohydrate sources in terms of honeydew from both hemipterans. In general, the results of this ant exclusion experiment using gluey barrier are agreed with previous studies done in fruit orchards; e.g. cherry (Stutz & Schmidt-Entling, 2011), apple (Stewart-Jones et al., 2008; Miñarro et al., 2010; Nagy et al., 2013); and in vineyards (Mgocheki & Addison, 2009).

In conclusion, mutualistic relationships between *M. bicolor* and honeydew–producing hemipterans were revealed. *M. bicolor* preferably collected honeydew of *A. gossypii* more than that of *F. virgata*, because it was not only easier to find, i.e. more abundant, but also more ingestible, due to the physical property of watery liquid. Other alluring properties of honeydew to ants could be honeydew composition in terms of sugars and amino acids, which needed further investigations. The ant–exclusion promoted an increment in predator densities, and thus leading to a tentatively conservation biological control of hemipterous pests in guava agroecosystem.

## Acknowledgment

We thank Horticulture Department of the Faculty of Agriculture at Kampkaeng Saen, Kasetsart University, particularly Unaroj Boonprakob and Kriengsak Thaipong, for their support in field experiment. Financial support was partially offered to junior author by the Graduate School, Kasetsart University.

## References

Billick, I., Hammer, S., Reithel, J.S. & Abbot, P. (2007). Ant–aphid interactions: Are ants friends, enemies, or both? Annals of the Entomological Society of America, 100: 887–892.

Blackman, R.L. & Eastop, V.F. (2000). Aphids on the World's Crops: An identification and formation guide (2nd edition). Chichester: John Wiley and Sons Ltd.

Blüthgen, N. & Fiedler, K. (2004). Preferences for sugars and amino acids and their conditionality in a diverse nectar-feeding ant community. Journal of Animal Ecology, 73: 155-166.

Buckley, R.C. (1987). Interactions involving plants, Homoptera, and ants. Annual Review of Ecology and Systematics, 18: 111-135.

Cushman, J. (1991). Host-plant mediation of insect mutualisms: variable outcomes in herbivore-ant interactions. Oikos 61:138-144.

da Silva-Torres, C.S.A., de Oliveira, M.D. & Torres, J.B. (2013). Host selection and establishment of striped mealybug, *Ferrisia virgata*, on cotton cultivars. Phytoparasitica 41: 31–40.

Daane, K.M., Sime, K.R., Fallon, J. & Cooper, M.L. (2007). Impacts of Argentine ants on mealybugs and their natural enemies in California's coastal vineyards. Ecological Entomology, 32: 583-596.

Delabie, J.H.C. (2001). Trophobiosis between Formicidae and Hemiptera (Sternorrhyncha and Auchenorrhyncha): a review. Neotropical Entomology, 30: 501-516.

Flatt, T. & Weisser, W.W. (2000). The effects of mutualistic ants on aphid life history traits. Ecology, 81: 3522-3529.

Gray, R.A. (1952). Composition of honeydew excreted by pineapple mealybugs. Science, 115: 129-133.

Grover, C.D., Kay, A.D., Monson, J.A., Marsh, T.C. & Holway, D.A. (2007). Linking nutrition and behavioural dominance: carbohydrate scarcity limits aggression and activity in Argentine ants. Proceedings of Biological Sciences, 274: 2951-2957.

Helms, K.R. & Vinson, S.B. (2008). Plant resources and colony growth in an invasive ant: the importance of honey-dew-producing hemiptera in carbohydrate transfer across trophic levels. Environmental Entomology, 37: 487–493.

Hölldobler, B. (1988). Chemical Communication in *Meranoplus* (Hymenoptera: Formicidae). Psyche, 95: 139-151.

Hölldobler, B. & Wilson, E.O. (1990). The Ants. Cambridge, MA: Harvard Univ. Press.

Kay, A.D., Zumbusch, T.B., Heinen, J.L., Marsh, T.C. & Holway, D.A. (2010). Nutrition and interference competition have interactive effects on the behavior and performance of Argentine ants. Ecology, 91: 57-64.

Lawo, N.C., Wäckers, F.L. & Romeis, J. (2009). Indian Bt cotton varieties do not affect the performance of cotton aphids. PLoS ONE 4(3): e4804. DOI:10.1371/journal.pone.0004804

Levine, T.R. & Hullett, C.R. (2002). Eta squared, partial eta squared, and misreporting of effect size in communication research. Human Communication Research, 28: 612-625.

Meyers, L.S., Gamst, G. & Guarino, A.J. (2013). Applied Multivariate Research (2nd edition). Los Angeles, CA: SAGE Publications, Inc.

Mgocheki, N. & Addison, P. (2009). Interference of ants (Hymenoptera: Formicidae) with biological control of the vine mealybug *Planococcus ficus* (Signoret) (Hemiptera: Pseudococcidae). Biological Control, 49: 180-185. DOI:10.1016/j.biocontrol.2009.02.001

Miñarro, M., Fernández-Mata, G., Medina, P., 2010. Role of ants in structuring the aphid community on apple. Ecological Entomology, 35, 206-215.

Nagy, C., Cross, J.V. & Markó, V. (2013). Sugar feeding of the common black ant, *Lasius niger* (L.), as a possible indirect method for reducing aphid populations on apple by disturbing ant-aphid mutualism. Biological Control, 65: 24–36. DOI:10.1016/j.biocontrol.2013.01.005

Nielsen, C., Agrawal, A.A. & Hajek, A.E. (2010). Ants defend aphids against lethal disease. Biological Letters, 6: 205-208.

Nixon, G.E.J. (1951). The Association of Ants with Aphids and Coccids. London: Commonwealth Inst. Entomol.

Pfeiffer, M. & Linsenmair, K.E. (2007). Trophobiosis in a tropical rainforest on Borneo: Giant ants *Camponotus gigas* (Hymenoptera: Formicidae) herd wax cicadas *Bythopsyrna circulata* (Auchenorrhyncha: Flatidae). Asian Myrmecology, 1: 105–119.

Rencher, A.C. & Christensen, W.F. (2012). Methods of Multivariate Analysis (3rd edition). Hoboken, NJ: John Wiley & Sons, Inc.

Salama, H.S. & Rizk, A.M. (1969). Composition of honey dew in the mealy bug, *Saccharicoccus sacchari*. Journal of Insect Physiology, 15: 1873-1875.

Schödlh, S. (1998). Taxonomic revision of Oriental *Meranoplus* F. Smith, 1853 (Insecta: Hymenoptera: Formicidae: Myrmicinae). Annalen des Naturhistorischen Museums in Wien 100B: 361-394.

Silverman, J. & Roulston, T.H. (2001). Acceptance and intake of gel and liquid sucrose compositions by the Argentine ant (Hymenoptera: Formicidae). Journal of Economic Entomology, 94: 511–515.

Stadler, B. & Dixon, A.F.G. (2005). Ecology and evolution of aphid–ant interactions. Annual Review of Ecology and Systematics, 36: 345-372.

Stewart-Jones, A., Pope, T.W., Fitzgerald, J.D. & Poppy, G.M. (2008). The effect of ant attendance on the success of rosy apple aphid populations, natural enemy abundance and apple damage in orchards. Agricultural and Forestry Entomology, 10: 37–43. DOI:10.1111/j.1461-9563.2007.00353.x

Styrsky, J.D. & Eubanks, M.D. (2010). A facultative mutualism between aphids and an invasive ant increases plant reproduction. Ecological Entomology, 35: 1-10.

Stutz, S. & Schmidt-Entling, M.H. (2011). Effects of the landscape context on aphid–ant–predator interactions on cherry trees. Biological Control, 57: 37-43.

Verma, J.P. (2013). Data Analysis in Management with SPSS Software. New Delhi: Springer India.

Völkl, W., Woodring, J., Fischer, M., Lorenz, M.W. & Hoffman, K.H. (1999). Ant-aphid mutualisms: the impact of honeydew production and honeydew sugar composition on ant preferences. Oecologia, 118: 483-491.

Way, M.J. (1963). Mutualism between ants and honeydew–producing homoptera. Annual Review of Entomology, 8: 307-344.

Wetterer, J.K. (2011). Worldwide spread of the tropical fire ant, *Solenopsis geminata* (Hymenoptera: Formicidae). Myrmecological News, 14: 21-35.

Wilder, S.M., Holway, D.A., Suarez, A.V. & Eubanks, M.D. (2011). Macronutrient content of plant-based food affects growth of a carnivorous arthropod. Ecology 92: 325–332.

Yao, I. (2014). Costs and constraints in aphid–ant mutualism. Ecological Research, 29:383–391. DOI:10.1007/s11284-014-1151-4

# 16

# Antipredator Behavior Produced by Heterosexual and Homosexual Tandem Running in the Termite *Reticulitermes chinensis* (Isoptera: Rhinotermitidae)

G Lɪ, X Zou, C Lᴇɪ, Q Hᴜᴀɴɢ

1 - Huazhong Agricultural University, Wuhan, China

**Keywords**

Predation risk, tandem running, *Reticulitermes chinensis*, dilution effect, sexual selection

**Corresponding author**

Qiuying Huang
Hubei Insect Resources Utilization and Sustainable Pest Management Key Laboratory
Huazhong Agricultural University
Wuhan, China 430070
E-Mail: qyhuang2006@mail.hzau.edu.cn

**Abstract**

Heterosexual and homosexual tandem running can be observed together in the alate pairings in some species of termites. This study examined the effect of heterosexual and homosexual tandem running in the termite *Reticulitermes chinensis* on the predation risk by a predatory ant, *Leptogenys kitteli*. Results showed that both heterosexual and homosexual tandem running reduced the predation risk of participants. When a male-male tandem encountered a female, the back male had a significant advantage over the front male in winning a female. Moreover, the back males were significantly heavier than the front males. These results indicated that the predation risk of dealates could be decreased by tandem running through the dilution effect. Furthermore, these data suggest that male-male tandem running could induce selection pressure in favor of vigorous males and may play an essential role in indirect sexual selection.

## Introduction

Many termite species generally reproduce by annual dispersal of alates that leave the parent colony and found new colonies from bisexual pairs (Bordereau *et al.*, 2002; Peppuy *et al.*, 2004). After the mating flight, individuals may exhibit calling behavior and tandem running (Hanus *et al.*, 2009; Hartke & Baer, 2011). Finally, pairs of dealates look for a suitable nesting site (Hartke & Baer, 2011). The other castes and the nest itself can not protect the imagoes while encountering predators during the period from swarming to colony foundation, which is when they are the most vulnerable (Deligne *et al.*, 1981). However, few studies focus on the role of tandem running in reducing predation risk during mating flights of termites.

Owing to the cryptic nesting habits and short swarming times in subterranean termites, it is very difficult to conduct extensive studies of their pairing behavior and antipreda-tor behavior in the field. Instead, researchers mainly focus on laboratory simulations and mathematical models to investigate social behavior in termites (Hayashi *et al.*, 2003; Huang *et al.*, 2008; Kenne *et al.*, 2000; Lee *et al.*, 2006; Matsuura & Kobayashi, 2007; Jeon & Lee, 2011). Matsuura *et al.* (2002) demonstrated that homosexual tandem running was an antipredator behavior in the Japanese subterranean termite, *Reticulitermes speratus*. However, how widespread this antipredator behavior is within the Isoptera is still unknown. Thus, it is necessary to further study the effect of tandem running on predation risk in other species of termites.

The termite *R. chinensis* is widely distributed in China, including Beijing, Tianjin, Shanxi and the Yangtze River drainage basin (Wei *et al.*, 2007). This termite species builds nests in the soil and wooden structures, and is an important pest of forest trees and urban buildings (Li *et al.*, 2010). However, knowledge about pairing behavior and antipredator behavior in *R. chinensis* is very limited currently. In this study, we examined the effect of heterosexual and homosexual tandem running in *R. chinensis* on the risk of predation by a sympatric predatory ant, *Leptogenys kittel*, to determine whether dealates of this species might also exhibit antipredator behavior.

## Methods and Materials

### Insects

The ant *Leptogenys kitteli* was chosen for this study because it is common in the habitat of *R. chinensis* and has been observed preying on it. On March 30, 2011, we collected a colony of *L. kitteli* from the decayed stump of a pine, *Pinus massoniana*, in Wuhan city, China. The *L. kitteli* colony was maintained in a plastic box ($75\times75\times60$ mm³) which was connected by a plastic tube to a clear plastic case ($75\times75\times60$ mm³) used as a foraging arena where the ants were fed on live *R. chinensis* workers every 3 days. On April 20, 2011, alates of *R. chinensis* were collected together with nest wood in Wuhan city just before the swarming season. They were housed in a plastic nest box ($670\times480\times410$ mm³) covered with nylon mesh and were held at 16 °C in a darkroom for 7 days to control the time of flight. Just before starting the experiments, the plastic nest box was transferred to a room with artificial light at 30 °C so that alates emerged from the nest wood and began to fly (Matsuura & Nishida, 2001; Matsuura *et al.*, 2002). The alates or dealates were anesthetized with $CO_2$ and separated by sex using configuration of the caudal sternites under a stereoscope (Roonwal, 1975). Then, the same-sex imagoes were put together in Petri dishes containing moist filter paper until they shed their wings. Each dealate was used only once, i.e. no dealate was re-used either within or between experiments.

### Effect of Unit Type on Post-encounter Risk

In this experiment, there were five treatments: single male, single female, male-male, female-female and male-female. Each unit type was selected randomly and placed in the foraging arena. After a single dealate was put in the foraging arena, the entrance to the foraging arena was opened and then shut after an ant entered. Once the ant encountered the single dealate, the capturing situations were recorded. When each pair of dealates began tandem running, the entrance was opened and then shut after one ant entered the foraging arena. Because an ant could only capture one dealate at a time, encounters will result in one of the following three situations: (1) the front dealate is captured, (2) the back dealate is captured, or (3) both dealates escape (Matsuura *et al.*, 2002). Each trial was run until the end of the ant's first attempt to capture a termite. Each treatment was replicated 50 times. Individual capture rates per predatory attack between single and tandem dealates, and between front and back dealates in tandems were compared. Statistical significance was analyzed using Fisher's exact probability test (SPSS Inc., 1989–2002).

### Effect of Unit Type on Encounter Risk

As tandem running increased the size of the prey unit, easier for ants to find prey. The actual change in the encounter risk could not be evaluated in a laboratory experiment, because dealates tend to run along the perimeter of a container as previously described in *R. speratus* (Matsuura *et al.*, 2002). Therefore, a mathematical model was needed to estimate the encounter risk of tandem dealates relative to a single dealate. The mathematical model of Matsuura *et al.* (2002) was used to estimate the frequency of encounters ($R$). The parameters $w$ and $l$ show the body width (the biggest diameter of abdomen) and length of dealates for the *R. chinensis*, respectively. The parameter $s$ represents the sensory range of an ant (the range between both antennae).

### Effect of Volatiles and Vision on Predator Behavior

If ants search for prey using visual or volatile cues, it is easier to detect tandem dealates than single dealate. Thus, a choice test was tested in a modified T-shaped box. The test is as follows: The apparatus ($200\times20\times20$ mm) was connected to the ant nest by a plastic tube. We used glass or a 60-mesh stainless-steel screen to separate a compartment ($20\times20\times20$ mm) at both ends of the apparatus. Two dealates were put at one end of the apparatus, but there were no dealates at the other end, as a control (Fig. 1). The entrance was shut after an ant entered. The ant was allowed to search for prey for 60 s. Then, the time spent on the dealates and control sides was recorded. Each treatment was replicated 20 times. New white paper was laid in the junction each time to remove the influence of ant trail pheromones. If visual or volatile cues from delalates can attract ants, the ants should spend more time searching on the dealate side than the control side. In addition, we put dealates wrapped in nylon mesh in the foraging arena and observed the reaction of ants, so that we could detect whether ants considered dealates as prey without direct antennal contact.

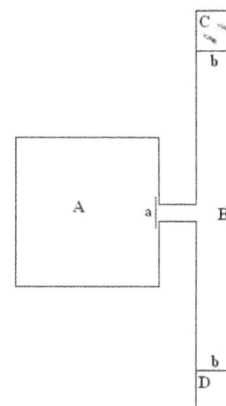

**Fig. 1.** The experimental apparatus used to detect whether ants were attracted to dealates by visual or volatile cues. A: ant nest; B: search area; C: dealate area; D: control area; a: entrance switch; b: glass

plate or a 60-mesh stainless-steel screen.

*Effect of Tandem Position on Pairing Opportunity*

An interesting phenomenon was found in our experiment as described previously in *R. speratus* (Matsuura *et al.*, 2002). When two males met, they would turn around in circles to compete for the back position, while such phenomenon did not occur in females. An important question was raised, whether the back position has dominance in subsequent pairing competition. Therefore, the following test was performed to examine this possibility. Two dealates of the same sex were chosen randomly and were placed in a 90 mm culture dish. A dealate of the opposite sex was chosen randomly and was put into the culture dish after the two dealates of the same sex began tandem running. There will be three results: (1) the front dealate successfully pairs, (2) the back dealate successfully pairs, or (3) triple tandem. Then, the front and back dealates were anesthetized with $CO_2$ and weighed. Each sex was replicated 20 times.

$p = 0.284$, Fisher's exact probability test).

## Results

*Effect of Unit Type on Post-encounter Risk*

The post-encounter risk of the back dealate in tandem running was significantly lower than that of a single individual (in male-male tandems: $p = 0.005$; in female-female tandems: $p = 0.003$; in female-male tandems: $p < 0.001$, Fisher's exact probability test). Furthermore, the post-encounter risk of the front dealate in tandem running was also significantly lower than that of a single individual, except in male-male tandems (in male-male tandems: $p = 0.070$; in female-female tandems: $p = 0.027$; in female-male tandems: $p = 0.016$, Fisher's exact probability test) (Table 1). Although the front individual is always captured at a higher rate than the one in the back, regardless of sex or pair type, there were no significant differences in the post-encounter risk between the back dealate and the front dealate (in male-male tandems: $p = 0.412$; in female-female tandems: $p = 0.532$; in female-male tandems:

Table 1. Comparison of the post-encounter risk in different unit types. Data in parentheses is the post-encounter risk relative to single dealates. ns, not significant; * $p < 0.05$; ** $p < 0.01$; *** $p < 0.001$.

| Unit type | Capture rate | | Escape rate |
|---|---|---|---|
| Single male | 0.64 | | 0.36 |
| Single female | 0.64 | | 0.36 |
| | Front captured | Back captured | Both escape |
| Male-male tandem | 0.44 (0.69)[ns] | 0.34 (0.53)** | 0.22 |
| Female-female tandem | 0.40 (0.63)* | 0.32 (0.50)** | 0.28 |
| Female-male tandem | 0.38 (0.59)* | 0.26 (0.41)*** | 0.36 |

*Effect of Unit Type on Encounter Risk*

The predation risk of tandem dealates relative to single dealates is yielded by multiplying the relative encounter risk and the relative post-encounter risk. The relative encounter risk is $R_2/R_1 = 1.341$ ($l = 4.92$, $w = 1.26$, $s = 5.27$ mm) (Table 2).

The relative predation risk of each position was as follows: male-male tandems, front males: 0.93, back males: 0.71; female-female tandems, front females: 0.84, back females: 0.67; female-male tandems, front females: 0.79, back males: 0.55. Because a value of 1 represents equal predation risk between tandem dealates and single dealates (Matsuura *et al.*, 2002), these results indicate that the total predation risk

was reduced by tandem running relative to single dealates.

Table 2. Body size of dealates and running speed of dealates and ants. [†] Data were the average from 20 dealates and 20 ants. [‡] Sensitive width was the interval between the tips of both antennae. [§] Running speed was determined according to the time required to run 20 cm on a white paper at 25 °C.

| | Termite alate[†] | | Predatory ant[†] |
|---|---|---|---|
| | Male | Female | |
| Body width $w$ (mm) | 1.220 ± 0.009 | 1.290 ± 0.005 | — |
| Body length $l$ (mm) | 4.827 ± 0.030 | 5.014 ± 0.033 | — |
| Sensitive width $s$ (mm)[‡] | — | — | 5.271 ± 0.072 |
| Running speed (mm/s)[§] | 29.048 ± 0.974 | 32.882 ± 0.641 | 50.956 ± 3.902 |

*Effect of Volatiles and Vision on Predator Behavior*

The differences were not significant in the residence times (visual: df = 19, $t$ = -0.471, $p$ = 0.643, volatile: df = 19, $t$ = 1.344, $p$ = 0.195, paired $t$-test) (Fig. 2), suggesting that the ants were not attracted to the dealates separated by glass or a steel screen. The supplementary experiment showed that the ants were not interested in dealates wrapped in nylon mesh when they appeared in the foraging arena. These results dem-

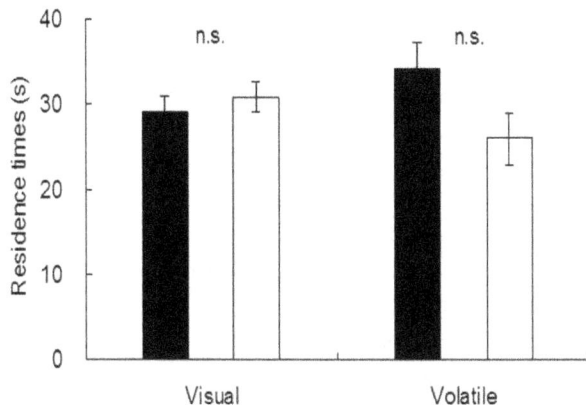

**Fig. 3.** Pairing success between front and back dealates in male-male tandems and female-female tandems. Two-tailed binomial test: (□) front, (■) back, (▨) triple-tandems.

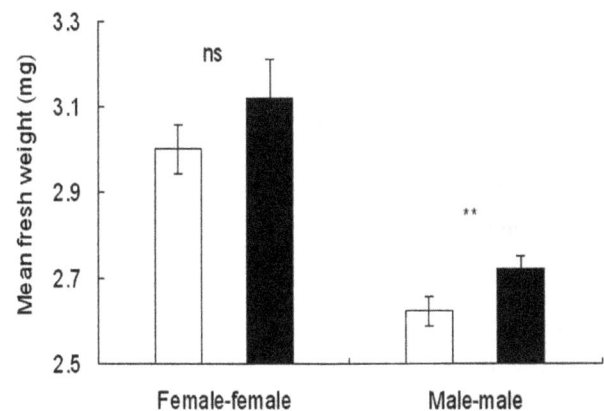

**Fig. 2.** Differences in the residence times of ants between arms of the apparatus having two dealates and lacking dealates. Error bars represent the standard error. ns, not significant; paired t-test: (■) two dealates, (□) no dealates.

*Effect of Tandem Position on Pairing Opportunity*

When a male-male tandem encountered a female, the back male had a significant advantage in winning the female over the front male ($p$ < 0.001, two-tailed binomial test) (Fig. 3). The back male won the female 12 times, while the front male only won 2 times. The "triple-tandems" occurred 6 times. These results clearly showed that back males had the superiority over front males in the pairing competition. This advantage was supported further by the weight results. The front males were significantly lighter than the back males in male-male tandems (df = 19, $t$ = -3.133, $p$ = 0.005, paired $t$-test) (Fig. 4). However, when a female-female tandem encountered a male, there was no significant difference between the front female and the back female in winning the male ($p$ = 0.155, two-tailed binomial test) (Fig. 3).

The front female paired with the male 11 times, and the back female paired 6 times. The "triple-tandems" occurred 3 times. In addition, there was no significant difference in the weights between the front females and the back females in female-female tandems (df = 19, $t$ = -1.371, $p$ = 0.186, paired $t$-test) (Fig. 4).

**Discussion**

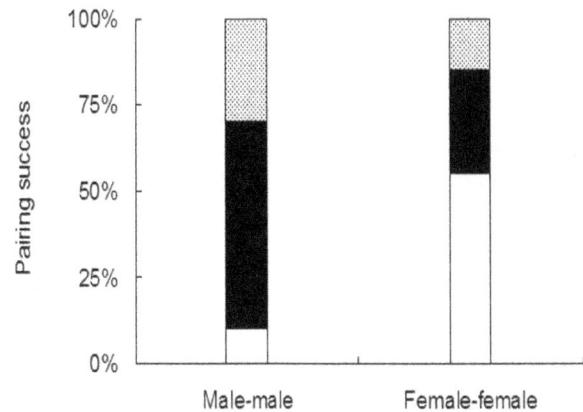

**Fig. 4.** Differences in body weight between front and back dealates in male-male and female-female tandems. Error bars represent the standard error. ns, not significant; **, $p$ < 0.01; paired t-test: (□) front, (■) back.

The dilution effect could reduce an individual's risk of predation in group-forming animals (Hamilton, 1971; Hall *et al.*, 2009; Marcoux, 2011; Rodgers *et al.*, 2011). Our results suggested that colonies of *R. chinensis* experienced the same dilution effect as described previously in *R. speratus* (Matsuura *et al.*, 2002). The alates of *R. chinensis* need to land and shed their wings in order to search for a mate, so it is easy for them to encounter potential predators during the period from swarming to colony foundation (Bordereau & Pasteels, 2011). Our results suggest that the probability of tandem dealates as the victim of a predator was reduced compared to single dealates. This can be explained by the fact that one ant cannot capture two dealates at the same time. In other words, the escape probability of an individual to each predation attack should be increased through tandem running, because one ant can only capture one dealate in each type of tandem (Matsuura *et al.*, 2002). In summary, the phenomenon of homosexual tandem before dealates encounter the opposite sex is an adaptive strategy to minimize predation risk. Moreover, heterosexual tandem running had also reduced the predation risk of two participants. This result was different from *R. speratus* in which the predation risk of the following male was larger than that of a single male in female-male tandems (Matsuura *et al.*, 2002).

The ant *L. kitteli* did not utilize visual or volatile cues to prey on the dealates in this study, indicating the encounter risk could not be increased by either the enlarged visual image or volatiles. Therefore, we could use the mathematic model of Matsuura *et al.* (2002) to estimate the encounter risk of *R. chinensis* by *L. kittel*. This model estimates that tandem running could reduce the total predation risk to an individual. That is to say, a dealate in a tandem run is safer than a single dealate. Reduced predation risk cannot be explained by reduced encounter rates or reduced post-encounter success of the ant against tandems. Average speed is the same between tandems and single dealates, the ant is apparently not utilizing termite pheromones as localization cues, and the escape rate is the same for both prey types (singles vs. tandems). Rather, ants can only handle one item of prey at a time, with the result that post-encounter predation risk is spread over more individuals. This suggests that the longer "trains" of dealates observed in the field are an extension of this adaptive strategy to reduce individual predation risk, although the limits to this tactic have not been explored.

When two males met, they turned round and round to compete for the back position. The likely interpretation of this phenomenon is that the post-encounter risk of the back dealate in male-male tandem running was significantly lower than that of a single individual. In fact, our results suggested that the ants less often captured back males than front males. In addition, the back males were significantly heavier than the front males in male-male tandems, and the back males have the superiority over front males in the pairing competition. These results suggested that male-male tandem running would induce selection pressure in favor of heavy vigorous males. Tandem running may therefor play a role in indirect sexual selection, if such "dominant" males contribute more to reproductive investment by both direct nutrient transfer and labor in the colony foundation stage (Shellman-Reeve 1990). In female-female tandem, we found that there were no differences in the number of pairing success and vulnerability between the positions in *R. chinensis*, consistent with the results in *R. speratus* (Matsuura *et al.*, 2002). In *R. speratus*, the cooperative colony foundation by female pairs was considered as one of the reasons that females do not compete for males as aggressively as males compete for females (Matsuura *et al.*, 2004; Matsuura *et al.*, 2002; Matsuura & Kobayashi, 2007). However, we need to further investigate whether there also is the cooperative colony foundation by female pairs in *R. chinensis*.

Antipredator effects of heterosexual and homosexual tandem running in termites has been tested only in *R. speratus* and *R. chinensis* until now, although homosexual tandem runs have been seen in many termite species. Austin *et al.* (2004) found that *R. speratus* and *R. chinensis* were close relatives within the genus *Reticulitermes*. Thus, whether the antipredator effect of tandem running exists in only these two species, throughout *Reticulitermes* or possibly beyond requires extensive studies in *Reticulitermes* and in closely related genera such as *Coptotermes* and *Heterotermes*. Our work extends the previous discovery of antipredator behavior in termites, however, the extent to which this is actually antipredator behavior needs to be measured under more realistic conditions. Moreover, the evolutionary significance of termite homosexual tandem runs remains further investigations. Also interesting is that the ants in this experiment were relatively smaller compared to the termites than in Matsuura's experiment, judging from the relative values of *l*, *w*, and *s*, suggesting position dependent predation risk may be generalizable to many termite-ant pairings regardless of the relative sizes of the interactants, although it still remains to be tested with ants that are much smaller than the termites they are preying on.

## Acknowledgments

We thank Dr. S.J. Tan for identifying the ant species. We also thank Drs. X.P. Wang and W.W. Zheng for revising the manuscript. We also thank the anonymous reviewers for providing valuable comments on earlier drafts of this manuscript. This work was supported by the National Natural Science Foundation of China (31000978) and the International Foundation for Science (D/4768-1).

## References

Austin, J.W., Szalanski, A.L. & Cabrera, B.J. (2004). Phylogenetic analysis of the subterranean termite family Rhinotermitidae (Isoptera) by using the mitochondrial cytochrome oxidase II gene. Ann. Entomol. Soc. Am. 97: 548-555. doi: 10.1603/0013-8746(2004)097[0548:PAOTST]2.0.CO;2)

Bordereau, C., Cancello, E., Sémon, E., Courrent, A. & Quennedey, B. (2002). Sex pheromone identified after solid phase microextraction from tergal glands of female alates in *Cornitermes bequaerti* (Isoptera, Nasutitermitinae). Insectes Soc., 49: 209-215. doi: 10.1007/s00040-002-8303-1)

Bordereau, C. & Pasteels, J.M. (2011). Pheromones and chemical ecology of dispersal and foraging in termites. In: Bignell D.E., Roisin Y. & N. Lo (eds.), Biology of Termites: A Modern Synthesis. Springer Verlag, Heidelberg, pp 279-320. doi: 10.1007/978-90-481-3977-4_11)

Deligne, J. (1981). The enemy and defense mechanism of termites. Soc. Insects, 2: 1-76.

Hall, S.R., Becker, C.R., Simonis, J.L., Duffy, M.A., Tessier,

A.J. & Cáceres, C.E. (2009). Friendly competition: evidence for a dilution effect among competitors in a planktonic host-parasite system. Ecology, 90: 791-801. doi: 10.1890/08-0838.1

Hamilton, W.D. (1971). Geometry for the selfish herd. J. Theor. Biol., 31: 295-311. doi: 10.1016/0022-5193(71)90189-5

Hanus, R., Luxová, A., Šobotník, J., Kalinovà, B., Jiroš, P. , Křeček, J., Bourguignon, T. & Bordereau, C. (2009). Sexual communication in the termite *Prorhinotermes simplex* (Isoptera, Rhinotermitidae) mediated by a pheromone from female tergal glands. Insectes Soc. 56: 111-118. doi: 10.1007/s00040-009-0005-5

Hartke, T. & Baer, B. (2011). The mating biology of termites: a comparative review. An. Behav., 82: 927-936. doi: 10.1016/j.anbehav.2011.07.022

Hayashi, Y., Kitade, O. & Kojima, J.-I. (2003). Parthenogenetic reproduction in neotenics of the subterranean termite *Reticulitermes speratus* (Isoptera: Rhinotermitidae). Entomol. Sci., 6: 253-257. doi: 10.1046/j.1343-8786.2003.00030.x

Huang, Q.Y., Wang, W.P., Mo, R.Y. & Lei, C.L. (2008). Studies on feeding and trophallaxis in the subterranean termite *Odontotermes formosanus* using rubidium chloride. Entomol. Exp. Appl., 129: 210-215. doi: /10.1111/j.1570-7458.2008.00764.x

Jeon, W. & Lee, S.H. (2011). Simulation study of territory size distributions in subterranean termites. J. Theor. Biol., 279: 1-8. doi: 10.1016/j.jtbi.2011.03.016

Kenne, M., Schatz, B., Durand, J.L. & Dejean, A. (2000). Hunting strategy of a generalist ant species proposed as a biological control agent against termites. Entomol. Exp. Appl., 94: 31-40. doi: 10.1046/j.1570-7458.2000.00601.x

Lee, S.H., Bardunias, P. & Su, N.Y. (2006). Food encounter rates of simulated termite tunnels with variable food size/distribution pattern and tunnel branch length. J. Theor. Biol., 243: 493-500. doi: 10.1016/j.jtbi.2006.07.026

Li ,W.Z., Tong, Y.Y., Xiong, Q. & Huang, Q.Y. (2010). Efficacy of three kinds of baits against the subterranean termite *Reticulitermes chinensis* (Isoptera: Rhinotermitidae) in rural houses in China. Sociobiology, 56: 209-222.

Marcoux, M. 2011. Narwhal communication and grouping behaviour: a case study in social cetacean research and monitoring. McGill University.

Matsuura, K., M. Fujimoto & K. Goka 2004. Sexual and asexual colony foundation and the mechanism of facultative parthenogenesis in the termite *Reticulitermes speratus* (Isoptera, Rhinotermitidae). Insectes Soc., 51: 325-332. doi: 10.1007/s00040-004-0746-0)

Matsuura, K., M. Fujimoto, K. Goka & T. Nishida 2002. Cooperative colony foundation by termite female pairs: altruism for survivorship in incipient colonies. An. Behav., 64: 167-173. doi: 10.1006/anbe.2002.3062

Matsuura, K. & Kobayashi, N. (2007). Size, hatching rate, and hatching period of sexually and asexually produced eggs in the facultatively parthenogenetic termite *Reticulitermes speratus* (Isoptera : Rhinotermitidae). Appl. Entomol. Zool., 42: 241-246. doi: 10.1303/aez.2007.241

Matsuura, K., Kuno, E. & Nishida, T. (2002). Homosexual tandem running as selfish herd in *Reticulitermes speratus*: Novel antipredatory behavior in termite. J. Theor. Biol., 214: 63-70. doi: 10.1006/jtbi.2001.2447

Matsuura, K. & Nishida, T. (2001). Comparison of colony foundation success between sexual pairs and female asexual units in the termite *Reticulitermes speratus* (Isoptera: Rhinotermitidae). Popul. Ecol., 43: 119-124. doi: 10.1007/PL00012022

Peppuy, A., Robert, A. & Bordereau, C. (2004). Species-specific sex pheromones secreted from new sexual glands in two sympatric fungus-growing termites from northern Vietnam, *Macrotermes annandalei* and *M. barneyi*. Insectes Soc., 51: 91-98. doi: 10.1007/s00040-003-0718-9

Rodgers, G.M., Ward, J.R., Askwith, B. & Morrell, L.J. (2011). Balancing the dilution and oddity effects: decisions depend on body size. PloS One, 6: e14819. doi: 10.1371/journal.pone.0014819

Roonwal, M.L. (1975). Sex ratio and sexual dimorphism in termites. J. Sci. Ind. Res. India, 34: 402-416.

Shellman-Reeve, J.S. (1990). Dynamics of biparental care in the dampwood termite, *Zootermopsis nevadensis* (Hagen): response to nitrogen availability. Behav. Ecol. Sociobiol., 26: 389-397. doi: 10.1007/BF00170895

Wei, J.Q., Mo, J.C., Wang, X.J. & Mao, W.G. (2007). Biology and ecology of *Reticulitermes chinensis* (Isoptera: Rhinotermitidae) in China. Sociobiology, 50: 553-555.

* The first two authors contributed equally to this work.

# 17

# Pollinator Sharing in Specialized Bee Pollination Systems: a Test with the Synchronopatric Lip Flowers of *Centrosema* Benth (Fabaceae)

M Ramalho[1], M Silva[1,2], G Carvalho[2]

1 - *Universidade Federal da Bahia (UFBA), Salvador, BA, Brazil.*
2 - *Faculdade de Tecnologia e Ciências, Salvador, BA, Brazil.*

**Keywords**
Nectar robber, size threshold , random interaction, bee pollinated flower

**Corresponding author**
Mauro Ramalho
Laboratório de Ecologia da Polinização
Instituto de Biologia
Universidade Federal da Bahia
Salvador, Bahia, Brazil
CEP: 40.210-730
E-mail: mrramauro@gmail.com

**Abstract**

Bee-pollinated lip flowers of two synchronopatric species of *Centrosema* were used as models to examine the influence of specialized pollination systems on the ecological mechanisms of pollinator sharing. Regression analysis of bee abundances in the habitat on bee abundances on *C. pubescens* flowers was significant (r = 0.69; $P$ = 0.001) and became very consistent and highly significant (r = 0.87; $P$ = 0.00001) using a size threshold of bee pollinators longer than 15mm. These same relationships were not significant ($P >$ 0.01), however, for *C. brasilianum* flowers. The structures of the two pollination systems also sustained the hypothesis of a size threshold for pollinators, although only the *C. pubescens*-bees interactions sustained the hypothesis of random interactions proportional to species abundances in the habitat. The flower visitor pools of the two plant species shared the same four main bee guilds: the pollinators Centridini, *Xylocopa*, and Euglossini and the primary nectar robber *Oxaea*. However, a significant divergence ($P$ < 0.01) was detected between the two systems when the abundances and behaviors (pollinators or cheaters) of the main shared flower visitors were incorporated into the overall quantitative analysis (NMDS). The flowers size differences are not significant ($P$ > 0.05) and could not explain these divergences. Particularly, the concentrations of the largest pollinators *Eulaema* and *Xylocopa* on *C. pubescens* flowers and the behavior shift of Centridini bees that act as legitimate pollinators in *C. pubescens* and as nectar robbers in *C. brasilianum* are better understood as functional foraging responses triggered by the synchronopatry and by nectar volume differences ($P$ = 0.001) between both lip flowers. Paradoxically, the robbery activity of Centridini bees arises as a supply side effect of smaller nectar volume in *C. brasilianum* flowers.

## Introduction

Floral attributes have been protagonists in ample and controversial debates concerning adaptive changes and the mechanisms subjacent to flower-visitor interactions (e.g., Herrera, 1996; Ollerton et al., 2007). The roots of these debates reside, in large part, in the understanding that floral characteristics are adaptively flexible (Endress, 1994), even when they involve presumably specialized pollination modes (Tripp & Manos 2008), as flower-visitor interactions are often scale and ecological context dependent (Herrera, 2005; Ollerton et al., 2007). In comparing extensive floral data (on floral traits and pollinators) from six communities around the world, Ollerton et al.

(2009) concluded that the "pollination syndrome hypothesis", as traditionally stated, did not properly describe the diversity of floral phenotypes and only poorly predicted plant species pollinators. These authors therefore recommended a 'fresh look at how the traits of flowers and pollinators relate to visitation and pollen transfer' in searching for appropriate descriptions of functional floral diversity.

A question central to this debate is if other variables can be as important as morphological restrictions in structuring pollination systems. Recent analyses of partnerships and the structures of flower-visitor webs have both indicated that morphological constraints continue to play a central role in the general theory of the organization of pollination systems.

The relationship between proboscis lengths and nectar chamber depth, for example, sustains the hypothesis that size thresholds are important in regulating those interactions (Stang et al., 2006). On the other hand, the pattern of "asymmetric specialization" that emerges from the webs (e.g., Vázquez, 2005) is also better explained when both size constraints and species abundance are incorporated into theoretical models (Stang et al., 2007). In this case, the following basic question directs the studies: are neutral mechanisms of random interactions proportional to species abundance sufficient to explain the asymmetric specialization (plants with specialized flowers frequently interacting with generalist visitors, and specialized floral visitors frequently interacting with generalized flowers) that has been observed in recent analyses of flower-visitor interaction webs (Vázquez & Aizen, 2004; Vázquez, 2005; Stang et al., 2007)?

Here we focus on the question of pollinator sharing in specialized pollination systems and have chosen as a model two synchronopatric species of *Centrosema* (Fabaceae) with specialized zygomorphic keel flowers. The highly zygomorphic keel flowers of Fabaceae are often pollinated by bees (Faegri & Van der Pijl, 1979; Westerkamp, 1996; Galloni et al., 2007). Bee-pollinated keel flowers have nectar guides and produce nectar as the main floral reward for those insects, but only large bees with appropriate nectar-searching behaviors are able to provoke the mechanical exposition of the fertile pollination organs (Van der Pijl, 1954; Gottsberger et al., 1988; Lopes & Machado 1996; Etcheverry et al., 2003; Ramalho & Rosa, 2010). A distinctive floral trait of bee pollinated keel flowers is protection of their pollen from bee consumption (which would presumably be more effective in lip-flowers, as their keel is dorsally positioned in relation to the landing petal) (Westerkamp, 1996).

*Centrosema pubescens* Benth and *Centrosema brasilianum* (L.) Benth have large and very similar lip-flowers with notable bee-pollination traits (Endress 1994). These two species show complete overlapping of their flowering periods and occur sympatrically along the tropical Atlantic coast of Brazil in herbaceous-shrub vegetation habitats (coastal "restingas" and sand dune areas). Their synchronopatric and specialized flowers should largely share pollinators according to the predictions of both the hypothesis of morphological constraints and that of random interactions proportional to species abundance. In cases such as this, pollinator sharing would favor the coexistence of species with subtle differences in their floral biology due to the potential loss of significant quantities of pollen to hetero-specific flowers (Jacobi et al., 2005). This study therefore assumed the following specific premises: first, the most shared groups of pollinators must be abundant in the habitat if divergences in floral biology of the two *Centrosema* species have only small influences on the functional differentiation of the two pollination systems; second, subtle morpho-functional differences in flower traits should produce greater differences among pollinating bees than among non-pollinating bees (flower visitors).

## Material and Methods

### Study area

The present study was undertaken in "restinga" vegetation along the eastern tropical Atlantic coast of Brazil. This coastal restinga vegetation develops under regional influences of the "The Tropical Atlantic Domain" (Por, 1992). The field study site was located in the Pituaçu Metropolitan Park (PMP) in the city of Salvador, Bahia State, Brazil (13°00' S; 38°30' W). The PMP is dominated by secondary growths of open shrubby-forests with patches of exposed sand dunes, covering an area of 450 ha, at approximately 50m above sea level. The regional climate is Af, according to Köppen classification system, with a mean annual rainfall of approximately 1500mm, and mean monthly temperatures vary between 18°C and 22°C. There is no marked dry season, although the vegetation, which grows on sandy soils, is exposed to sporadic water deficits throughout the year, and there are many areas with only thin vegetation covers.

Plant species. *Centrosema pubescens* Benth. and *Centrosema brasilianum* (L.) Benth. occupy open vegetation areas (dominated by herbs and small shrubs) in tropical restinga and coastal sand dune sites. *C. brasilianum* is a procumbent herb with glabrous inflorescences; its flowers are bluish-purple with white-yellowish pale stripes (nectar guides) on the banner petal. *C. pubescens* is a climbing vine with velutinous inflorescences; its flowers are bluish-pale lilac with pink and creamy white stripes on the banner. The flag (or banner) serves as a large landing platform for the bees on both lip flowers and the corolla base is completely surrounded by a robust tubular calyx. The robust keel (in a dorsal position) surrounds and protects the fertile verticils and it must be displaced upwards by bees moving towards the corolla base. Nectar is the main floral reward, and is well-protected at the base of the corolla by the tight juxtaposition of the petal bases (which form a rigid and very narrow passage to the nectar chamber). The nectar can be reached through the interiors of the flowers by robust bees with long proboscis (personal observation).

The two *Centrosema* species have complete overlapping flowering periods: *C. brasilianum* blooms year-round and *C. pubescens* from March to July (M. Ramalho & M. Silva unpublished). Permanent 20m X 20m plots were delimited in high density patches for field experiments, sampling, and bee visitation observations. The observed numbers of open flowers/day were relatively low (5-10 flowers/m²/day and 10-15 flowers/m²/day of *C.pubescens* and *C. brasilianum*, respectively; M.Silva & M. Ramalho unpublished) even in these dense patches where the flowers of both species were sometimes intermingled.

Floral Biology and morphology measures. The basic analyses of floral biology were performed following Dafni (1992). Stigma receptivity was tested with hydrogen peroxide, and nectar volumes were measured in ten flowers from five different individuals of each *Centrosema* species. Flowers (n = 10) were bagged one day before opening and the nectar was withdrawn using micropipettes soon after 08:00 h. This time was standardized based on a posteriori characterization of the period of highest floral visitor activity. Ten flowers from ten different individuals of each *Centrosema* species were measured with an electronic digital caliper (accuracy 0.1mm): the length and width of the banner, the length of the keel and the depth of the corolla were measured in fresh flowers. In this latter case, it was measured the distance between the point of insertion of the banner's spur and the internal base of the nectar chamber.

Bee Sampling Data. Floral visitors were captured using hand-held insect nets during the periods of overlapping flowering of the two *Centrosema* species (March to July) in two successive years. Based on observational field data on floral biology (principally anthesis and anther dehiscence), the bees were intensively sampled on the flowers of both species from 07:00 h to 12:00 h (most of the flowers are senescent after 12:00 h) on a daily basis for 15 minutes/species every hour for 24 days (30 hours of sampling efforts) along the overlapping flowering period. Visitor behavior on the flowers was observed simultaneously (and photographed for posterior analysis) during 15 minutes/h (totaling 15 hours) to characterize pollinators and non-pollinators. Bees that behaved in discordance with the keel morphology (i.e., making holes in the corolla and accessing nectar from the outside of the flower or biting the anthers to collect pollen) were considered flower robbers (see Inouye, 1980) and were separated into two categories (Inouye, 1980): primary or secondary. Primary flower robbers make perforations in the corolla in order to gain access to the nectar (or pollen), while secondary flower robbers take advantage of the perforations made by primary flower robbers. In general, visitors are called "cheaters" if they consume floral resources without entering into contact with the fertile verticils (Inouye, 1980).

The bee morpho-species were determined by consulting published keys and bee references in the Pollination Ecology Laboratory (ECOPOL) and the Bee Biology and Ecology Laboratory (LABEA) at the Biology Institute of the Universidade Federal da Bahia (UFBA). The identifications of the Centridini and Euglossini bee species were confirmed by Dr. Fernando Zanella (Universidade Federal da Integração Latino-Americana UNILA, Paraná, Brasil) and Dr. Ednaldo L. das Neves (Faculdade Jorge Amado, Bahia) respectively; the scientific names follow Moure et al. (2007). All specimens were deposited in the ECOPOL. The *Centrosema* species were identified by Dr. Luciano Paganucci de Queiroz of the Universidade Estadual de Feira de Santana - Bahia (UEFS), and those specimens were deposited in the Alexandre Leal Costa Herbarium at the Universidade Federalda Bahia (UFBA-ALCH).

Data Analysis. The Mann-Whitney non-parametric test (α = 0.05) was used to compare the quantities of nectar produced by *C. pubescens* and *C. brasilianum*. The t-test was used to compare flower measures between both species (Gotelli & Ellison 2004).

In the global comparative analyses of the two pollination systems pseudo-species were created corresponding to two behavioral categories: robbers and pollinators. Data concerning the composition and abundance of 'pseudo-species' visiting the flowers of six individual plants of *C. pubescens* and *C. brasilianum* were used in non-metric multidimensional scaling (NMDS) based on a Bray-Curtis similarity matrix. Analyses of similarity (ANOSIM) of each plant species were performed to test the null hypothesis of equality in the composition and abundance of floral visitors of these pollination systems. Detailed analyses of the contributions of each bee group to the observed dissimilarity (SIMPER) were also performed. Rare pseudo-species with frequencies equal or less than one were excluded from the analysis.

The hypothesis of random interactions being proportional to species abundances (Vázquez & Aizen, 2004; Vázquez, 2005;) was evaluated comparing bee abundances in the local habitat. The bee abundances in the local habitat (PMP) were estimated by bee sampling during one year period. Using hand insect nets, the bees were captured on the flowers of all detected bee plants (Sakagami et al., 1967) in a transect of 2.5km length, from 07:00 h to 17:00 h, totalizing 240h of sampling effort. Linear regressions between the absolute abundance values of each bee species in the habitat and their respective 'relative abundance' on the flowers were estimated using GraphPad Instat 3 software (at a significance level of 0.01). The relative abundance of each bee species were estimated as follows: the abundance was transformed into a value between 0 and 1 (with 1= the maximum abundance observed on the flowers) by dividing the number of individuals on each flower by the total number of individuals of the most abundant bee species observed on both flowers. Using values of relative abundance between 0 and 1 facilitate interpretations of the graphs in terms of the probability of interactions being proportional to species abundance. A significance level of 0.01 was used.

As the flowers of *Centrosema* are quite large and their nectar is well-protected at the base of the corolla, we assumed an a posteriori size threshold of floral visitors to test for size constraints on random interactions proportional to species abundances (Stang et al., 2007). Body size was used as a surrogate for proboscis length, as all of the pollinators observed belonged to the general category of long-tongued bees (Michener, 2000). It is important to note that all members of the orchid bees or Euglossini have proboscis longer than the other bee groups with similar body size. The orchid bees have very long proboscis (longer than ¾ of their body length) that are more or less proportional to their body size (with several exceptions): being, for example, up to 10mm long in species whose bodies are approximately 10mm long, and up to 40mm long in species larger than 20mm (Roubik & Hanson

2004). Five individuals of each bee species were measured to estimate their body lengths using an electronic digital caliper (accuracy 0.1mm). The distance between the top of the head (at the height of middle ocellus) and the end extremity of abdomen was measured in order to obtain a rough estimate of relative size for insertion of bees in the following body length categories: small bees <9.9mm; medium bees $\geq 10 < 15$mm; large bees $\geq 15 < 20$mm; very large bees $\geq 20$mm.

## Results

The very similar zygomorphic lip-flowers of *C. brasilianum* and *C. pubescens* both offer nectar as the main reward to flower visitors. Nectar is produced from the start of flower anthesis (during the night) until pre-senescence (near 12:00 h) in both species, thus being available from sunrise until noon. In synchrony with nectar availability, the stamen is receptive from early morning until floral senescence in both species. *C. pubescens* produced greater quantities of nectar ($P < 0.01$) than *C. brasilianum* (Table 1), so that the availability of this resource to foragers is potentially greater on the flowers of the former species. Active pollen harvests were made by very few non-pollinators that often visited the flowers after the peak of activity of the pollinators – so that pollen protection by the keel structure appears to be quite effective considering the legitimate visitors.

The sizes of the flowers are very similar in the two species of *Centrosema* and there are no significant differences

Table 1. Floral biology of two synchronopatric species of *Centrosema* (Fabaceae) in an coastal tropical restinga (Brazil). Flower measures (N = 10) are described in methods.

| Character | | *Centrosema pubescens* | *Centrosema brasilianum* |
|---|---|---|---|
| Anthesis (start-end) | | 00:00 h – 05:00 h | 00:00 h – 02:00 h |
| Stigma receptivity (start-end) | | 05:00 h until senescence | 02:00 h until senescence |
| Nectar volume (ml) | | 26 ± 4.20 | 14 ± 5.50 |
| Floral Reward | | Nectar | Nectar |
| Flower color | | bluish-pale lilac, with magenta and creamy-white stripes on the banner (nectar guide) | bluish-purple, with whitish-yellow pale stripes on the banner (nectar guide) |
| Flower Measures (mm) | Banner length | 33.71 ± 2.16 | 36.4 ± 3.54 |
| | Banner width | 37.66 ± 2.79 | 38.85 ± 3.75 |
| | Keel length | 20.41 ± 1.75 | 16.34 ± 0.79 |
| | Corolla depth | 5.34 ± 0.31 | 5.46 ± 0.36 |

in width ($P = 0.45$) and length ($P = 0.67$) of the banner and, mainly, in the depth of the corolla ($P = 0.45$). The difference between the flowers is observed only on the dimensions of the keel ($P < 0.001$), more robust in *C. pubescens*. The keel size probably does not modify the nectar accessibility by large bees ($\geq 15 < 19.9$ mm) or very large bees ($\geq 20$ mm), however, it could affect the behavior of small and medium sized bees on the flowers.

A total of 489 flower visitors were sampled on the flowers of the two *Centrosema* species, of which almost 98% and 27 species were bees (principally robust bees, with body sizes > 10mm; Table 2). Most of the bees collected nectar in the flowers of both *Centrosema* species, with the exception of few pollen robbers. Eighty-six percent of the observed bee species made the legitimate nectar harvests on *C. pubescens* flowers, as compared to only 46% on *C. brasilianum*. During legitimate visits in both flowers, pollinators typically landed on the ventral lip (banner) of the flower and forced their head and thorax towards the corolla base, displacing the keel upwards and triggering pollen deposition on their backs (nototribic pollination). They accessed the nectar by inserting their long tongues into the nectar chamber through a rigid and very narrow passage at the base of the corolla. This corolla structure therefore impedes legitimate nectar access by small bees (e.g., bees < 10mm).

In terms of both species richness and abundances on flowers, the major groups of pollinators were large (body length > 15mm < 20mm) or very large bees (body length $\geq$ 20mm), all with long (e.g., *Xylocopa*) or very long proboscis (*Eulaema*, Euglossini), as well as some medium-sized Euglossini bees with very long proboscis (*Euglossa*, Euglossini). The non-pollinators were medium-sized nectar robber bees (*Oxaea*) and small pollen robber bees (*Ceratina* and *Augochloropsis*). The roles of medium-size Centridini bees varied with *Centrosema* species (Table 2).

In contrast to legitimate visitors, nectar robbers always moved along the outside of the flower to the base of the perianth, where they would pierce the calyx to gain access to the nectar chamber. This type of behavior was often displayed by individuals of *Oxaea* species on *C. brasilianum* flowers (14% of total flower visitors) and on *C. pubescens* flowers (17% of total flower visitors), and by *Centris* and *Epicharis* bees on *C. brasilianum* flowers. *Oxaea* usually acted as a primary nectar robber, making holes in the calyx that could be used by secondary nectar robbers (*Centris*, *Epicharis*, *Ceratina*, *Pseudaugochlora*, and *Augochloropsis*). *Ceratina*, *Pseudaugochlora*, and *Augochloropsis* also acted as pollen robbers, harvesting it with their mouth parts directly from the anthers; *Ceratina* bees were often the primary pollen robbers, punching holes in the anthers inside the keel that the other two groups would later take advantage of.

A high similarity was seen between the two pollination systems, considering their sharing of higher taxa and functional bee groups. By contrast, considering the actual num-

**Table 2.** Abundance distributions (on the flowers and in the PMP habitat), size categories, and behaviors of the floral visitors to *Centrosema* flowers: *Centrosema pubescens* and *Centrosema brasilianum*. The behavioral categories follow Inouye (1980): R = Primary flower robbers; Rs = Secondary flower robbers; P = pollinators; n.r. = not recorded. The size categories of the bees were based on body lengths: (•) Small bees <9.9mm; (••) Medium sized bees ≥10 <14.9mm; (••••) Large bees ≥15 <19.9mm; (•••••••) Very Large bees ≥ 20mm.

| Bee Groups | Flower Visitor Abundance | | | Bee Size Categories |
|---|---|---|---|---|
| | Habitat PMP | *C.pubescens* | *C.brasilianum* | |
| **Euglossini** | | | | |
| *Euglossa cordata* (Linnaeus, 1758) | 87 | 17(P) | 57(P) | •• |
| *Euglossa ignita* Smith, 1874 | 0 | 1(P) | 0 | •• |
| *Euglossa securigera* Dressler, 1982 | 0 | 2(P) | 0 | •• |
| *Eufriesia* cf. *mussitans* Fabricius, 1787 | 26 | 0 | 1(P) | •••• |
| *Eulaema cingulata* Moure 1950 | 13 | 12(P) | 3(P) | •••• |
| *Eulaema flavescens* Friese, 1899 | 6 | 6(P) | 0 | ••••••• |
| *Eulaema nigrita* (Lepeletier, 1841) | 71 | 45(P) | 10(P) | •••• |
| *Eulaema bombiformis niveofasciata* Friese, 1899 | 4 | 17(P) | 0 | ••••••• |
| **Bombini** | | | | |
| *Bombus brevivillus* Franklin,1913 | 26 | 10(P) | 3(P) | •••• |
| **Centridini** | | | | |
| *Centris (Hemisiella) tarsata* (Smith, 1874) | 15 | 1(P) | 1(Rs) | •• |
| *Centris (Centris) flavifrons* (Fabricius, 1775) | 14 | 1(P) | 1(Rs) | •••• |
| *Centris (Centris) leprieuri* (Spinola, 1841) | 48 | 0 | 8(Rs) | •• |
| *Centris (Trachina) fuscata* (Lepeletier, 1841) | 65 | 2(P) | 2(Rs) | •• |
| *Epicharis (Xanthepicharis) bicolor* Smith, 1854 | 2 | 1(P) | 1(Rs) | •••• |
| *Epicharis (Epicharis) flava* (Friese, 1900) | 11 | 15(P) | 1(Rs) | •••• |
| **Xylocopini** | | | | |
| *Xylocopa (Megaxylocopa) frontalis* (Olivier, 1789) | 131 | 85(P) | 1(P) | ••••••• |
| *Xylocopa (Neoxylocopa) nigrocincta* Smith,1854 | 15 | 3(P) | 0 | •••• |
| *Xylocopa (Neoxylocopa) suspecta* Moure & Camargo, 1988 | 14 | 2(P) | 1(P) | •••• |
| *Xylocopa (Neoxylocopa) cearensis* Ducke, 1910 | 1 | 0 | 1(P) | •••• |
| *Xylocopa (Neoxylocopa) grisescens* Lepeletier,1841 | 4 | 1(P) | 0 | ••••••• |
| **Oxaeini** | | | | |
| *Oxaea flavescens* Klung, 1807 | 18 | 2(R) | 13(R) | •• |
| *Oxaea* sp.1 | 48 | 57(R) | 10(R) | •• |
| **Ceratinini** | | | | |
| *Ceratina (Crewella)* sp.1 | | 32(R) | 24(R) | • |
| **Augochlorini** | | | | |
| *Pseudaugochlora pandora* Smith, 1853 | 50 | 0(Rs) | 6(Rs) | •• |
| *Augochloropsis callichroa* Cockerell, 1900 | 37 | 4(Rs) | 6(Rs) | • |
| **Exomalopsini** | | | | |
| *Exomalopsis* sp.1 | 13 | 0 | 2(R) | • |
| **Ericrocidini** | | | | |
| *Acanthopus* sp.1 | 2 | 3(P) | 0 | • |
| **Other Insects** | (n.r.) | 8 (n.r.) | 10 (n.r.) | |

bers of species, only 46% of the pollinators (12/26) and 67% of the non-pollinators (4/6) were shared by both *Centrosema* species. The percentage of shared bee species between the two flowers partially reflected sampling artifacts (the small numbers of sampled individuals of several species; Table 2). The main differences between the systems, however, are related to the distributions of abundances of the large pollinators and the shifting behavior of a shared bee group (Centridini).

Apparent size constraints can explain basic differences in the behaviors of bee visitors to the specialized lip flowers of *Centrosema* (Table 2) – with pollinators usually being larger than 15mm; non-pollinators were consistently less than 15mm long. Among the medium-sized bees (10-15mm), only some *Euglossa* species with very large proboscis (> 10mm) were abundant on flower and would legitimately access the nectar while performing pollination. The primary nectar flower robbers were medium-sized bees *Oxaea* (< 15mm). The behaviors of medium-size Centridini bees on the flowers were less predictable as they acted as pollinators of *C. pubescens* and as secondary nectar robbers of *C. brasilianum*. In this latter case, flower robbing was probably related to the lower nectar volume in *C. brasilianum* flowers. The nectar robber behavior of some larger Centridini bees (two species of *Epicharis*) probably was stimulated by easy access to nectar in *C. brasilianum* flowers that had been previously perforated by primary flower robbers (*Oxaea*).

The relationship between the relative abundance of bees on *Centrosema* flowers and their abundance in the habitat (Fig 1; see also Table 2) was very consistent and extremely significant (r = 0.87; $P < 0.00001$) when the analyses were restricted to large bees (≥ 15mm) visiting *C. pubescens* flowers (Fig 1a). Although significant, a loss of consistency was seen in the relationship when all floral visitors to *C. pubescens* flowers were included, independent of their sizes (r = 0.69; $P = 0.001$). The abundance of bees in the habitat, however, was not predictive of their abundance on *C. brasilianum* flowers under any circumstances – whether considering a minimum size constraint of 15mm (Fig 1b; $P = 0.49$; $P = 0.04$) or including all floral visitors in the analyses (r = 0.42; $P = 0.034$). The smaller nectar rewards and the activities of secondary flower robbers on *C. brasilianum* flowers seem to affect mainly large pollinators (e.g., *Eulaema nigrita* and *Xylocopa frontalis*) that probably shift to *C. pubescens* flowers with large nectar volume.

Considering the abundances and behaviors of floral visitors (pollinator or non-pollinator), the NMDS analysis revealed two distinct structural and functional organizations of the pollination systems of *C. brasilianum* and *C. pubescens* (Fig 2). Analysis of similarity (ANOSIM) confirmed significant differences between the pooled visiting bees (average dissimilarity = 83%; $P = 0.011$). Quantitative analyses of the contributions of particular bee groups to the observed dissimilarity indicated that the abundance distributions of the very large *Eulaema* and *Xylocopa* bees and the medium-sized *Euglossa* bees (with very large proboscides) were particularly

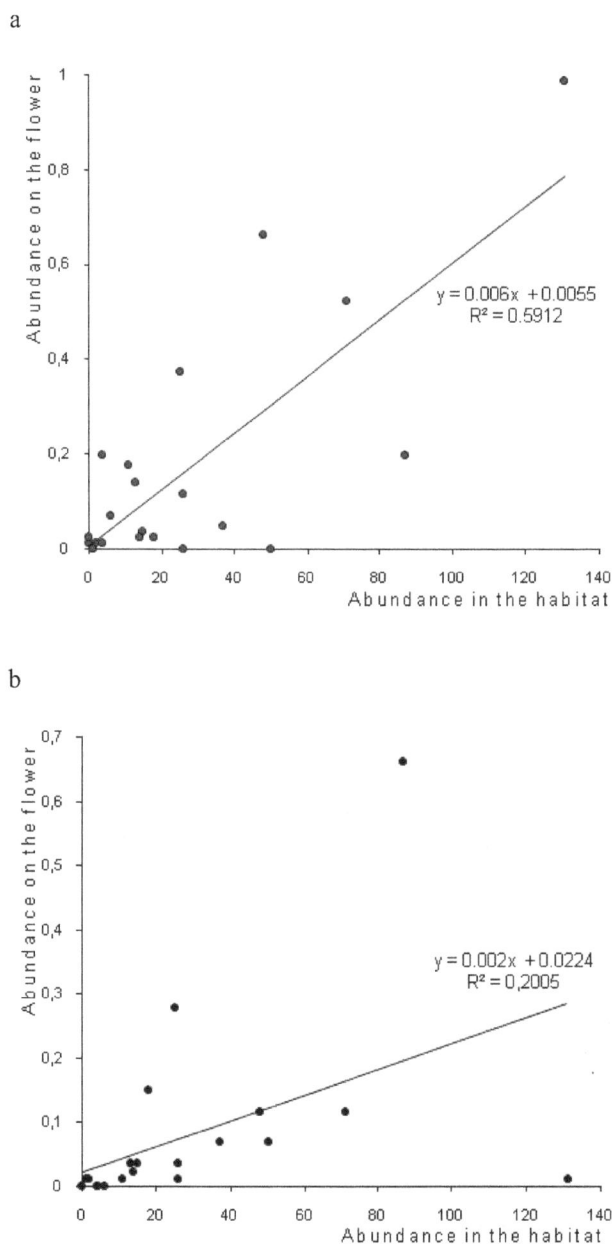

**Fig. 1.** Relationships between abundance of bees (pollinators + non-pollinators) in the habitat (PMP) and on the flowers of *Centrosema*: a) *Centrosema pubescens* (r = 0.87; $P < 0.00001$); b) *Centrosema brasilianum* (r = 0.49; $P = 0.04$). It is presented only the regression curves for the large bees + very large bees + Euglossini (bees with very long proboscis).

important to the ecological differentiation between the two pollination systems. The pollinating bees *Xylocopa frontalis*, *Eulaema nigrita* and *Eulaema meriana* together, for example, were responsible for 27% of the observed dissimilarity between the two systems, while medium-sized pollinating bees *Euglossa* contributed 25%; medium-sized *Centris* bees contributed 4.2%, mainly due to the fact that they behave as pollinators or robbers depending on the *Centrosema* species. Altogether, the primary nectar or pollen robber bees *Oxaea* and *Ceratina* respectively, contributed for only 10% of the dissimilarity.

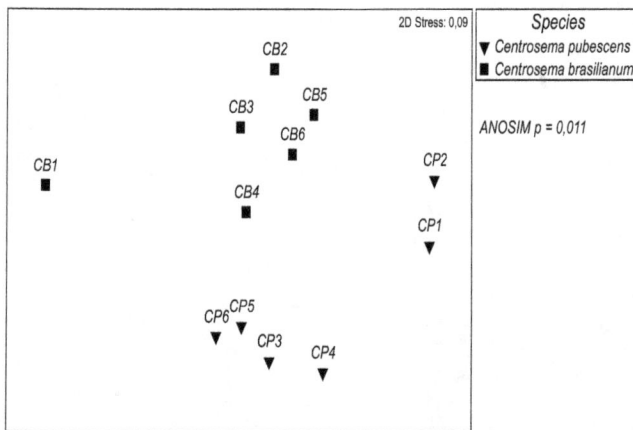

**Fig. 2**. Ordination diagram of the two pollination systems by nonmetric multidimensional scaling (NMDS): CP = *Centrosema pubescens* and CB = *Centrosema brasilianum*.

## Discussion

Similarity in floral biology and mainly synchronopatry expose the two *Centrosema* species to share the main visiting bees and functional bee groups (Camargo et al., 1984; Roubik, 1989; Michener, 2000): the huge carpenter bees of the genus *Xylocopa* and the very large-tongued Euglossini bees as the main pollinators; *Oxaea* primary nectar robbers; *Ceratina* primary pollen robbers; and the floral oil bees Centridini whose role is dependent on *Centrosema* species. In particular, the primary nectar and pollen robbers were bees evenly shared by both flowers, as expected, suggesting they are weakly responding to subtle floral differences between the *Centrosema* species (e.g., flower color and keel size). On the other hand, the significant differences in the abundances of shared pivotal pollinators, mainly the large bees *Xylocopa* and *Eulaema*, and secondary nectar robbers, cannot be attributed to random effects only.

The hypothesis of a size threshold modulating the interactions between nectar-flowers and consumers (Stang et al., 2006) is mainly supported by the predictable interactions of *C. pubescens* with large and very large bees that have large or very large proboscis (*Xylocopa* and *Eulaema* species). These relationships also corroborate the influence of random interactions proportional to bee species abundances in the habitat (Vázquez & Aizen, 2004, Vázquez, 2005; Stang et al., 2007).

The observed partnerships between pollinators and the very specialized *Centrosema* lip-flowers should therefore be attributed to the action of two basic mechanisms: neutral random interactions driven by species' abundances in the habitat, and interactions modulated by morphological constraints (i.e., bee size threshold). On the other hand, neither morphological constraints (e.g., proboscis x corolla length) nor neutral interactions (Vázquez, 2005; Stang et al., 2006) fully encompass the mechanisms responsible for the high concentrations of very large bees on *C. pubescens* flowers and the observed shifting of the behavior of Centridini bees from legitimate *C.*

*pubescens* pollinators to *C. brasilianum* nectar robbers. Significant difference in keel size between both flowers doesn't explain adequately the exploitation modes and sharing of these two nectar sources by the bees. If the keel provides resistance to visitors, it was expected some difference in the frequency distribution of mid-sized bees between the two *Centrosema* species and a higher frequency of nectar thefts in *C. pubescens* flowers with the largest keel: only the first prediction is partially supported by the high abundance of *Euglossa cordata* on *C. brasilianum* flowers. The abundance distributions of primary and secondary nectar robbers on both *Centrosema* species are therefore not affected by the keel size.

Bees are notable for their ability to choose among nectar sources comparing foraging cost/benefit ratios (Waddington, 1980; Pyke, 1984). For instance, among coexisting *Bombus* species, the largest bees with the largest proboscis tend to choose flowers with the longest corollas, from which they can collect more nectar more efficiently than smaller bees (e.g., Harder, 1985; Heinrich, 2004). If the bees were mainly responding to size restrictions, therefore, the largest bees should visit the flowers of both *Centrosema* species with similar frequencies, in light of their similar conditions of nectar access (i.e., similar flower sizes and corolla lengths) and similar flower densities in the habitat (see methods). The same would not be true, however, in terms of floral nectar volume, and some large and very large-sized bees (e.g., *Xylocopa frontalis* and *Eulaema* species) likely choose *C. pubescens* flowers simply because they can obtain more nectar per visit. The difference in flower color between the *Centrosema* species must be used for recognition and choice of *C. pubescens* flowers, with more nectar, by these large bees.

Despite the low contribution of secondary nectar robbers to the structural divergences between *Centrosema* pollination systems (i.e., 4% dissimilarity), the Centridini bees deserve attention because of the distinctive roles of "cheaters" on structuring flower-visitor webs (Genini et al., 2010). Centridini is one of the most abundant bee group in tropical coastal restinga and sand dunes (Ramalho & Silva, 2002; Viana & Kleinert, 2006; Oliveira-Rebouças & Gimenes, 2011; Rosa & Ramalho, 2011), and therefore its low abundance on *Centrosema* flowers would not sustain the hypothesis of random interactions proportional to abundances in the habitat (Vázquez, 2005; Vázquez & Aizen, 2004). Behaving as secondary nectar robbers in *Centrosema* flowers is probably a response of Centridini bees to a contingent relationship: encounters facilitated by the abundance of these bees in the habitat (Ramalho & Silva, 2002; Rosa & Ramalho, 2011) and by the long flowering periods of *Centrosema* species.

In light of the intense activity of primary flower robbers (*Oxaea* spp) on both *Centrosema* species (14% to 17% of total flower visitors on flowers), and particularly on *C. pubescens*, if secondary flower robbers were responding to access opportunities to nectar by preexisting perforations in the corolla, it would be expected that they would rob the flowers of both species or that the secondary robbery activity should be slightly higher in *C. pubescens* flowers. As such, it must not be by chance that changes from legitimate (pollinator) to illegitimate (secondary nectar robbery) visiting behavior mainly involve medium-sized Centridini bees (=15mm) and *C. brasilianum* flowers with smaller nectar volumes.

The high observed frequency of medium-sized *Euglossa* bees, with very long tongues (proboscides >10mm) on *C. brasilianum* flowers provides indirect evidence that the body size/proboscis length ratio play a role in this relationship – as to maximize returns from the exploitation of the smaller nectar volumes in *C. brasilianum* flowers, medium-sized bees must have very long tongues (*Euglossa* species) or be primary flower robbers (*Oxaea*); being a secondary flower robber (Centridini) would be the "best thing to do with a worst thing", as they would be highly exposed to visit depleted flowers.

In some circumstances, nectar robbing (even from specialized flowers) can be a very rewarding strategy for bees, depending on their ability to make adjustments in their foraging behaviors (Zhang et al., 2011), and that is probably why cheaters are ubiquitous in mutualistic flower-visitors networks (Genini et al., 2010). Nectar robbing behavior could be stimulated by size restrictions triggered by subtle differences in corolla lengths among very similar zoophilous flowers and size differences between individuals visiting the same flower (Urcelay et al., 2006; Zhang et al., 2011) or by difference in nectar volume in *Centrosema* species. Paradoxically, in the relationship between the two *Centrosema* species and Centridini bees, secondary nectar robber seems to be more advantageous in the flowers with smaller nectar volume.

In synthesis, pollinators sharing by the two *Centrosema* species is potentialized by synchronopatry and modulated by pollinator choices between flower sources with different nectar volume and, apparently, by direct or indirect interactions among floral visitors, including flower robbers. The concentrations of the largest bee pollinators on *C. pubescens* flowers and the secondary nectar robber activity of medium-sized Centridini bees on *C. brasilianum* flowers are foraging responses better understood by nectar volume differences than by differences in floral morphology per se.

From the point of view of plant reproduction, nectar robbery has detrimental effects on maternal functions (e.g. seed set), depending on the species' reproductive system (e.g., Irwin et al., 2001). As such, by reducing the nectar volume, probably *C. brasilianum* is selecting for medium-sized bees with the largest proboscis (e.g. *Eulgossa*) as its major pollinators, and therefore it should present some reproductive adjustment to compensate for the loss of large-bodied pollinators and the parallel-paradoxical increase in robbers activities. Both *Centrosema* species invest principally in cross-pollination, although the ratios of seeds/ovules and seeds/fruits are both significantly smaller ($P = 0.0001$) in *C. brasilianum* than in *C. pubescens* (M. Ramalho & M. Silva unpublished), suggesting the first species is more adjusted to being visited and cross-pollinated mainly by a smaller number of large or very large bee species.

## References

Camargo, J.M.F., Gottsberger, G. & Silberbauer-Gottsberger, I. 1984.On the phenology and flower visiting behavior of *Oxaea flavescens* (Klug) (Oxaeinae, Andrenidade, Hymenoptera) in São Paulo, Brazil. Beitrage zur Biologie der Pflanzen, 59: 159-179.

Dafni, A. (1992). Pollination Ecology: The practical Approach series. Ed. IRL Press. Oxford: Oxford University Press. 250p.

Endress, P.K. (1994). Diversity and evolutionary biology of flowers. Cambridge: Cambridge Univ. Press. 511p.

Etcheverry, A.V, Protomastro, J.J. & Westerkamp, C. (2003). Delayed autonomous self-pollination in the colonizer *Crotalaria micans* (Fabaceae: Papilionoideae): structural and functional aspects. Plant Systematics and Evolution, 239: 15-28. doi: 10.1007/s00606-002-0244-7

Faegri, K. & Van Der Pijl, L. (1979). The principles of pollination ecology. Oxford: Pergamon Press.

Galloni, M., Podda, L., Vivarelli & Cristofolini, G. (2007). Pollen presentation, pollen-ovule rations, and other reprodutive traits in Mediterranean Legumes (Fam. Fabaceae – Subfam.Faboideae). Plant Systematics and Evolution, 266: 147-164. doi: 10.1007/s00606-007-0526-1.

Genini, J., Morellato, P.C., Guimarães, P.R. & Olesen, J.M. (2010). Cheaters in mutualism networks. Biology Letters, 6: 494-497. doi: 10.1098/rsbl.2009.1021

Gotelli, N.J. & Ellison, A.M. (2004). A primer of ecological statistics. Massachusetts: Sinauer Associates, Inc.510p.

Gottsberger, G., Camargo, J.M.F. & Silberbauer-Gottsberger, I. (1988). A bee-pollinated tropical community: the beach dune vegetation of Ilha de São Luis, Maranhão, Brazil. Botanische Jahrbücher für Systematik, 109: 469-500.

Herrera, C.M. (1996). Floral traits and plant adaptation to insect pollinators: a devil advocate approach. In. Lloyd, D.G. & Barrett, S.C.H.(eds) Floral Biology (pp. 65-87). New York: Chapman and Hall.

Herrera, C.M. (2005). Plant generalization on pollinators: species property or local phenomenon? American Journal of Botany, 92: 13-20.

Harder L.D. (1985). Morphology as a predictor of flower choice by bumble bees. Ecology, 66: 198-209.

Heinrich, B. (2004). Bumblebee economics. Cambridge: Harvard University Press.

Inouye, D.W. (1980). The terminology of floral larceny. Ecology, 61: 1251-1253.

Irwin, R.E., Brody, A.K. & Waser, N.M. (2001). The impact of floral larceny on individuals, populations and communities.

Oecologia: 129, 161-168. doi: 10.1007/s004420100739

Jacobi, C,M., Ramalho, M. & Silva, M. (2005). Pollination biology of the exotic ratlleweed *Crotalaria retusa* L. (Fabaceae) in NE Brazil. Biotropica, 37: 357-363. doi: 10.1111/j.1744-7429.2005.00047.x

Lopes, A.V. De F. & Machado, I.C.S. (1996). Biologia floral de *Swartzia pickelii* Killipex Ducke (Leguminosae-Papilionoideae) e sua polinização por *Eulaema* spp. (Apidae-Euglossini). Revista Brasileira de Botânica, 19: 17-24.

Michener, C.D. (2000). The bees of the World. London: The Johns Hopkins Univ. Press. 912p.

Moure, J.S., Urban, D. & Melo, G.A.R. 2007. Catalogue of Bees (Hymenoptera, Apoidea) in the Neotropical Region. Sociedade Brasileira de Entomologia. 1058p.

Ollerton J., Killick, A., Lamborn, E., Watts, S. & Whiston, M. (2007). Multiple meanings and modes: on the many ways to a generalist flower. Taxon, 56: 717–728.

Ollerton, J., Alarco, R., Waser, N.M., Price, M.V., Watts, S., Cranmer, L., Hingston, A., Peter, C. I. & Rotenberry, J. (2009). A global test of the pollination syndrome hypothesis. Annals of Botany, 103: 1471-1480. doi:10.1093/aob/mcp031.

Por, F.D. (1992). Sooretama. The Atlantic Rain Forest of Brazil. The Nehterlands: SPB Academic Publish.

Pyke, G.H. (1984). Optimal foraging theory: a critical review. Annual Review of Entomology, 15: 523-75

Ramalho, M. & Silva, M. (2002). Flora oleífera e sua guilda de abelhas em uma comunidade de restinga tropical. Sitientibus: Série Ciências Biológicas, 2: 34-43.

Ramalho, M. & Rosa, J.F. (2010). Ecologia da Interação entre as pequenas flores de quilha de *Stylosanthes viscosa* (Faboideae) e as grandes abelhas *Xylocopa cearensis* (Apoidea, Hymenoptera), em duna tropical. Biota Neotropica, 10: 93-100.

Oliveira-Rebouças, P & Gimenes, M. (2011). Polinizadores potenciais de *Comolia ovalifolia* DC Triana (Melastomataceae) e *Chamaecrista ramosa* (Vog.) H.S. Irwin e Barneby var. ramosa (Leguminosae-Caesalpinioideae), na restinga, Bahia, Brasil. Brazilian Journal of Biology, 71: 343-351. doi: 10.1590/S1519-69842011000300002.

Rosa, J.F. & Ramalho, M. (2011). Spatial Dynamics of Diversity in Centridini Bees: The Abundance of Oil Producing Flowers as a Measure of Habitat Quality. Apidologie, 42: 150-158. doi: 10.1007/s13592-011-0075-z

Roubik, D.W. (1989). Ecology and natural history of tropical bees. Cambridge: Cambridge Univ. Press. 503p.

Roubik, D.W. & Hanson, P.E. (2004). Orchid bees of tropical America. Costa Rica: Editorial INBio. 370p.

Sakagami, S.F, Laroca, S. & Moure, J.S. (1967). Wild bee Biocoenotics in São José dos Pinhais (PR) South Brazil. Preliminary Report. Journal of the Faculty of Sciences. Hokkaido University. Ser.VI, Zool., 16: 253-291.

Stang, M., Klinkhamer, P.G.L. & van der Meijden, E. (2006). Size constraints and flower abundance determine the number of interactions in a plant–flower visitor web. Oikos, 112: 111-121. DOI: 10.1111/j.0030-1299.2006.14199.x.

Stang, M., Klinkhamer, P.G.L. & van der Meijden, E. (2007). Asymmetric specialization and extinction risk in plant–flower visitor webs: a matter of morphology or abundance? Oecologia, 151: 442–453. doi: 10.1007/s00442-006-0585-y

Tripp, E.A. & Manos, P.S. (2008). Is floral specialization an evolutionary dead-end? Pollination system transitions in *Ruella* (Acanthaceae). Evolution, 62: 1712-1737. doi: 10.1111/j.1558-5646.2008.00398.x

Urcelay, C., Morales, C. & Chalcoff, V.R. (2006). Relationship between corolla length and floral larceny in the South American hummingbird-pollinated *Campsidium valdivianum* (Bignoniaceae). Annales Botanici Fennici, 43: 205-201

Van Der Pijl, L. (1954). *Xylocopa* and flowers in the tropics. I. The bees as pollinators. Lists of flowers visited. Botany serie C. 57, 413-423.

Vázquez, D.P. (2005). Degree distribution in plant–animal mutualistic networks: forbidden links or random interactions? Oikos, 108: 421-426. doi: 10.1111/j.0030-1299.2005.13619.x

Vázquez, D.P. & Aizen, M.A. (2004). Asymmetric specialization: a pervasive feature of plant–pollinator interactions. Ecology, 85: 1251-1257. doi: 10.1890/03-3112.

Viana, B.F. & Kleinert, A.M.P. (2006). Structure of bee-flower system in the coastal sand dune of Abaeté, northeastern Brazil. Revista Brasileira de Entomologia, 50: 53-63. doi: 10.1590/S0085-56262006000100008.

Waddington, K.D. (1980). Flight patterns of foraging bees relative to density of artificial flowers and distribution of nectar. Oecologia, 44: 199-204.

Westerkamp, C. (1996). Pollen in bee-flower relations. Botanica Acta, 109: 325-332.

Zhang, Y. W., Zhao, J. M., Yang, C. F. & Gituru, W. R. (2011). Behavioural differences between male and female carpenter bees in nectar robbing and its effect on reproductive success in *Glechoma longituba* (Lamiaceae). Plant Biology, 13 (Suppl. 1): 25-32. doi: 10.1111/j.1438-8677.2009.00279.x.

# 18

# Seasonality of Epigaeic Ant Communities in a Brazilian Atlantic Rainforest

PSM Montine[1], NF Viana[1], FS Almeida[2], W Dáttilo[3], AS Santanna[1], L Martins[†], AB Vargas[1]

1 - Centro Universitário de Volta Redonda (UniFOA), Volta Redonda, Rio de Janeiro, Brazil.
2 - Universidade Federal Rural do Rio de Janeiro, Três Rios, Rio de Janeiro, Brazil.
3 - Universidad Veracruzana, Xalapa, Veracruz, Mexico.
† In memorian.

**Keywords**
Myrmecology, richness, diversity community structure

**Corresponding author**
André Barbosa Vargas
Centro Universitário de Volta Redonda
Av. Paulo Erlei Alves Abrantes, 1325
Volta Redonda, Rio de Janeiro, Brazil
27240-560
E-Mail: andrebvargas@yahoo.com.br

**Abstract**
In this study we assessed the leaf-litter ant community in the Cicuta Forest, a semideciduous forest located in the state of Rio de Janeiro, southeastern Brazil. Specifically, we tested the following hypotheses: (1) ant richness and diversity are higher in the rainy season, due to higher resource availability and better temperature and humidity conditions; and that (2) the structure of the ant community is influenced by climate seasonality. We collected 83 ant species of 35 genera and eight subfamilies. In total, 64 species were collected in the dry season and 73 species in the rainy season. Based on rarefaction curves with confidence intervals, we observed that species richness in the dry and rainy seasons did not differ significantly from each other. Shannon diversity did not differ significantly (t = -1.20; $P$ = 0.23) between the dry (3.43) and rainy seasons (3.52). We did not observe a significant effect of climate seasonality neither on ant species composition, richness, and diversity, nor on community structure. These results may be explained by the degree of isolation and degradation of this forest remnant. In short, our study contributes to knowledge on how seasonal variations affects ant communities.

## Introduction

Ants are extremely abundant and diverse in tropical rainforests (Hölldobler & Wilson, 1990; Ward, 2000; Lach et al., 2010). Previous studies showed that ant diversity is influenced by several biotic factors, such as plant richness and density (Gomes et al., 2010a; Tews et al., 2004), as well as by abiotic factors, such as leaf litter depth (Nakamura et al., 2003; Vargas et al., 2007), temperature (Almeida et al., 2007), rainfall (Kaspari, 2000; Speight et al., 2008), and physical and chemical properties of the soil (Gomes et al., 2010b). Moreover, due to the high spatial abundance of ants in tropical environments, they play several ecological roles (Hölldobler & Wilson, 1990).

In the soil-litter interface, ants also are related with essential ecological processes, such as decomposition of organic matter, soil aeration, seed dispersal, and population control of other arthropods (Hölldobler & Wilson, 1990; Folgarait, 1998; Passos & Oliveira, 2004; Dattilo et al., 2009).

The structure of ant assemblages varies in time and space, following environmental heterogeneity (Lassau & Hochuli, 2004; Jankowski et al., 2009). It is well known that variations in rainfall intensity affect temperature and humidity, and so influence ecosystem productivity (Kaspari, 2001). Therefore, rainfall seasonality is expected to lead to variations in the activity, abundance, and species richness and composition of ants (Coelho & Ribeiro, 2006; Almeida et al., 2007; Castro et al., 2012). Indeed, rainfall may affect ant assemblages in two ways, by altering microclimate at a local scale or ecosystem productivity at a regional scale (Castilho et al., 2011; Castro et al., 2012).

On the other hand, seasonal variations may not affect ant richness and composition due to environmental simplification and changes in habitat structure, either at a local or at a regional scale, which makes these variations less pronounced (Vasconcelos & Laurance, 2005; Silva et al., 2011).

For example, Castro et al., (2012) did not observe variations in the structure of ant assemblages in an impacted area, but only between different areas. In this study, we assessed temporal variations in ant assemblages that live in the soil-litter interface in the Cicuta Forest. This area is a rainforest remnant located in an urban matrix in southeastern Brazil. Specifically, we hypothesized that, due to higher resource availability and better temperature and humidity conditions during the warm and rainy season, ant richness and composition should differ between dry and rainy seasons.

## Material and Methods

### Study area

The Cicuta Forest has 125.14 ha and is located within the municipalities of Volta Redonda and Barra Mansa, in the southern region of Rio de Janeiro State, southeastern Brazil. In the 18th and 19th centuries this region was economically very important due to coffee production (Dean, 1996). Classified as an area of relevant ecological interest, according to the Brazilian environmental law, the reserve was created by the CONAMA resolution #05 of June 5th, 1984 and by the Decree #90,792 of January 9th, 1985. The local vegetation is classified as submontane semi-deciduous seasonal forest (IBGE, 1992). The surrounding landscape is classified as an urban matrix, with small forest fragments at different successional stages, *Eucalyptus* plantations, and mainly pastures.

The altitude ranges between 300 and 500 m a.s.l. (Monsores et al., 1982) and the climate is Cwa (dry winter and warm and rainy summer) in Koeppen classification. The average annual rainfall is 1,300mm; February is the warmest month (24°C) and July is the coldest (17°C). The local geomorphology is characterized by flattened hilly terraces, isolated structural hills, and dissected tablet-shaped terraces. The Red-Yellow Podzolic soil, whose constitution may facilitate the occurrence of erosion, predominates in the region (Brasil, 1983; Dias et al., 2001). The Cicuta Forest, which is considered an important Atlantic Forest remnant, harbors large tree species such as jequitibá (*Cariniana estrellensis*), chichá (*Sterculia chicha*) and figueira branca (*Ficus guaranítica*).

### Sampling

For sampling ants we used pitfall traps at three equidistant sites, 500 m apart from one another. At each site, we placed 20 pitfalls at every 10 m in four parallel transects. Pitfall traps were made of 500-mL plastic cups (8 cm diameter) containing approximately 150 ml of ethanol 70% as preservative fluid, which remained in the field for seven days. During the exposure period, the traps were constantly checked so that they always contained the preservative fluid. We repeated sampling four times: two times in the rainy season and two times in the dry season (June 2001 - dry season, March - rainy season and June - dry season in 2002, and March 2003- rainy season).

### Data analysis

We used the Mao Tau moment-based rarefaction (Gotelli & Colwell, 2001) to build rarefaction curves of ant species for the dry and rainy seasons. In spite of our sampling effort, this technique eliminates the need for replicates, as it allows a direct comparison of richness between seasons of the year (Colwell et al., 2004). In addition, we used the first order Jackknife non-parametric species richness estimator for extrapolating species richness in the study area. Both analyses (rarefaction and richness estimation) were made in EstimateS 7.5.2 (Colwell, 2005). We calculated the Shannon diversity index (H') for each season (winter and summer) and compared them with a t test in PAST (Hammer et al., 2001).

We also tested for a turnover in the composition of dominant ant species between the dry and rainy seasons. We considered dominant the species that were present in more than 25% of the samples.

To summarize the structure of the ant community in dry and rainy seasons, we ordered the samples with non-metric multidimensional scaling (NMDS). This type of ordination is one of the most robust, as it summarizes more information in fewer axes compared with other techniques (Legendre & Legendre, 1998). The NMDS was performed based on a distance matrix calculated with the Bray-Curtis dissimilarity index. Next, we tested for differences in ant species composition between samples collected in the dry and rainy seasons, using a permutation test (10,000 permutations) based on an analysis of similarities (ANOSIM) (Clarke, 1993). Both the ordination and similarity analysis were performed in R 2.13.1 (R Development Core Team). In order to not overestimate the ant species with more efficient systems for recruiting and / or those whose colonies are closer to the bait (Gotelli et al. 2011), all analyses used in this study were calculated based on the frequency of species occurrence in the pitfall traps and not based on the number of workers.

## Results

We collected 83 ant species of 35 genera and eight subfamilies (Table 1). The subfamily Myrmicinae had the largest number of species (42 - 51%), followed by Formicinae (17 - 20%), Ponerinae (12 - 14%), and Ecitoninae (5 - 6%). In total, 64 species were collected in the dry season and 73 in the rainy season.

Although the rarefaction curves evidenced a fast increase in the number of species in both seasons (dry and rainy), no curve reached an asymptote. This suggests that more species could have been added with a larger sampling effort or with the addition of other sampling methods (Figure 1). According to the Jackknife 1 richness estimator, the sampling efficiency was 71.2 % in the dry season (observed richness: 64 species; estimated richness: 89.7 species) and 73.9 % in the rainy season (observed richness: 73 species; estimated richness: 98.78

species). Diversity (H') was 3.43 in the dry season and 3.52 in the rainy season, which did not differ significantly from each other (t = -1.20; $P$ = 0.23).

Although some species occurred only in one season, the structure of the ant assemblage did not differ between the dry and rainy seasons (Figure 2) (NMDS followed by ANOSIM: r = 0.048; P < 0.001). In addition, we observed that 10 species (15.6 %) were particularly dominant in the dry season and only five species (6.8 %) in the rainy season. We recorded 27 (33%) rare species (Table 1).

**Table 1** - List of species by subfamilies sampled in two different periods (dry and wet) in Cicuta Forest, Volta Redonda, Rio de Janeiro.

| Subfamilies/Species | Sampling period | | Functional groups |
|---|---|---|---|
| | Dry | Wet | |
| **Dolichoderinae** | 14 | 2 | |
| Linepithema sp. 1 | | 1 | omn |
| Linepithema sp. 2 | 14 | 1 | omn |
| **Ecitoninae** | 32 | 12 | |
| Eciton cf. vagans Olivier | | 3 | arm |
| Labidus praedator (Fr. Smith) | 28 | 8 | arm |
| Neivamyrmex sp. 1 | 2 | 1 | arm |
| Neivamyrmex sp. 2 | 1 | | arm |
| Neivamyrmex sp. 3 | 1 | | arm |
| **Ectatomminae** | 11 | 2 | |
| Ectatomma edentatum Roger | 7 | 1 | lit-dom |
| Gnamptogenys minuta (Emery) | 1 | 1 | gen-pred |
| Gnamptogenys sp. 3 | 3 | | gen-pred |
| **Formicinae** | 43 | 43 | |
| Acropyga sp. 1 | 1 | 1 | sub |
| Brachymyrmex sp. 1 | 1 | 11 | omn |
| Brachymyrmex sp. 2 | 5 | 2 | omn |
| Brachymyrmex sp. 3 | 5 | 2 | omn |
| Brachymyrmex sp. 4 | 4 | 1 | omn |
| Brachymyrmex sp. 5 | | 1 | omn |
| Camponotus crassus Mayr | | 1 | omn |
| Camponotus sericeiventris Guérin | 1 | 1 | omn |
| Camponotus sp. 1 | 5 | 5 | omn |
| Camponotus sp. 2 | 1 | 3 | omn |
| Camponotus sp. 3 | | 1 | omn |
| Camponotus sp. 4 | 16 | 11 | omn |
| Camponotus sp. 5 | | 1 | omn |
| Camponotus sp. 6 | 1 | 1 | omn |
| Camponotus sp. 7 | 1 | | omn |
| Camponotus sp. 8 | 1 | | omn |
| Nylanderia sp. 1 | 1 | 1 | omn |
| **Heteroponerinae** | | 1 | |
| Heteroponera sp. 1 | | 1 | gen-pred |
| **Myrmicinae** | 263 | 291 | |
| Acanthognathus brevicornis Smith | 6 | 2 | lit-pred |
| Acromyrmex sp. 1 | 5 | 11 | fung |
| Apterostigma gr. pilosum | 5 | 2 | fung |
| Atta sp. 1 | | 1 | fung |
| Carebara urichi (Wheeler) | 3 | 1 | omn |
| Cephalotes pallens (Klug) | | 1 | omn |
| Crematogaster sp. 1 | | 2 | omn |
| Crematogaster sp. 2 | 1 | 2 | omn |
| Cyphomyrmex sp. 2 | 1 | 7 | fung |
| Cyphomyrmex sp. 3 | | 1 | fung |
| Cyphomyrmex sp. 4 | 1 | | fung |
| Hylomyrma balzani (Emery) | 4 | 4 | lit-pred |
| Leptothorax sp. 1 | 1 | 2 | lit-pred |
| Megalomyrmex sp. 1 | 6 | 6 | lit-pred |
| Mycetarotes carinatus Mahyé-Nunes | | 3 | fung |
| Octostruma sp. 1 | 3 | 7 | omn |
| Pheidole gertrudae Forel | | 1 | omn |
| Pheidole sp. 1 | 1 | 16 | omn |
| Pheidole sp. 10 | 1 | 3 | omn |
| Pheidole sp. 11 | 2 | 3 | omn |
| Pheidole sp. 2 | 23 | 21 | omn |
| Pheidole sp. 3 | 14 | 13 | omn |
| Pheidole sp. 4 | 24 | 40 | omn |
| Pheidole sp. 5 | 15 | 10 | omn |
| Pheidole sp. 6 | 28 | 11 | omn |
| Pheidole sp. 7 | 21 | 16 | omn |
| Pheidole sp. 8 | 17 | 7 | omn |
| Pheidole sp. 9 | | 1 | omn |
| Procryptocerus sp. 1 | 1 | | omn |
| Solenopsis sp. 1 | 1 | 4 | lit-omn |
| Solenopsis sp. 2 | 24 | 40 | lit-omn |
| Solenopsis sp. 3 | 39 | 19 | lit-omn |
| Solenopsis sp. 4 | 4 | 10 | lit-omn |
| Solenopsis sp. 5 | | 2 | lit-omn |
| Solenopsis sp. 6 | | 2 | lit-omn |
| Solenopsis sp. 7 | 1 | 7 | lit-omn |
| Strumigenys apretiata | 3 | 5 | lit-pred |
| Strumigenys sp. 1 | 2 | 3 | lit-pred |
| Strumigenys sp. 2 | 4 | 3 | lit-pred |
| Strumigenys sp. 3 | 1 | | lit-pred |
| Strumigenys sp. 4 | 1 | | lit-pred |
| Trachymyrmex sp. 1 | | 2 | fung |
| **Ponerinae** | 116 | 110 | |
| Hypoponera sp. 1 | 3 | 8 | gen-pred |
| Hypoponera sp. 2 | | 1 | gen-pred |
| Hypoponera sp. 3 | 6 | 2 | gen-pred |
| Hypoponera sp. 4 | 8 | 1 | gen-pred |
| Leptogenys sp. 1 | 1 | 1 | lit-pred |
| Leptogenys sp. 2 | | 1 | lit-pred |
| Odontomachus chelifer (Latreille) | 7 | 8 | gen-pred |
| Odontomachus meinerti Forel | 22 | 20 | gen-pred |
| Pachycondyla harpax (Fabricius) | 26 | 15 | gen-pred |
| Pachycondyla sp. 1 | 1 | | gen-pred |
| Pachycondyla sp. 2 | 3 | 1 | gen-pred |
| Pachycondyla striata Fr. Smith | 39 | 52 | gen-pred |
| **Pseudomyrmecinae** | 1 | 2 | |
| Pseudomyrmex sp. 1 | 1 | 2 | omn |
| Total Richness | 64 | 73 | |

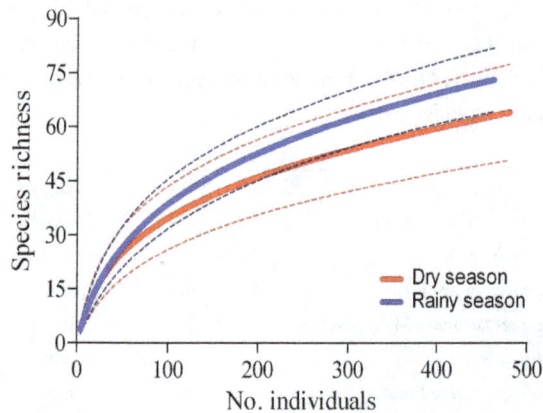

**Fig 1**. Rarefaction curves of ant species richness (Mao Tao) for dry and wet seasons based on the number of individuals collected. The thinner lines represent the confidence interval of 95%.

## Discussion

Several studies have shown that abiotic factors, such as rainfall and temperature, are directly related to the availability of food and nesting sites for insects (Speight et al. 2008). These factors may also influence the foraging activity of ants (Levings, 1983; Almeida et al., 2007). Some previous studies carried out in semi-deciduous areas of the Brazilian Atlantic Forest pointed to an effect of climate seasonality on ant assemblages (Vargas et al., 2007; Castilho et al., 2011). In this study, despite the highest ant richness found in the rainy period, we did not observe an effect of climate seasonality on species richness, diversity, or assemblage structure. However, the expressive number of exclusive species and their frequency in each season suggest that some species may be influenced by seasonality.

Our results corroborate Castro et al., (2012) who also did not found a relationship between ant species richness and seasonality in a degraded area. The lack of correlation observed in the present study may be explained by environmental degradation, which makes the environments simpler (Vasconcelos et al., 2006; Sobrinho & Schoereder, 2007), and increases the competition and abundance of generalistic species (Schoereder et al., 2004). We observed the same pattern in the hypogaeic fauna, as also observed by Figueiredo et al., (2013).

The Cicuta Forest has faced different impacts throughout the years, such as fire, hunting, cattle farming, and logging, mainly in its surroundings, which influence vegetation structure. These impacts certainly restrict the occurrence of more demanding species and contribute to the dominance of a smaller number of generalist species. Currently access to the forest is prohibited, except for scientific research, in order to minimize human impacts and help its conservation. This forest remnant is extremely important to biodiversity maintenance in the region, mainly because it is located within an urban matrix. One example of its importance is the occurrence of a population of *Alouatta guariba clamitans* Cabrera, 1940 (red howler monkey), which may be considered one of the last populations in the Paraíba Valley (Alves & Zaú, 2005).

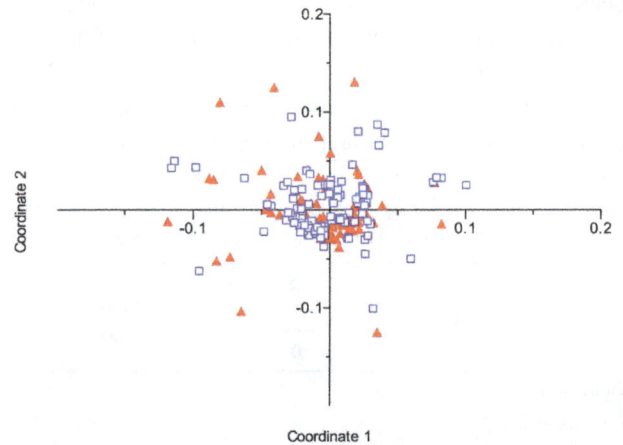

**Fig 2**. Non-metric multidimensional scaling (NMDS) composition of ants collected in dry (triangles) and rainy seasons (squares) between June 2001 and March 2003 in the hemlock forest, State of Rio de Janeiro, Brazil. This ordination analysis was calculated from Bray-Curtis dissimilarity index's (Stress = 0.615; Axis Axis 1 + 2 = 39.3% of explanation).

The ant richness found in the Cicuta Forest is high compared to other forest remnants (Veiga-Ferreira et al., 2005; Castro et al., 2012; Vargas et al., 2013). The assemblage composition of ants is highly generalist, but also we can found a high occurrence of rare species (33%). In addition, it is worth to mention the record of a well-documented pattern for tropical forest fragments with Myrmicinae, Formicinae, and Ponerinae as the subfamilies with highest diversity. The same is true for the genera *Pheidole*, *Camponotus*, and *Solenopsis*, which stood out as the most diverse (Ward, 2000 Ward, 2010; Castilho et al., 2011; Miranda et al., 2013; Dattilo et al., 2011).

However, subfamilies as Amblyoponinae, Cerapachyinae, and Proceratiinae, which are usually recorded in forests, were not represented. This has possibly happened due to the sampling technique used here (see Vargas et al., 2009), as most of their species present cryptic behavior, and reduced size and they rarely forage in the leaf litter (Hölldobler & Wilson, 1990; see Longino et al., 2002; Figueiredo et al., 2013).

Hence, biodiversity conservation in the Atlantic Forest is related to the maintenance of its forest remnants, even if they are relatively small. Therefore, the Cicuta Forest is important for the conservation of biodiversity in semi-deciduous seasonal forests, especially if we consider that variations in rainfall and temperature are important to regulate the ant community. In short, our study contributes to knowledge of how seasonal variations affects ant communities.

## Acknowledgements

To the University Center of Volta Redonda (UniFOA), FAPERJ, and CNPq. We thank the Brazilian Institute for the Environment and Natural Resources (IBAMA) for permission to collect ants in the Cicuta Forest. We also thank two anonymous reviewers for valuable comments that improved this article.

# References

Almeida, F.S., Queiroz, J.M. & Mayhe-Nunes, A.J. (2007). Distribuição e abundância de ninhos de *Solenopsis invicta* Buren (Hymenoptera: Formicidae) em um agroecossistema diversificado sob manejo orgânico. Floresta e Ambiente., 14: 33-43.

Alves, S.L., & Zaú S.A. (2005). A importância da área de relevante interesse ecológico da floresta da Cicuta (RJ) na conservação do Bugio-Ruivo (*Alouatta guariba clamitans* Cabrera, 1940). Revista Universidade Rural, 25: 41-48.

Brasil. (1983). Levantamento de recursos naturais, Rio de Janeiro/Vitória. RADAMBRASIL., 31: 23-24.

Castilho, G.A., Noll, F.B., Silva, E.R. & Santos, E.F. (2011). Diversidade de Formicidae (Hymenoptera) em um fragmento de Floresta Estacional Semidecídua no Noroeste do estado de São Paulo, Brasil. Revista Brasıleira de Biociências, 9: 224-230.

Castro, S.F., Gontijo, A.B., Castro, P.T.A. & Ribeiro, S.P. (2012). Annual and Seasonal Changes in the Structure of Litter-Dwelling Ant Assemblages (Hymenoptera: Formicidae) in Atlantic Semideciduous Forests. Psyche., 2012: 95971-12. doi: 10.1155/2012/959715.

Clarke, K.R., Warwick, R.M. & Brown, B.E. (1993). An index showing breakdown of seriation, related to disturbance, in a coral reef assemblage. Marine Ecology Progress Series, 102: 153-160.

Colwell, R.K., Mao, C.X. & Chang, J. (2004). Interpolating, extrapolating, and comparing incidence-based species accumulation curves. Ecology, 85: 2717-2727.

Colwell, R.K. (2005). EstimateS: Statistical estimation of species richness and shared species from samples. version 7.5. http://purl.oclc.org/ estimates (último acesso em 21/05/2013).

Dáttilo, W., Marques, E.C., Falcão, J.C.F., & Moreira, D.D.O. (2009). Interações mutualísticas entre formigas e plantas. EntomoBrasilis, 2: 32-36.

Dáttilo, W., Sibinel, N., Falcão, J.C.F., & Nunes, R.V. (2011). Mirmecofauna em um fragmento de Floresta Atlântica urbana no município de Marília, Brasil. Bioscience Journal, 27: 494-504.

Dean, W. (1996). A Ferro e Fogo: a história e a devastação da Mata Atlântica brasileira. São Paulo: Companhia das Letras, 484p.

Dias, J.E., Gomes, O.V.O. & Goes, M.H.B. (2001). Áreas de risco de erosão do solo: uma aplicação por geoprocessamento. Floresta e Ambiente, 8: 1-10.

Figueiredo, C.J., Silva, R.R., Munae, C.B., & Morini, S.C. (2013). Fauna de formigas (Hymenoptera: Formicidae) atraídas a armadilhas subterrâneas em áreas de Mata Atlântica. Biota Neotropica, 13: 176-182.

Folgarait, P.J. (1998). Ant biodiversity and its relationship to ecosystem functioning: a review. Biodiversity and Conservation, 7: 1221-1244.

Gomes, J.P., Iannuzzi, L. & Leal, I.R. (2010a). Resposta da Comunidade de Formigas aos Atributos dos Fragmentos e da Vegetação em uma Paisagem da Floresta Atlântica Nordestina. Neotropical Entomology, 39: 898-905. doi: 10.1590/S1519-566X2010000600008.

Gomes, J.B.V., Barreto, A.C., Michereff, M.F., Vidal, W.C.L., Costa, J.L.S., Oliveira-Filho, A.T., & Curi, N. (2010b). Relações entre atributos do solo e atividade de formigas em restingas. Revista Brasileira de Ciências do Solo, 34: 67-78.

Gotelli, N.J & Colwell, R.K. (2001). Quantifying biodiversity: Procedures and pitfalls in the measurement and comparison of species richness. Ecology Letters, 4: 379-391.

Gotelli, N.J., Ellison, A.M., Dunn, R.R., Sanders, N.J. (2011). Counting ants (Hymenoptera: Formicidae): biodiversity sampling and statistical analysis for myrmecologists. Myrmecological News, 15: 13-19.

Hammer, Q., Harper, D.A.T., Ryan, P.D. (2001). PAST: Paleontological Statistics Software Package for Education and Data Analysis. Palaeontologia Electronica, 4: 1-9. http://palaeo-electronica. org/2001_1/past/issue1_01.htm.

Holldobler, B. & Wilson, E.O. (1990). The Ants. Cambrige: Belknap of Harvard University Press. 732 p.

IBGE. (1992). Instituto Brasileiro de Geografia e Estatística. Manual técnico da vegetação brasileira. Departamento de Recursos Naturais e Estudos Ambientais. Rio de Janeiro.

Jankowski, J.E., Ciecka, A.L., Meyer, N.Y. & Rabenold, K.N. (2009). Beta diversity along environmental gradients: implications of habitat specialization in tropical montane landscapes. Journal of Animal Ecology, 78: 315-327. doi: 10.1111/j.1365-2656.2008.01487.

Kaspari, M. (2001). Taxonomic level, trophic biology and the regulation of local abundance. Global Ecology and Biogeography, 10: 229-244.

Kaspari, M. (2000). A primer of ant ecology. In: D. Agosti, J.D. Majer, L.E. Alonso & T.R. Schultz (Eds.). Ants standard methods for measuring and monitoring biodiversity. Washington: Smithsonian Institute Press. p. 9-24.

Lach, L., Parr, C.L. & Abbott, K.L. (2010). Ant ecology. Oxford: Oxford University Press. 402p.

Lassau, S.A., & Hochuli, D.F. (2004). Effects of habitat complexity on ant assemblages. Ecography, 27: 157–164. doi: 10.1111/j.0906-7590.2004.03675.x.

Legendre, P., & L. Legendre. (1998). Numerical ecology. Third English edition. Amsterdam: Elsevier Press 870 pp.

Longino, J. T.; Coddington, J.; Colwell, R. K. (2002). The ant fauna of a tropical rain forest: estimating species richness in three different ways, Ecology, 83: 689-702.

Miranda, T.A. ; Santanna, A.S.; Almeida, F.S.; Vargas, A.B. (2013). Aspectos estruturais do ambiente e seus efeitos nas assembléias de formigas em ambientes de floresta e bosque. Cadernos UniFOA (Online), 21: 63-72.

Monsores, D.W., Bustamante, J.G.G., Fedullo, L.P.L., & Gouveia, M.T.J. (1982). Relato da situação ambiental com vistas à preservação da área da Floresta da Cicuta. Relatório técnico, 17 p.

Nakamura, A., Proctor, H., & Catterall, C.P. (2003). Using soil and litter arthropods to assess the state of rainforest restoration. Ecological Management and Restoration, 4: 20–28. doi: 10.1046/j.1442-8903.4.s.3.x

Passos, L. & Oliveira, P.S. (2004). Interactions between ants and fruits of *Guapira opposita* (Nyctaginaceae) in a Brazilian sand plain rain forest: ant effects on seeds and seedling. Oecologia, 139: 376-382. doi: 10.1007/s00442-004-1531-5.

Silva, P.S.D., Bieber, A.G.D., Corrêa, M.M., & Leal, I.R. (2011). Do Leaf-litter Attributes Affect the Richness of Leaf-litter Ants? Neotropical Entomology (Impresso), 45: 542-547. doi: 10.1590/S1519-566X2011000500004.

Sobrinho, T.G. & Schoereder, J.H. (2007). Edge and shape effects on ant (Hymenoptera: Formicidae) species richness and composition in forest fragments Biodiversity and Conservation, 16: 1459-1470. doi: 10.1007/s10531-006-9011-3.

Schoereder JH, Sobrinho TG, Ribas CR, Campos RBF. (2004). The colonization and extinction of ant communities in a fragmented landscape. Austral Ecology, 29: 391-398. doi: 10.1111/j.1442-9993.2004.01378.x.

Speight, R., Hunter, M.D. & Watt, A.D. (2008). Ecology of Insects: concepts and applications. Oxford, Blackwell Publishing, 628 p.

Tews, J., Brose, U., Grimm, V., Tielbörger, K., Wichmann, M.C., Schwager, M. & Jeltsch, F. (2004). Animal species diversity driven by habitat heterogeneity/diversity: The importance of keystone structures, Journal of Biogeography, 31: 79-92. doi: 10.1046/j.0305-0270.2003.00994.x.

Vargas, A.B., Mayhe-Nunes, A.J., Queiroz, J.M., Orsolon, G.S., Folly-Ramos, E. (2007). Efeito de fatores ambientais sobre a mirmecofauna em comunidade de restinga no Rio de Janeiro, RJ. Neotropical Entomology, 36: 28-37. doi: 10.1590/S1519-566X2007000100004.

Vargas, A.B., Queiroz, J.M., Mayhe-Nunes, A.J., Orsolon, G.S. & Folly-Ramos, E. (2009). Teste da regra de equivalência energética para formigas de serapilheira: efeitos de diferentes métodos de estimativa de abundância em floresta ombrófila. Neotropical Entomology, 38: 867-870. doi: 10.1590/S1519-566X2009000600023.

Vargas, A.B., Mayhe-Nunes, A.J., Queiroz, J.M. (2013). Riqueza e composição de formigas de serapilheira na Reserva Florestal da Vista Chinesa, Rio de Janeiro, Brasil. Cadernos UniFOA. Vol. esp. 85-94.

Vasconcelos H.L., Laurance W.F. (2005). Influence of habitat, litter type, and soil invertebrates on leaf-litter decomposition in a fragmented Amazonian landscape. Oecologia, 144: 456-462.

Vasconcelos, H.L., Vilhena, J.S.M, Magnusson W.E. & Albernaz, A.L.K.M. (2006). Long-term effects of forest fragmentation on Amazonian ant communities. Journal of Biogeography, 33: 1348–1356. doi: 10.1111/j.1365-2699.2006.01516.x.

Veiga-Ferreira, S., Mayhé-Nunes, A.J., Queiroz, J.M. (2005). Formigas de serapilheira na Reserva Biológica do Tinguá, Estado do Rio de Janeiro, Brasil (Hymenoptera: Formicidae). Revista Universidade Rural, 25: 49-54.

Ward, P. S. (2000). Broad-scale patterns of diversity in leaf litter ant communities. In: D.J.D. Agosti, L. Majer, E. Alonso & T.R. Schultz (Eds.). Ants: Standard methods for measuring and monitoring biodiversity. Washington: Smithsonian Institution.

Ward, P. S. (2010). Taxonomy, phylogenetics and evolution. In: Lori Lach, Catherine L. Parr, and Kirsti L. Abbott (Ed.). Ant ecology, 429p.

# 19

# Evidences of Batesian Mimicry and Parabiosis in Ants of the Brazilian Savanna

MC Gallego-Ropero[1], RM Feitosa[2]

1 - University of Cauca, Department of Biology, Popayán, Colombia.
2 - Departamento de Zoologia, Universidade Federal do Paraná, Curitiba, Brazil.

**Keywords**
*Camponotus blandus, Pseudomyrmex termitarius,* termites, mimicry evolution, Parabiosis, Brazilian Cerrado.

**Corresponding author**
María Cristina Gallego Ropero
University of Cauca
Department of Biology
Calle 5 No. 4 – 70, Popayán, Colombia.
E-Mail: macrisgaro@yahoo.es

**Abstract**
Despite the numerous records of ant-mimicking arthropods, reports of ant species that are mimics among themselves are still rare. In the savanna of central Brazil we found two ant species that are remarkably similar in color pattern and body size, *Pseudomyrmex termitarius* and *Camponotus blandus*. Both species are widely distributed in the Neotropical Region, but the cases of mimicry between them are apparently restricted to populations inhabiting nests of the termite *Cornitermes cumulans* in the Brazilian Cerrado. Field observations and excavation of the termitaries revealed that *Camponotus blandus* shares nest chambers and foraging trails with *P. termitarius,* and workers of both species are mutually tolerant. Our observations suggest that the morphological and behavioral similarities between these species represent a Batesian mimicry relationship in which the relatively palatable *Camponotus blandus* mimics the unpalatable *P. termitarius* for predator avoidance. The pacific association between the termitophilous colonies of these species may also suggest some level of parabiotic interaction.

## Introduction

Mimicry is considered a conspicuous demonstration of Darwinian selection (Fisher 1930) and can incorporate a wide range of sensory modalities, including visual, auditory, vibrational and chemical (Pasteur 1982). Batesian mimicry has been described as the mechanism by which a palatable species look similar to an unpalatable one to avoid predation. Mimetic systems have been recorded in a vast range of invertebrates and despite the fact that the majority of quantitative studies of Batesian mimicry have examined defensive visual mimicry among lepidopterans (e.g. Joron & Mallet 1998), social insects are one of the most common "models" of Batesian mimics (Hölldobler &Wilson 1990; McIver & Stonedahl 1993; Cushing 1997; Ceccarelli & Crozier 2006). It occurs because social insects may have strong defense mechanisms that are effective against predators. Nevertheless, records of social insects mimicking themselves are relatively rare.

Batesian mimicry has been suggested in few ant species (Ward 1984; Hölldobler & Wilson 1992; Gobin

et al. 1998; Merrill & Elgar 2000; Ito et al. 2004), despite the vast diversity of arthropods that mimic ants (e.g. Pie & Del-Claro 2002; Taniguchi et al. 2005; Nelson & Jackson 2009). Here we provide a novel account of a remarkable similarity between two species of unrelated ants. We found workers of the formicine ant *Camponotus blandus* (Fr. Smith) reproducing the body coloration and sharing nest chambers and foraging trails of the pseudomyrmecine *Pseudomyrmex termitarius* (Fr. Smith). The two species are exclusively Neotropical, occurring from Central America to northern Argentina. However, the records of Batesian mimicry in these species are apparently restricted to the populations inhabiting termite nests in the savanna of central Brazil.

Like most members of Pseudomyrmecinae, *P. termitarius* is a very aggressive ant with precise mandibles and a painful, venomous sting which serves as a solid deterrent to many potential predators. As its name suggests, *P. termitarius* has a preference for establishing its colonies inside termite nests, although subterranean nests can also be found (Mill 1981; Jaffé et al.1986; Pulgarín 2004). This species is an opportunistic predator and, when inhabiting termitaries, it can prey upon termite brood or inquiline arthropods that also occupy the termite nests. Populations of *P. termitarius* are widespread in the Brazilian Cerrado, where they can they can most commonly be found inside nests of the termite *Cornitermes cumulans* (Kollar) (Redford 1984), although colonies can also be found in open soil and in association with other termite species (Kempf 1960).

In contrast, the defense mechanisms of formicine ants are limited to a formic acid spray, since these ants do not possess a functional sting. This is the case with *Camponotus blandus*. The formic acid of these ants renders some of them unpalatable to some predators. However, lacking more substantial defenses, many formicines may be relatively more vulnerable to predation (Lamon & Topoff 1981; Montgomery 1985; Merrill & Elgar 2000). For example, ants of the genus *Camponotus* are among the most commonly preyed upon by Australian birds (Barker & Vestjens 1990). Like many species of this genus, *C. blandus* is a polymorphic, generalist ant (Fernández 2003). The nests are found in the soil, in preexisting cavities under rocks, decaying logs or termite mounts. The foraging is largely arboreal but the workers can be efficient ground predators, preying on terrestrial arthropods and even raiding termite galleries (Fowler & Crestana 1987; Mendonça & Resende 1996).

Morphologically, *Camponotus blandus* can be highly variable regarding the color patterns along its wide geographical distribution, but as far as we know, only in the Brazilian Cerrado this species assumes the reddish and black pattern typical of *P. termitarius*. Both species can share nests of *Cornitermes cumulans* in central Brazil. Our morphological, geographical and behavioral lines of evidence indicate that one species is a visual model of the other. We argue that *C. blandus*, a member of a normally highly predated genus, mimics an aggressive and venomous ant, *P. termitarius*, in order to reduce the risk of predation in the open areas of central Brazil. This is the first formal report of mimicry of an ant by another ant in the Neotropical Region and may also represent a new case of parabiosis between ant species.

## Materials and Methods

Collecting and observations on interactions between *Camponotus blandus* and *Pseudomyrmex termitarius* were carried out in two field stations. The first one, Estação Experimental Embrapa Cerrados, is located in the Planaltina municipality, near Brasília, in Brazil's Federal District (15,36'35,5"S; 47,44'09,5"W). It is a 3,500 hectares site with permanent ecological reserves and 10 different Cerrado phytophysiognomies (Embrapa 2010). The second station, Reserva Acangaú, is situated in the Paracatu municipality, State of Minas Gerais (17,12'08,2"S; 47,4'19,6"W). It is classified as a Private Natural Heritage Reserve of 3,000 hectares, where the Cerrado *sensu strictu* is the prevalent physiognomy (INMET 2011).

We made qualitative behavioral observations of workers from five colonies of *Camponotus blandus* and *P. termitarius* cohabiting in termitaries of *Cornitermes cumulans* (Fig. 1a). Termitaries containing colonies of both species were excavated to determine the position of the ant nests and internal organization of the colonies. Then, eight and three additional colonies were collected for *P. termitarius* and *Camponotus blandus*, respectively, from termitaries of both study sites. These additional colonies were found to be isolated, without the presence of the second species. Our observations were made throughout the day, primarily to establish the periods when the ants were most active and abundant above the ground. When possible, all the individuals present in the colonies were collected, including sexuals and brood, and fixed in 70% ethanol. Voucher specimens are deposited at the insect collection of the Instituto de Biologia, Universidade de Brasília (UnB), DF, Brazil.

**Fig. 1** A typical termitary of *Cornitermes cumulans* in the Brazilian Cerrado (**a**), worker of *Pseudomyrmex termitarius* foraging solitarily on a termitary (**b**), mixed sample from a nest chamber containing individuals of *Camponotus blandus* (Cb) and *Pseudomyrmex termitarius* (Pt) (**c**).

## Results

The foraging behavior of *Pseudomyrmex termitarius* is predominantly solitary (Fig. 1b); however, workers may follow disperse trails on the surface of termitaries and in the adjacent areas. We found workers of the mimic *Camponotus blandus* sharing the trails of *P. termitarius* in colonies studied from both study sites, though workers of *Camponotus blandus* could be seen foraging independently.

Termitaries excavation revealed that *P. termitarius* forms polydomous colonies, occupying several chambers inside the termite nests. These chambers were normally found on the hypogaeic portion of the termitaries, from 40 cm high to 15 cm below the soil surface and, in many cases, were shared with colonies of *Camponotus blandus* (Fig. 1c). It was not possible to observe if brood of both species was kept together, but no agonistic interaction was observed between workers of the two species. The presence of the termitary builder

(*Cornitermes cumulans*) is not a requirement for the presence of the ants, since ant nests were found in both abandoned and occupied termite nests. In general, colonies of *P. termitarius* were larger than those of *Camponotus blandus*, with an average number of individuals of 45 and 23, respectively, including brood and sexuals.

Morphologically, the most conspicuous similarity between these two species is their color pattern. Both share the reddish antennae, mesosoma, waist, anterior segments of gaster, and distal portion of legs, and the black head, coxae and apical segments of gaster. Additionally, the total body length is also very similar in both species, about 7 mm (Fig. 2). Extensive collecting was performed in other areas where the species co-occur and the study of specimens of both species deposited in myrmecological collections indicates that the morphological similarity only occurs in populations of termitaries in central Brazil. Most of the *Camponotus blandus* specimens found without the association with *P. termitarius* are predominantly blackish in color. Still, regarding all the species phylogenetically related (subgenus *Myrmaphaenus*) to *Camponotus blandus*, the termitophilous populations of this species are the only ones to exhibit the color pattern observed here.

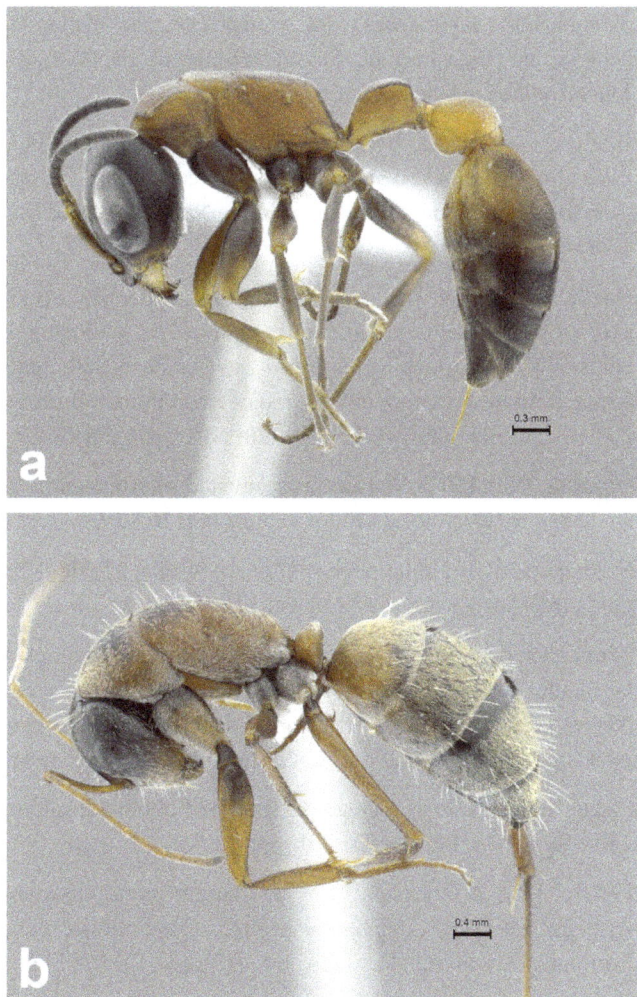

**Fig. 2** Lateral view of workers: *Pseudomyrmex termitarius* (a) and *Camponotus blandus* (b).

## Discussion

The color combination exhibited by *P. termitarius* and *Camponotus blandus* is most likely aposematic. The black and reddish coloring can frequently be seen as a warning pattern in unpalatable plants and animals of different groups and regions (Lythgoe 1979). Many ichneumonoid wasps and butterflies of the "tiger-complex" combine these colors, apparently to advertise mechanical or chemical defenses to potential predators (Quicke et al. 1992; Beccaloni 1997). Similar reddish and black color patterns can be found within a variety of species in the Australian ant genus *Myrmecia*. Species of these ants are among the most aggressive ants in the World (Haskins & Haskins 1950; Hölldobler & Wilson 1990).

Populations of *Camponotus blandus* occupying termite nests in the Brazilian Cerrado differ in worker caste from the majority of the conspecific populations with arboreal habits and also from those of different biomes and geographic regions. While most species of *Camponotus* have a wide range of size classes, termitophilous *Camponotus blandus* exhibits relatively little variation in the size of the foragers. This conservative caste system helps to create a more consistent congruence in size between *Camponotus blandus* and *P. termitarius* foragers, and thus increases the efficacy of mimicry, given that *P. termitarius* is a monomorphic species.

Mimicry will be most effective when the ranges of the mimic and its model overlap broadly, such that predators encountering the mimic are likely to have had some experience with the model (Pough 1994). The range of *Camponotus blandus* lies entirely within that of *P. termitarius*, although the mimicry cases between both species are apparently restricted to the termitaries in central Brazilian Cerrado. The geographic sympatry of these species is even tighter at the ecological level. Both *Camponotus blandus* and *P. termitarius* share the same nesting and foraging strategies in the open areas of the Cerrado. More significantly, however, is the fact that workers of both species can share nest chambers inside the termitaries, showing an almost complete ecological overlapping.

Theory also predicts that mimicry will be most effective when the model is relatively abundant (Lindström et al. 1997), and in all our observations *P. termitarius* is relatively more common than *Camponotus blandus* within their range, with more individuals per colony. Museum collections corroborate this finding; *P. termitarius* is well represented in the ant collection of the Museu de Zoologia da USP (MZSP), with more specimens than for most of its congeners (RMF, pers. obs.).

The striking similarities between *P. termitarius* and *Camponotus blandus* are obviously unlikely to be derived from a common phylogenetic ancestry given that the formicines and pseudomyrmecines are very divergent phylogenetic lineages (Brady et al. 2006). Instead, our observations suggest that these similarities represent a Batesian mimicry relationship in which the relatively palatable *Camponotus blandus* mimics the unpalatable *P. termitarius* in order to escape predation.

Batesian mimicry in ants has been reported in *Pheidole nasutoides*, *Camponotus bendigensis*, *Polyrhachis rufipes* and *Camponotus* sp. The major workers of *Pheidole nasutoides* mimic soldiers of nasutitermitinae termites, which have formidable chemical defenses against predation in the Costa Rican tropical forest (Hölldobler & Wilson 1992). *Camponotus bendigensis* shares body size and color patterns with *Myrmecia fulvipes*, a very aggressive Australian ant with a painful venomous sting (Merrill & Elgar 2000). In the Oriental tropics, the species *Polyrhachis rufipes* is often found on trails of *Gnamptogenys menadensis*. Workers of *Polyrhachis rufipes* can follow the trails of *G. menadensis* and thus reach sugar sources (Gobin *et al.* 1998). Finally, workers of an undescribed species of an arboreal *Camponotus* were exclusively observed on foraging trails of the myrmicine ant *Crematogaster inflata* in western Malaysia. The bright yellow and black color pattern, as well as the walking behavior, are very similar in both species (Ito *et al.* 2004).

The most remarkable result of our study is the finding of *Camponotus blandus* sharing nest chambers and foraging trails with *Pseudomyrmex termitarius,* and the mutual tolerance between workers of both species during foraging and intranidal activities. It must be confirmed in the future whether *Camponotus* workers can recognize signals from *Pseudomyrmex termitarius* and *vice versa*, but our observations may suggest some level of parabiotic interaction between these species. Parabiosis is defined as a particular form of facultative or obligatory symbiosis in which two or more species utilize the same nest structure and sometimes even the same odor trails, but normally keep their broods separate (Hölldobler & Wilson, 1990). It is important to emphasize the facultative nature of this particular association, since neither of the participating species is dependent on the other. Parabiotic associations have been described for ant species from all over the globe, but most records report facultative relations concentrated in the Neotropics, where several cases, generally involving members of different subfamilies, have been recorded (e.g. Adams 1990; Ipinza-Regla *et al.* 2005; Sanhudo *et al.* 2008).

It remains to be explained why this tight association between the wide distributed *Camponotus blandus* and *Pseudomyrmex termitarius* involves only the termitophilous populations of these species in the Brazilian savanna. The variables favoring this mimetic and probably parabiotic system may include the predation of *Camponotus blandus* by birds and terrestrial reptiles, both are extremely common in the savanna of central Brazil. It is likely that these predators have sufficient spectral sensitivity to detect the color pattern exhibited by these ants, especially in the open lands of the Cerrado, where foragers are more exposed than in the forests and woodlands where *Camponotus blandus* also occurs.

The efficient defensive mechanisms of *Pseudomyrmex termitarius* and the solid protection of the termitaries of *Cornitermes cumulans* have apparently resulted in the evolution of the mimicry and parabiotic syndromes in *Camponotus blandus*.

## Acknowledgements

We thank the staff of the termitology laboratory of Universidade de Brasília for their assistance in the field and the colleagues of Embrapa and Reserva Acangaú for the collecting permits. Thanks to the PEC-PG program for the grant offered and to Universidad del Cauca, Colombia. This study was funded by CAPES, PRONEX-FAPDF-CNPQ project (proc.193.000563-2009). We acknowledge the research grant received from Fundacão de Amparo à Pesquisa do Estado de São Paulo to R.M. Feitosa (no. 11/24160–1).

## References

Adams, ES. (1990) Interaction between the ants *Zacryptocerus maculatus* and *Azteca trigona*: interspecific parasitization of information. Biotropica, 22: 200-206.

Barker, RD. & Vestjens, WJM. (1990) The food of Australian birds. II Passerines. CSIRO, Melbourne, 557 pp.

Beccaloni, GW. (1997) Vertical stratification of ithomiine butterfly (Nymphalidae: Ithomiinae) mimicry complexes: the relationship between adult flight height and larval host-plant height. Biological Journal of the Linnean Society, 62: 313-341.

Brady, SG., Schultz, TR., Fisher, BL. &Ward, PS. (2006) Evaluating alternative hypotheses for the early evolution and diversification of ants. Proceedings of the National Academy of Sciences USA, 13 (48): 18172-18177.

Ceccarelli, FS. & Crozier, RH. (2006) Dynamics of the evolution of Batesian mimicry: molecular phylogenetic analysis of ant-mimicking *Myrmarachne* (Araneae: Salticidae) species and their ant models. Journal of Evolutionary Biology, 20 (1): 286-295. DOI: 10.1111/j.1420-9101.2006.01199.x

Cushing, PE. (1997) Myrmecomorphy and myrmecophily in spiders: a review. Florida Entomologist, 80: 165-193.

Embrapa (2010) Home page.< http://www.cpac.embrapa.br/unidade/historia>. Acessed August, 12th 2011.

Fernández, F. (2003) Capítulo 21: Subfamilia Formicinae. Fernández F. (ed.). Introducción a las Hormigas de la Región Neotropical. Instituto de Investigación de Recursos Biológicos Alexander von Humboldt, Bogotá, Colombia. XXVI + 398 p.

Fisher, RA. (1930) The genetical theory of natural selection, 2nd edn. Dover, London. X+ 195 pp.

Fowler, HG. & Crestana, L. (1987) Group recruitment and its organization in *Camponotus blandus* (Fr. Smith) (Hym.: Formicidae). Revista Brasileira de Entomologia, 31: 55-60.

Gobin, B., Peeters, C., Billen, J. & Morgan, ED. (1998) Interspecific trail following and commensalism between the ponerine ant *Gnamptogenys menadensis* and the formicine ant *Polyrhachis rufipes*. Journal of Insect Behaviour, 11: 361-368.

Haskins, CP. & Haskins, EF. (1950) Notes on the biology and social behaviour of the archaic ponerine ants of the genus *Myrmecia* and *Promyrmecia*. Annals of the Entomological Society of America, 43: 461-491.

Hölldobler, B. & Wilson, EO. (1990) The Ants. Cambridge: Harvard University Press. 732 pp.

Hölldobler, B. & Wilson, EO. (1992) *Pheidole nasutoides*, a new species of Costa Rican ant that apparently mimics termites. Psyche, 99: 15-22.

Instituto Nacional de Meteorologia – INMET. Normais climatológicas. Home page <http://www.inmet.gov.br/html/clima.php>. Acessed in February 20th 2011.

Ipinza-Regla, J., Fernández, A. & Morales, AM. (2005) Hermetismo entre *Solenopsis gayi* Spinola, 1851 y *Brachymyrmex giardii* Emery, 1894 (Hymenoptera, Formicidae). Gayana, 69: 27-35. doi.org/10.4067/S0717-65382005000100005

Ito, F., Hashim, R., Huei, YS., Kaufmann, E., Akino, T. & Billen, J. (2004) Spectacular Batesian mimicry in ants. Naturwissenschaften, 91: 481-484.

Jaffé, K., López, ME. & Aragot, W. (1986) On the communication systems of the ants *Pseudomyrmex termitarius* and *P. triplarinus*. Insectes Sociaux, 33: 105-117.

Joron, M. & Mallett, JLB. (1998) Diversity in mimicry: paradox or paradigm? Trends in Ecology and Evolution, 13: 461-466.

Kempf, WW. (1960) Estudo sôbre *Pseudomyrmex I*. (Hymenoptera: Formicidae). Revista Brasileira de Entomologia, 9: 5-32.

Lamon, B. & Topoff, H. (1981) Avoiding predation by army ants: Defensive behaviours of three ant species of the genus *Camponotus*. Animal Behaviour, 29 (4): 1070-1081.

Lindström, L., Alatalo,. RV. & Mappes, J. (1997) Imperfect Batesian mimicry – the effects of the frequency and the distastefulness of the model. Proceedings of the Royal Society of London B, 264: 149-153.

Lythgoe, JN. (1979) Ecology and vision. Clarendon, Oxford. xi 244 pp.

McClure, M., Chouteau, M. & Dejean, A. (2008) Territorial aggressiveness on the arboreal ant *Azteca alfari* by *Camponotus blandus* in French Guiana due to behavioural constraints. Comptes Rendus Biologies, 331: 663-667. Doi : 10.1016/j.crvi.2008.06.008

McIver, JD. & Stonedahl, G. (1993) Myrmecomorphy: morphological and behavioral mimicry of ants. Annual Review of Entomology, 38: 351-379.

Mendonça, GM. & Resende, JJ. (1996) Predação de *Syntermes molestus* (Burmeister, 1839) (Isoptera-Termitidae) por *Camponotus blandus* (Fr. Smith, 1858) (Hymenoptera-Formicidae) em Feira de Santana-Ba. Sitientibus, 15: 175-182.

Merrill, DN. & Elgar, MA. (2000) Red legs and golden gasters: Batesian mimicry in Australian ants. Naturwissenschaften, 87: 212-215. 10.1007/s001140050705

Mill, AE. (1981) Observations on the ecology of *Pseudomyrmex termitarius* (F. Smith) (Hymenoptera, Formicidae) in Brazilian Savannas. Revista Brasileira de Entomologia, 25: 271-274.

Montgomery, GG. (1985) Movements, foraging and food habits of the four extant species of neotropical vermilinguas (Mammalia; Myrmecophagidae). In: Montgomery, GG. (ed) The Evolution and ecology of armadillos, sloths, and vermilinguas. Smithsonian Institution Press. Washington. D.C. 365-377.

Nelson, XJ. & Jackson, RR. (2009) Collective Batesian mimicry of ant groups by aggregating spiders. Animal Behaviour, 78: 123-129. doi:10.1016/j.anbehav.2009.04.005

Pasteur, G. (1982) A classificatory review of mimicry systems. Annual Review of Ecology and Systematics, 13: 169-199.

Pie, MR. & Del-Claro, K. (2002) Male-Male agonistic behavior and ant-mimicry in a Neotropical richardiid (Diptera: Richardiidae). Studies on Neotropical Fauna and Environment, 37: 19-22.

Pough, FH. (1994) Mimicry and related phenomena. In: Gans, C. & Huey, R.B. (eds) Biology of the reptilia, vol 16. Branta, Ann Arbor. 153-254.

Pulgarín, JA. (2004) Algunas observaciones sobre la estructura del nido y composición de colonias de *Pseudomyrmex termitarius* (Hymenoptera: Formicidae) en una localidad de Bello (Antioquia, Colombia). Revista de la Facultad Nacional de Agronomía, Medellín.

Quicke, DLJ., Ingram, SN., Proctor, J. & Huddleston, T. (1992) Batesian and Müllerian mimicry between species with connected life histories, with a new example involving braconid wasp parasites of *Phoracantha* beetles. Journal of Natural History, 26: 1013-1034.

Redford, KH. (1984) The termitaria of *Cornitermes cumulans* (Isoptera: Termitidae) and their role in determining a potencial keystone species. Biotropica 16,: 112-119.

Sanhudo, CED., Izzo, TJ. & Brandão, CRF. (2008) Parabiosis between basal fungus-growing ants (Formicidae, Attini). Insectes Sociaux, 55: 296-300. DOI:10.1007/s00040-008-1005-6

Taniguchi, K., Maruyama, M., Ichikawa, T. & Ito, F. (2005) A case of Batesian mimicry between a myrmecophilous staphylinid beetle, *Pella comes*, and its host ant, *Lasius* (*Dendrolasius*) *spathepus*: an experiment using the Japanese treefrog, *Hyla japonica* as a real predator. Insectes Sociaux, 52: 320-322. DOI: 10.1007/s00040-005-0813-1

Ward, PS. (1984) A revision of the ant genus *Rhytidoponera* (Hymenoptera: Formicidae) in New Caledonia. Australian Journal of Zoology, 32: 131-175.

# 20

# Competitive Interactions in Ant Assemblage in a Rocky Field Environment: Is Being Fast and Attacking the Best Strategy?

TA Sales, IN Hastenreiter, LF Ribeiro, JFS Lopes

*PPGCB: Comportamento e Biologia Animal, Universidade Federal de Juiz de Fora, Juiz de Fora, Minas Gerais, Brazil*

**Keywords**

behavior, competition, discovery ability, ecologically dominant

**Corresponding author**

Tatiane Archanjo de Sales
PPGCB, Comportamento e Biologia Animal
Universidade Federal de Juiz de Fora
Instituto de Ciências Biológicas
Campus Universitário s/n
Juiz de Fora, Minas Gerais, Brazil
36036-300
E-Mail: tatiane.archanjo@gmail.com

**Abstract**

The ant assemblage structure can be molded by mechanisms such as competition and discovery-dominance trade-off. In harsh circumstances it is likely that ant species that control the food resource are the first to arrive at the food source, the most aggressive (behavioral dominance), abundant (numerical dominance) and, thus, ecologically dominant. By these characteristics combination, the discovery-dominance should not be a trade-off, but a positive relationship. Here, we examined the interactions among nine ant species in a rocky field area, in the Ibitipoca State Park, Minas Gerais, Brazil. By offering attractive baits at field, we determined the discovery-dominance ability and the frequency of attack, avoidance and coexistence behaviors of each species. We showed that *Crematogaster sericea*, *Pheidole obscurithorax* and *Pheidole radoszkowskii* abundance has a positive and significant correlation with their discovery ability. They were the first to arrive at the baits (best discoverers) and were numerically dominant, being thus considered ecologically dominant. Despite *P. radoszkowskii* being part of this relationship, this interpretation should be taken cautiously. Its dominance was assured by their high discovery ability and abundance, but the behavioral strategy exhibited was avoidance, not attacking as *C. sericea* and *P. obscurithorax*. The discovery-dominance trade-off could be broken by the linked characteristics that define the ecological dominant status of the ant species studied. Also, *P. radoszkowskii* demonstrates that other strategies could surpass the combination of being fast and attacking, and thus this is not the best strategy for all. In harsh circumstances each species has its own best strategy.

## Introduction

The structure of an ant assemblage can be influenced by various factors, the main ones being mutualism, competition, parasitism and predation. Competition plays an important role in the structure of the assemblage, since competitive interactions could control the access of different species to resources (food and nesting sites), thus determining the coexistence among species within an assemblage. Evidence which supports the role of competition as a structuring factor of ant assemblages comes from observation of physical and chemical aggression among species to protect resources and territorial limits (Parr & Gibb, 2010). The high competitive potential of each ant species arises from the fact that most of them are omnivorous. This increases the competition effect and its importance in structuring ant assemblages (Benson & Harada,1988), which leads ants to exhibit various strategies to obtain food resources (Carroll & Janzen, 1973; Detrain & Deneubourg, 1997).

The competitive strategy of some species to control the food source is through numerical dominance, with the predominance of particular species in numbers, biomass and/or frequency of occurrence (Davidson, 1998). Others exhibit aggressive behaviors which forces their competitors to avoid them (behavioural dominance) (Bestelmeyer, 2000; Davidson, 1998; Fellers, 1987). However, this dominant status is not immutable. A study conducted by Markó and Kiss (2002) indicates that *Myrmica ruginodis* changes its behavior from aggressive to submissive in the presence of a stronger competitor (*Manica rubida*). Espírito-Santo et al. (2012) showed that the presence of a strong competitor (*Camponotus sericeiventris*) leads *C. rufipes* to show different degrees of aggressiveness. When conspecific from different origins meet, the behavior of *Camponotus sericeiventris* workers also changes from the simple inspection to foreigner-chasing (Yamamoto & Del-Claro, 2008).

Another mechanism widely investigated related to competition and coexistence in ant assemblages is the discovery-dominance trade-off. Various authors have cited that there is a balance between discovery ability and resource dominance (Vepsäläinen & Pisarski, 1982; Fellers, 1987; LeBrun & Feener, 2007; Pearce-Duvet et al., 2011), wherein dominat species which are slower to find food have greater capacity to defend it. On the other hand, species that are good at finding food can be classified as subordinate and their strategy for success is to find food quickly so as to exploit it partially before being dislodged by a dominant species (Vepsäläinen & Pisarski, 1982; Fellers, 1987; LeBrun & Feener, 2007). However, this trade-off can be broken by ecological dominant ant species (Davidson, 1998), invasive ant species (Holway, 1999) and the presence of parasitoids (Lebrun & Feener, 2007), in structurally complex habitats (Gibb & Parr, 2010) or at high temperatures (Bestelmeyer, 2000).

In highly diverse ant assemblages high variation of the competitive interactions is likely to exist (Andersen, 2008). The rocky field area, located at State Park of Ibitipoca, Minas Gerais, Brazil, besides having an impressive richness of ant species (Lopes et al., 2012), is an environment considered extreme and hostile, where food resources become scarce in certain periods (Fowler et al., 1991). Matching up all these characteristics is likely to find at rocky field-type habitats a more enhanced competition among ant species, wherein species that will control the food resource will be very aggressive (behavioral dominance), abundant (numerical dominance) and, wherefore, ecologically dominants (Davidson, 1998). If this is true, the discovery-dominance will be a positive relationship and not a trade-off.

Here, we examine the interactions among nine ant species in a rock field, considered a harsh environment due to extreme daily variation of temperature and humidity; patches with different levels of sun incidence daily; and swallow soils (Rodela & Tarifa, 2002). By offering baits, we aimed to verify which kind of relationship between dominance and food discovery is found for ants in such environment. Also we evaluated if the aggressiveness level is related to the dominance status of ant species.

## Material and Methods

### Study site

The study was conducted in a rocky field area, at State Park of Ibitipoca (Parque Estadual do Ibitipoca-PEIb), Minas Gerais, Brazil (S 21°42.493', W 043°53.738'). The region

has a humid mesothermal climate, with a mild summer and dry winter seasons and at 1500m average elevation (Rodela, 1998). Mean annual precipitation is 1532 mm (being high between months December and January). The average summer-maximum and winter-minimum temperatures are 36°C and -4°C, respectively, with extremes daily fluctuations of temperatures (Rodela & Tarifa, 2002). The studied rocky field is characterized by grassland vegetation consisting of grass, herbs and shrubs on outcrops of quartzitic rocks associated with shallow soils and high solar incidence (Rodela, 1998). Vegetations composition includes Velloziaceae, Compositae, Melastomataceae, Orchidaceae, Gramineae, Asclepiadaceae, Eriocaulaceae, Bromeliaceae and Cyperaceae (Rodela, 1998).

### Experimental design

We carried out the observations between June 2010 and February 2011. The experimental system consisted of six contiguous plots measuring 8 x 8 m, each one included 25 points set out 2m from each other in a grid pattern. Of these 25 points, the 16 edge points were not utilized, and the 9 internal points constituted the bait stations (Delsinne et al., 2007 adapted) (Fig. 1).

The plots order for baits offer was randomly chosen, so that each one of the six plots was sampled 10 times. At each of the nine bait stations, we placed 3g of sardine with honey (1:1; g:g) over a square PVC plate (10x10 cm). The nine baits were monitored until the appearance of the first forager ant at one of the baits, which lasted at maximum 15 minutes. Then we removed the other eight baits not visited, in order to reduce the excess of resources available, thus avoiding the distribution of potential competitors over the baits spread through the experimental system. In the first discovered bait, we started recording by filming for 40 minutes. The temperature and relative humidity were measured at the beginning of each recording session. This procedure was repeated until we obtained 60 recording sessions, totalizing 40 hours of records. The experiment was always carried out between 8:00 AM and 3:00 PM. Voucher specimens were collected for later identification in the laboratory. At the end of the observations we removed the bait with a plastic bag.

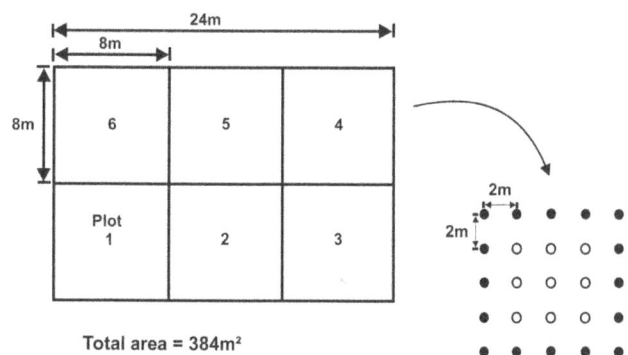

**Fig 1.** Diagram of the plots with the bait stations. Empty circles represent sites where the baits were offered.

The video allowed us to identify the species that was the first to find the food source, the number of species that visited each bait and its respective abundance. Interactions at baits were also observed and registered by frequency of occurrence into three categories, following Fellers (1987): attack, avoidance and coexistence. An attack constituted of when one ant bit, turned its gaster toward another ant or there was an outright fight. We considered it to be avoidance when ants fleed or avoided direct confrontation with an individual of another species. Lastly, coexistence was considered when two or more workers of different species fed at the same bait without any interaction. Using this classification, we registered the frequency of each behavioral category for ant species.

*Data analysis*

We restricted our analyses to species which occurred in at least 10% of the baits in order to obtain sufficient numbers of behavioral interactions to reliably assess their dominance. Using this method, we sorted the nine most common species, coincidentally the same number of species used in other studies (Fellers, 1987; Lebrun, 2005; Delsinne et al., 2007).

For data analysis, we divided the 40 minutes of observation into 5-minute intervals, in which we registered the abundance of each species, which enables the calculation of specific average abundance.

The discovery ability of each species (DA) was the number of baits at which the species was the first to arrive (NF) divided by the total number of baits in which that species was observed (NO): (DA = NF / NO). Values near 1 indicate higher discovery ability (Pearce-Duvet et al., 2011). We analyzed the relationship between the specific average abundance (log10x+1 transformed) and their respective discovery ability by fitting a linear regression model.

To verify whether there was a dependence between the frequency of occurrence of each behavior category (attack, avoidance and coexistence) and the species, we subjected the data to the independence Pearson's chi-squared ($\chi^2$) test for contingency tables, in order to access if a species exhibits more often one of the behavioral categories. When attacks occurred more than expected, the species was considered aggressive. We used the R program for all the analyses, at 5% significance in all cases (R Development Core Team, 2013).

## Results

We sampled a total of 20 ant species, which belong to 11 genera, distributed into 6 subfamilies: Ectatomminae, Ponerinae, Formicinae, Dolichoderinae, Pseudomyrmicinae and Myrmicinae. Nine of these twenty species were observed at more than 10% of the baits (Table 1). The majority of the baits were visited by more than one species, while only 7% attracted only one species (Fig. 2). *C. crassus* and *C. renggeri*

visited more baits than the other species, while *Pheidole* sp.1 exploited the smallest number (Fig. 2).

The specific average abundance calculated for each species showed a positive and significant correlation with their discovery ability (df=7, $P<0.001$, $R^2=0.91$), showing that the best discoverer species were also numerically dominant. Conversely, for the species that were not the first ones to arrive at the baits, we registered lower values of abundance (Fig. 3).

During the experimental period, the average temperature was $29.2 \pm 4.0$ °C ($39.9 - 19.5$ °C) and the relative humidity was $50.5 \pm 10.5\%$ ($75.4 - 28.6\%$). Through the results of the regression analyses it was not possible to show a significant effect of these abiotic variables on the abundance for any of the nine species.

**Table 1.** Frequency of occurrence of the ant species sampled at 60 baits in Ibitipoca State Park, Minas Gerais, Brazil. * Indicate the species used in the analyses (observed at more than 10% of the baits).

| Subfamily | Species | Baits (%) |
|---|---|---|
| Formicinae | *Camponotus crassus* Mayr, 1862* | 65.0 |
| Formicinae | *Camponotus renggeri* Emery, 1894* | 48.0 |
| Ectatominae | *Ectatomma edentatum* Roger, 1863* | 38.3 |
| Formicinae | *Camponotus genatus* Santschi, 1922* | 35.0 |
| Myrmicinae | *Pheidole obscurithorax* Naves, 1985* | 35.0 |
| Ponerinae | *Pachycondyla striata* Smith, 1858* | 23.3 |
| Myrmicinae | *Pheidole radoszkowskii* Mayr, 1884* | 21.7 |
| Myrmicinae | *Pheidole* sp1* | 18.3 |
| Myrmicinae | *Crematogaster sericea* Forel, 1912* | 16.7 |
| Dolichoderinae | *Linepithema cerradense* Wild, 2007 | 8.3 |
| Myrmicinae | *Pheidole* sp3 | 8.3 |
| Myrmicinae | *Pheidole* sp2 | 6.7 |
| Myrmicinae | *Cephalotes pavonii* (Latreille, 1809) | 5.0 |
| Myrmicinae | *Cephalotes pusillus* (Klug, 1824) | 5.0 |
| Ectatomminae | *Ectatomma* sp1 | 5.0 |
| Ponerinae | *Odontomachus* sp1 | 3.3 |
| Myrmicinae | *Solenopsis* sp1 | 3.3 |
| Formicinae | *Myrmelachista* sp1 | 1.7 |
| Myrmicinae | *Pheidole* sp4 | 1.7 |
| Pseudomyrmicinae | *Pseudomyrmex* sp1 | 1.7 |

When two or more species were recorded foraging at the same bait, approximately 53% of the behaviors exhibited were avoidance. The chi-square test indicates dependence between the behavior category and species (Pearson's chi-squared test: df = 16, $\chi^2 = 356.84$, $P < 0.001$), supplying further evidence that the behavioral strategy shown by a

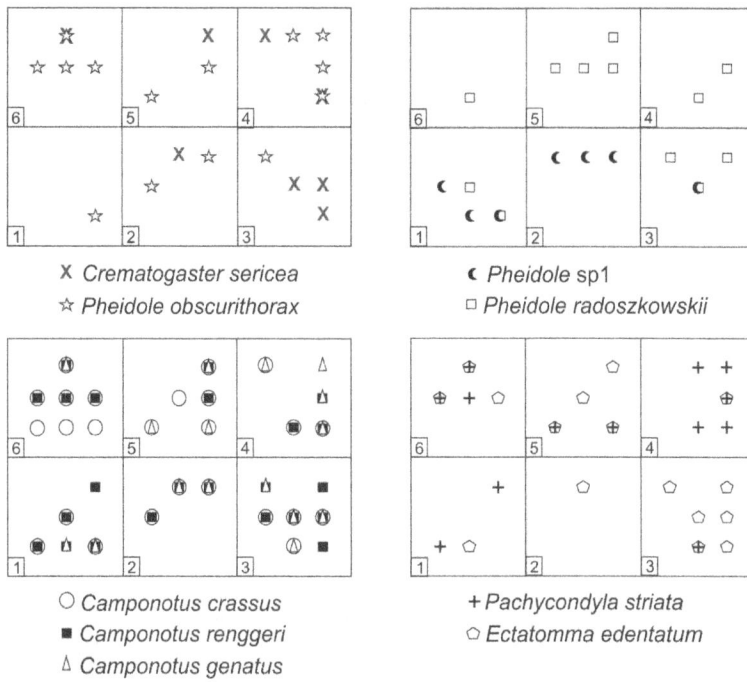

**Fig 2.** Bait occupancy by the nine species. The species are separated only for illustrative purposes.

X *Crematogaster sericea*
☆ *Pheidole obscurithorax*

( *Pheidole* sp1
□ *Pheidole radoszkowskii*

○ *Camponotus crassus*
■ *Camponotus renggeri*
△ *Camponotus genatus*

+ *Pachycondyla striata*
⬠ *Ectatomma edentatum*

species varies according to the species with which it interacts.

　*C. sericea*, *P. obscurithorax* and *C. renggeri* attacked more than expected. However, such aggressiveness only assured dominance of the bait for the first two species, not for *C. renggeri*. In the present study, the aggression exhibited by *C. renggeri* did not assure dominance at the baits, probably due to its low abundance per bait ($N_{max} = 2$). In turn, *C. sericea* stood out for the absence of avoidance behaviors (Table 2), which along with its abundance and discovery ability characterizes it as an aggressive species that can potentially exert a limiting effect of resource use by subordinate species.

　*P. radoszkowskii* also presented a high abundance, but the workers exhibited avoidance behaviors more often than expected, suggesting is the avoidance strategy that assures its ability to remain at the resource (Table 2).

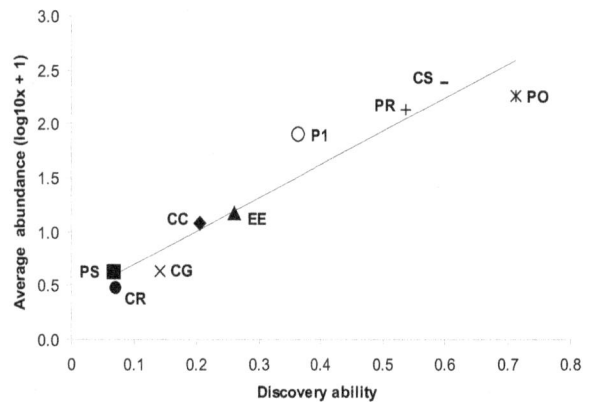

**Fig 3.** Relationship between the average abundance of the species at the baits and respective discovery ability. CC: *Camponotus crassus*, CG: *Camponotus genatus*, CR: *Camponotus renggeri*, CS: *Crematogaster sericea*, EE: *Ectatomma edentatum*, PS: *Pachycondyla striata*, PO: *Pheidole obscurithorax*, PR: *Pheidole radoszkowskii*, P1: *Pheidole* sp1.

**Table 2.** Relative frequency for the attack, avoidance and coexistence behaviors and standardized residuals in contingency tables (Pearson's $\chi^2$ test for standardized residuals in contingency tables). In boldface behaviors that occurred above the expected (>+1.96) and behaviors that occurred below the expected (<-1.96).

| Species | Relative Frequency of Behaviors (%) | | | *Standardized residuals in contingency tables (Z) | | |
|---|---|---|---|---|---|---|
| | Attack | Avoidance | Coexistence | Attack | Avoidance | Coexistence |
| *Camponotus crassus* | 27.43 | 51.43 | 21.14 | -2.09 | **3.21** | -1.79 |
| *Camponotus genatus* | 16.85 | 47.19 | 35.96 | -2.37 | 0.95 | 1.60 |
| *Camponotus renggeri* | 73.63 | 2.75 | 23.63 | **7.72** | **-6.52** | -0.61 |
| *Crematogaster sericea* | 69.64 | 0.00 | 30.36 | **3.68** | **-3.72** | 0.54 |
| *Ectatomma edentatum* | 29.00 | 31.00 | 40.00 | -0.79 | -1.07 | **2.41** |
| *Pachycondyla striata* | 3.85 | 46.15 | 50.00 | -2.17 | 0.44 | **2.06** |
| *Pheidole obscurithorax* | 41.42 | 29.75 | 28.83 | **2.33** | **-2.93** | 1.11 |
| *Pheidole radoszkowskii* | 10.13 | 73.00 | 16.88 | **-5.63** | **7.01** | **-2.61** |
| *Pheidole* sp1 | 22.06 | 50.00 | 27.94 | -1.45 | 1.12 | 0.25 |

Species with low abundances (*C. renggeri*, *C. genatus* and *P. striata*) can be considered submissive. However, this does not necessarily imply absence of aggressiveness, only a lower competitive ability, as observed for *C. renggeri* (Table 2).

Attack behavior frequencies lower than expected were registered for *P. radoszkowskii*, *C. crassus*, *C. genatus* and *P. striata*. We believe that for these *Camponotus* this result can reflect their low abundance and discovery ability. In contrast, for *P. striata* the most probable explanation is its solitary foraging strategy, characteristic of the species. In the case of *E. edentatum*, the solitary foraging strategy also explains the lower than expected occurrence of avoidance behaviors (Table 2).

## Discussion

The relationship between dominance and food discovery was a positive one and not a trade-off, at the rocky field studied. The discovery-dominance trade-off predicts that good discoverers are subordinate species, which maximize their rates of finding resources to access them before being dislodged by a behaviorally dominant species (Vepsäläinen & Pisarski, 1982; Fellers, 1987; LeBrun & Feener, 2007). Inversely, the ant assemblage here studied presents a set of ecologically dominant ant species that arrived first and controlled baits through the combination of numerical dominance and aggressive behavior, with the exception of one species that was not aggressive.

Species with the greatest discovery ability were those with the highest abundance at the baits. On the other hand, species with low discovery ability (bad discoverers) who are less abundant, visited a larger number of baits. One can therefore assume that the low discovery ability of these species could have been offset by exploitation of a broader area. Since these species have low capacity to defend resources, they probably search for food over a larger area. In contrast, for the species with high discovery ability and abundance, we can hypothesize that the baits exploited by them were within their territory and thus were more quickly located and successfully defended.

This positive relationship can also be a reflex of the typical environmental conditions of rocky fields. These areas are characterized by extreme daily fluctuations of temperature and relative humidity, plus the effects of winds and strong sunlight (Pirani et al., 1994). Besides, they are located at high altitudes (above 1,000 meters) and have soils with outcrops of quartzite rocks (Guedes & Orge, 1998). In extreme and hostile environments such as rocky fields, where food resources become scarce in certain periods (Fowler et al., 1991), it is reasonable to assume that after spending energy to find resources, an ant species must take maximum advantage of it. Also, we must consider that the relationships in rocky fields are more specialized, as shown for wasps, due to limitations in resource collection (Clemente et al, 2013). Therefore,

after locating the resource, we noted that dominant species present a particular behavioral strategy to remain at the food source. This might also have been the reason why the average abundance of the species at the baits was not influenced by abiotic factors.

*Pheidole radoszkowskii* used the avoidance as behavioral strategy to remain at the baits. When faced with aggressive ants from other species, most of the time they responded with avoidance. Actually, *P. radoszkowskii* exhibits asymmetry in the competitive relationship with other dominant species (Perfecto, 1994). Our data does not exclude their potential for aggression or coexistence. Rather, they demonstrate its behavioral plasticity towards the species with which it co-occurs.

Further, *P. radoszkowskii* forms colonies with small populations (Perfecto, 1994), which can explain the avoidance strategy observed in the present study. In species with small colonies, direct confrontation represents a greater cost than it does for species with more numerous colonies (Carroll & Jansen, 1973). Therefore, avoidance - a typical interference behavior of ants (Fellers, 1987; Yanoviak & Kaspari, 2000; Delsinne et al., 2007) - seems to be more efficient to allow their use of the bait than attacking or giving up and seeking another resource.

Similarly, *P. obscurithorax* exhibited avoidance behaviors soon after locating the resource but became more aggressive after recruitment and the arrival of soldiers. Storz and Tschinkel (2004) reported a combination of foraging tactics of this species in function of the resource size. For small resources, the scouts carried the food back to the colony alone and only recruited when there was a larger resource. In the present study, the bait offered was hard to transport, thus requiring workers and soldiers recruitment to assure its use for a longer period.

Another strategy employed by *P. obscurithorax* to exploit the food resource was the use of tools. Workers took small pebbles and pieces of leaves onto the plate and placed them in contact with the bait, then removed these materials and transported them back to the nest altogether with the food. The use of tools to carry resources that are not represented by discrete units assures obtaining approximately 10 times more food than direct transport (Fellers & Fellers, 1976). In the case of a subordinate species, this behavior is also utilized to assure later use of the food, since the parts of the resource covered become unavailable to dominant species (Fellers & Fellers, 1976).

Mass recruitment was the strategy presented by *Pheidole* sp1 and *C. sericea* to dominate the bait. Such strategy assured *Pheidole* sp.1 the use of the resource towards *C. crassus* and *C. renggeri*, species with which it co-occurred most and which consequently presented low abundance. In general, the *Pheidole* soldiers are recruited to carry the resource (Mertl et al., 2010), but in the case of *Pheidole* sp.1, we observed soldiers acting only for defense.

The attacks exhibited by *C. sericea* combined with its lack of avoidance confirms this species' high aggressiveness (Longino, 2003). A considerable proportion of the attacks only consisted of turning the gaster up, in order to release small droplets of venom, which can be related to both offensive and defensive behavior (Buren, 1959). Therefore, its high abundance is due to a combination of aggressiveness and fast recruitment (Longino, 2003).

*C. sericea*, *P. obscurithorax* and *P. radoszkowskii* were considered ecologically dominants. This terminology is proposed by Davidson (1998) who named as ecologically dominant ant species those which are behavioral (due to superior fight and/or recruitment abilities) and numerically dominant. We can suppose that the discovery-dominance trade-off was broken by the linked characteristics that define the ecological dominant status of these species. Despite *P. radoszkowskii* being part of this relationship, this interpretation should be taken cautiously. Its dominance at the baits was assured by their high discovery ability and abundance, but the behavioral strategy exhibited was avoidance, not aggressiveness. *P. radoszkowskii* demonstrates that others strategies could surpass the set of being fast and attack, and thus this is not the best strategy for all. In harsh circumstances each species has its own best strategy, which is also illustrated by seed disperser ant species at Caatinga. In this case, high-quality disperser ant species showed a strong preference for diaspores with highest elaiosome mass, transporting the seeds for longest distances, until their nests, whereas to the low-quality disperser ants, the best strategy was fed on elaiosomes in situ, and never transporting the seeds to their nests (Leal et al., 2014).

The competitive interactions recorded show a range of foraging strategies employed by different ant species composing an assemblage that guarantees exploitation of food resources for all of them. The nature of the competition and the ant behavioral strategies have interesting implications in understanding the species' richness and composition of assemblages, especially in an environment where resources are scarce and ephemeral.

## Acknowledgements

The authors are grateful to R. S. Camargo and K. Del-Claro for their constructive comments. Also we are grateful to the Fundação de Apoio e Amparo a Pesquisa do Estado de Minas Gerais (FAPEMIG grant No. 00411/08), the National Concil for Scientific and Technological Development (CNPq) for a fellowship granted to the last author (process No. 307335/2009-7) and Instituto Estadual de Florestas (IEF) from Minas Gerais in special to Clarice Nascimento Lantelme Silva for technical support at Ibitipoca Park..

## References

Andersen, A.N. (2008). Not enough niches: non-equilibrial processes promoting species coexistence in diverse ant communities. Austral Ecology, 33: 211-220. doi: 10.1111/j.1442-9993.2007.01810.x

Benson, W. & Harada, A.Y. (1988). Local diversity of tropical and temperature ant faunas (Hymenoptera: Formicidae). Acta Amazonica, 18:275-289.

Bestelmeyer, B.T. (2000). The trade-off between thermal tolerance and behavioural dominance in a subtropical South American ant community. Journal of Animal Ecology, 69: 998-1009. doi: 10.1111/j.1365-2656.2000.00455.x

Buren, W.F. (1959). A review of the species of *Crematogaster*, sensu stricto, in North America (Hymenoptera: Formicidae). Part I, Journal of the New York Entomological Society, 66: 119-134.

Carroll, C.R. & Janzen, D.H. (1973). Ecology of foraging by ants. Annual Review of Ecology and Systematics, 4: 231-257.

Clemente, M.A., Lange, D., Dátillo, W.; Del-Claro, K & Prezoto, F. (2013). Social Wasp-Flower Visiting Guild Interactions in Less Structurally Complex Habitats are More Susceptible to Local Extinction. Sociobiology, 60: 337-344. doi:10.13102/sociobiology.v60i3.337-344.

Davidson, D.W. (1998). Resource discovery versus resource domination in ants: a functional mechanism for breaking the trade-off. Ecological Entomology, 23: 484-490.

Delsinne, T., Roisin, Y. & Leponce, M. (2007). Spatial and temporal foraging overlaps in a Chacoan ground-foraging ant assemblage. Journal of Arid Environments, 71: 29-44. doi: 10.1016/j.jaridenv.2007.02.007

Detrain, C. & Deneubourg, J.L. (1997). Scavenging by *Pheidole pallidula*: a key for understanding decision making systems in ants. Animal Behavior, 53: 537–547.

Espírito-Santo, N.B., Ribeiro, S.P. & Lopes, J.F.S. (2012). Evidence of competition between two canopy ant species: Is aggressive behaviour innate or shaped up by a competitive environment? Psyche. doi: 10.1155/2012/609106

Fellers, J.H. (1987). Interference and exploitation in a guild of woodland ants. Ecology, 68: 1466-1478

Fellers, J.H. & Fellers, G.M. (1976). Tool use in a social insect and its implications for competitive interactions. Science, 192: 70-72.

Fowler, H.G., Forti, L.C., Brandão, C.R.F., Delabie, J.H.C. & Vasconcelos, H.L. (1991). Ecologia nutricional de formigas. In: A.R. Panizzi & J.R.P. Parra (Eds.), Ecologia nutricional de insetos e suas implicações no manejo de pragas (pp 131-223). Manole, São Paulo.

Guedes, M.L.S. & Orge, M.D.R. (1998). Checklist das

espécies vasculares de Morro do Pai Inácio (Palmeiras) e Serra da chapadinha (Lençóis). Chapada Diamantina, Bahia. Brasil. Projeto Diversidade florística e distribuição das plantas da Chapada Diamantina, Bahia. Salvador, BA: Instituto de Biologia da Universidade Federal da Bahia.

Holway, D.A. (1999). Competitive mechanism underlying the displacement of native ants by the invasive Argentine Ant. Ecology, 80: 238-251.

Leal, L.C.; Neto, M.C. L.; Oliveira, A.F.M.; Andersen, A.N.; Leal, I.R. (2014). Myrmecochores can target high-quality disperser ants: variation in elaiosome traits and ant preferences for myrmecochorous Euphorbiaceae in Brazilian Caatinga. Oecologia, 174: 493-500. doi: 10.1007/s00442-013-2789-2.

Lebrun, E.G. (2005). Who is the top dog in ant communities? Resources, parasitoids, and multiple competitive hierarchies. Oecologia, 142: 643-652. doi: 10.1007/s00442-004-1763-4.

Lebrun, E.G. & Feener, D.H. (2007). When trade-offs interact: Balance of terror enforces dominance discovery trade-off in a local ant assemblage. Journal of Animal Ecology, 76: 58-64. doi: 10.1111/j.1365-2656.2006.01173.x.

Longino, J.T. (2003). The Crematogaster (Hymenoptera, Formicidae, Myrmicinae) of Costa Rica. Zootaxa, 151: 1-150.

Lopes, J.F.S., Hallack, N.M.R., Sales, T.A., Brugger, M.S., Ribeiro, L.F., Hastenreiter, I.N. & Camargo, R.S. (2012). Comparison of the ant assemblages in three phytophysionomies: Rocky Field, Secondary Forest and Riparian Forest—A case study in the State Park of Ibitipoca, Brazil. Psyche, 1-7. doi: 10.1155/2012/928371.

Markó, B. & Kiss, K. (2002). Searching for food in the ant Myrmica rubra (L.) (Hymenoptera: Formicidae) – How to optimize? In: Tomescu N, Popa V (szerk.) In: Memoriam "Professor Dr. Doc. Vasile Gh. Radu" Corresponding member of Romanian Academy of Sciences, Cluj University Press, Kolozsvár (Románia).

Mertl, A.L., Sorenson, M.D. & Traniello, J.F.A. (2010). Community-level interactions and functional ecology of major workers in the hyperdiverse ground-foraging Pheidole (Hymenoptera, Formicidae) of Amazonian Ecuador. Insectes Sociaux, 57: 441–452. doi:10.1007/s00040-010-0102-5.

Parr CL, Gibb H (2010) Competition and the Role of Dominant

Ants. In: L. Lach, C.L. Parr & K.L Abbott. Ant Ecology (pp 77 – 96). Oxford University Press Inc, New York.

Pearce-Duvet, J.M.C., Moyano, M., Adler, F.R. & Feener Jr, D.H. (2011). Fast food in ant communities: How competing species find resources. Oecologia, 167: 229-240. doi: 10.1007/s00442-011-1982-4.

Perfecto, I. (1994). Foraging behavior as a determinant of asymmetric competitive interaction between two ant species in a tropical agroecosystem. Oecologia, 98: 184-192.

Pirani, J.R., Giulietti, A.M., Mello-Silva, R.E. & Meguro, M. (1994). Checklist and patterns of geographic distribution of vegetation of Serra do Ambrósio, Minas Gerais, Brazil. Revista Brasileira Botânica, 17: 133-147.

R Development Core Team. (2013). R: A language and environment for statistical computing. R Foundation for Statistical Computing, Vienna, Austria. ISBN 3-900051-07-0, URL http://www.R-project.org/.

Rodela, L.G. (1998). Cerrados de altitude e campos rupestres do Parque Estadual do Ibitipoca, sudeste de Minas Gerais: distribuição e florística por subfisionomias da vegetação. Revista do Departamento de Geografia da Universidade Federal de Juiz de Fora, 12: 163-189.

Rodela, L.G. & Tarifa, J.R. (2002). O clima da serra do Ibitipoca, sudeste de Minas Gerais. Espaço e Tempo, 11:101-113.

Storz, S.R. & Tschinkel, W.R. (2004). Distribution, spread, and ecological associations of the introduced ant Pheidole obscurithorax in the southeastern United States. Journal of Insect Science, 4: 1-11.

Vepsäläinen, K. & Pisarski, B. (1982). Assembly of island ant communities. Annales Zoologici Fennici., 19: 327–335.

Yamamoto, M. & Del-Claro, K. (2008). Natural history and foraging behavior of the carpenter ant Camponotus sericeiventris Guérin, 1838 (Formicinae, Campotonini) in the Brazilian tropical savanna. Acta Ethologica, 11: 55-65. doi: 10.1007/s10211-008-0041-6.

Yanoviak, S.P. & Kaspari, M. (2000). Community structure and the habitat template: Ants in the tropical forest canopy and litter. Oikos, 89: 259-266.

# 21

# Tree species used for nesting by stingless bees (Hymenoptera: Apidae: Meliponini) in the Atlantic Rain Forest (Brazil): Availability or Selectivity

MD Silva[1,2], M Ramalho[1]

1 - Universidade Federal da Bahia, Salvador, BA, Brazil

2 - Instituto Federal de Educação, Ciência e Tecnologia Baiano, Governador Mangabeira, BA, Brazil

**Keywords**

stingless bees, preexisting cavities, arboreal substrates.

**Corresponding author**

Marilia Dantas Silva
Rua Waldemar Mascarenhas, s/nº
Portão (Estrada Velha da Chesf)
Governador Mangabeira, Bahia, Brazil
44.350-000
E-Mail: ailirambio@hotmail.com

**Abstract**

The stingless bees (Meliponini) are numerically dominant in tropical forests and most species depend on preexisting cavities for nesting, mainly tree hollows. However, it is still incipient the knowledge about basic characteristics of forest trees used for nesting. The basic questions addressed in this study include: would appropriate hollows be restricted to a few tree species? Would there be selectivity in the use of tree hollows in the forest? These issues are addressed from the comparison of usage patterns among forest trees in different stages of forest regeneration in the Atlantic Forest (Michelin Reserve in northeastern Brazil). Among 89 nests (from six species) found in tree hollows, in a sampled area of 32 ha of forests, 78.7% were associated with live plants and 21.3% to dead trees. This result does not support the hypothesis of selectivity for living trees, considering the high rate of living trees: dead trees (40:1). Nests were sampled from 41 tree species of 31 genera and 22 plant families. Meliponini species showed no differential association with any tree species. The absence of selectivity of tree species as nesting site is probably due to the high diversity of trees per hectare of Atlantic rainforest. The stingless bees also showed no selectivity for wood hardness, therefore the potential durability of tree hollows probably exerts weak selective pressure on bees, or at least the hardness variation range of trees used for nesting has no important influence on reproductive success of the colonies of stingless bees.

## Introduction

Probably, the abundance of stingless bees (Meliponini) is constrained by floral resources (Hubbell & Johnson, 1977; Eltz et al., 2002) and/or by nesting sites (Oliveira et al., 1995; Samejima et al., 2004; Teixeira & Viana, 2005). In these eusocial bees, the choice of nesting site has profound influence on the longevity and reproductive success of colonies (Hubbell & Johnson, 1977) due to the limited flight range (Araújo et al., 2004) and progressive swarm (Nogueira-Neto, 1997).

The stingless bees use several substrates for nesting (Nogueira-Neto, 1997); however, the majority depends on preexisting cavities such as tree hollows, and few species build exposed nests (Roubik, 1989; Michener, 2000; Antonini & Martins, 2003; Batista et al., 2003; Martins et al., 2004; Silva et al., 2013). The knowledge about tree species used by stingless bees for nesting is still incipient (Moreno & Cardoso, 1997; Aguilar-Monge, 1999; Martins et al., 2004; Kleinert, 2006) and it is not known if these bees have shown some preference for nesting trees. According to Nogueira-Neto et al. (1986), some species of stingless bees may have specificity in the use of trees for nesting, which is corroborated by data sampled under low diversity of nesting trees (Batista et al., 2003; Martins et al., 2004; Teixeira & Viana, 2005; Serra et al., 2009). However, the very high diversity and very low population densities of trees in tropical forests should favor no selectivity of nesting sites: for instance in the Atlantic rainforest, the tree diversity can reach over 400 species of trees per hectare (see Guedes et al., 2005).

As an example, stingless bees would not present specificities when choosing trees for nesting in tropical dipterocarp forests and they would be opportunistic, although the tree hollows could be selected by their characteristics, such as size (Eltz et al., 2002). In the tropical forests of Central America, Hubbell and Johnson (1977) observed low selectivity in the using of tree hollows and argued that the availability of nesting substrates would be higher than the density of stingless bee nests in the forest habitats. The hypothesis that tree hollow availability exceed the demand for nesting by stingless bees is supported by the low density of nests (2.8 nests/ha) compared to the high density of trees potentially suitable for nesting in the Atlantic rain Forest (Silva et al., 2013).

Estimates of availability and use of nesting substrates are necessary for understanding the role of choice or preferences in structuring communities of stingless bees in forests (Kleinert, 2006; Silva et al., 2013). The basic questions addressed in this study include: would availability of suitable hollows for nesting be restricted to a few tree species? Would there be selectivity in the use of tree hollows in the forest? This study compares the tree species used by the stingless bee community at different stages of forest regeneration in the Atlantic rainforest. It is assumed that the availability of tree hollows with suitable sizes varies between stages of forest regeneration, affecting choice opportunities, according to the species-specific needs of stingless bees. If the availability and selectivity of cavities are important, the communities of stingless bees must change between the stages of forest regeneration.

## Material and Methods

The present study was conducted in the Michelin Ecological Reserve - MER (13°50'S, 39°15'W) in the Brazilian Atlantic rainforest in northeastern Brazil. The MER encompasses 3,096 ha of tropical rainforest at altitudes from 160 to 327 masl. The native forest forms a mosaic with rubber (*Hevea brasiliensis* Muell. Arg., Euphorbiaceae) plantations. The MER forest areas experienced severe anthropogenic impacts prior to 2004, generating a mosaic of forest fragments at different stages of regeneration. At the present time, the preserved nuclear areas of the largest fragments have attained a mature old growth stage of regeneration, with canopy heights of more than 20 m and many trees with circumferences >190 cm at breast height, as well as numerous old growth trees >300 cm in circumference reaching more than 30 m height. There are also extensive patches of forest at early stages of regeneration (with lower canopies and no old growth trees and with shrub and herbaceous plant cover in the understory), mainly at the edges of the largest fragments (Flesher, 2006).

Two categories of forested habitats were discriminated to verify and compare the nest density and Meliponini species richness: mature old growth forest and early stages of regeneration (IS = initial state and AS = advanced state). Four replicates of the two forest categories were sampled in each of the four largest MER forest fragments. A total of 64 25x25 m plots was established and sampled in each of the four replicates (total area of 4 ha/replicate), for an overall total of 16 ha for each of the two forest categories (Silva et al., 2013). All the trees

within each plot were visually inspected in search of nests, with special attention being paid to large trees with circumferences at breast height (CBH) >190 cm – in which stingless bee nests tend to be concentrated in forest habitats (Eltz et al., 2002; Batista et al., 2003). The botanical material was collected for preparation of exsiccates and identification of tree species.

To determine the wood density (hardness) of nesting trees (with Meliponini nests), samples of 2.0 x 2.0 cm were collected from the tree trunk at breast height (CBH = 1.30 m): one block near to the core and the other close to the bark. The mercury porosimetry technique was used to estimate the wood dry mass, according to the method of Vital (1984).

The vegetation structure was measured using the T-square method (Sutherland, 2006). Twenty random points were drawn/replica and the distance from the point to the nearest individual (x) and its distance to the nearest neighbor on orthogonal were used to estimate the density of living trees (all trees, with or without nests) in the following perimeter categories (circumference at breast height-CBH = 1.30 cm): (1) 21-50 cm; (2) 51-80 cm; (3) 81-110 cm and (4) above 110 cm. The T-square technique was not used to estimate the density of dead trees because it is not suitable for events with very low frequency (Sutherland, 2006). In this case, ten plots of 25x25 m (randomly chosen) were established for density measurement in each of the four replicates/forest stage and all the dead trees were counted (according to the perimeter categories used in the samplings of living trees). The density of living trees was calculated using the program Ecological Methodology, 2nd Ed. (Kenney & Krebs, 2000), while the density of dead trees was obtained by dividing the number of trees by the area sampled in each habitat category. In addition to the density and size of nesting trees, both stages were compared in relation to the hardness of the wood. The "Permutational Multivariate Analysis of Variance" test - PERMANOVA (Anderson, 2005) was used for data analysis because the assumptions of homoscedasticity (Levene test) and normality (Kolmogorov-Smirnov test) were not satisfied. The tests were run on Graphpad Instat and SPSS (SPSS 13.0 for Windows, SPSS Inc., Chicago, IL, USA) software.

Three dependent variables (total abundance, richness and wood hardness of nesting trees) were tested in relation to a single factor: stage of forest regeneration in two levels (IS = initial state and AS = advanced state). The Bray-Curtis measure was used with untransformed data for comparing both stages of forest regeneration at a significance level of 0.05 (Anderson 2005).

The nonparametric correlation test (Spearman) was applied in two situations: 1) to estimate the relationship between the number of trees with nests and the number of largest trees (CBH> 80cm) per unit area of each habitat category (IS and AS); 2) to assess the relationship between the size of trees-CBH (or tree hardness) and richness and abundance of Meliponini nests. The tests were run on Graphpad Instat 3.05 software (GraphPad-Software, 1998) at a significance level of 0.05.

Selectivity was used here as synonymous for preference that would be detected when a used category of nesting trees (size or hardness) is higher than its availability in local forest habitat.

The similarity of the nesting trees between species of stingless bees was estimated by cluster analysis (Bray-Curtis coefficient) and data were run on PAST program - (PAlaeontological STatistics, see. 1.81).

## Results and Discussion

Of the total of 118 nests of stingless bees found in MER, 75.4% were in trees, 9.4% amid the rocky substrate, 7.6% inside termite nests and 7.6% in soil and slopes. Among 89 nests found in tree hollows, 78.7% were found in living trees and 21.3% in dead trees. Nests of five Meliponini species were observed in hollows of living trees, of which only *Scaptotrigona xanthotricha* Moure was not found in dead trees (Table 1). Batista et al. (2003) found 5.4% of the nests of stingless bees in dead trees in a disturbed area of Atlantic rainforest.

In a recent review, Cortopassi-Laurino et al. (2009) found that living trees predominate largely over dead trees as nesting substrates to stingless bees. Roubik (1989) has also argued that the stingless bees should occupy most durable substrates, which provide good physical protection in the forest, therefore living trees. Although these data and arguments have internal coherence, they often lack a measure of availability of cavities in dead and living trees for testing preference or selectivity. In the Atlantic rainforest of MER, the density of living trees is about 40 times higher than the density of dead trees and, therefore, these data do not support the hypothesis that stingless bees would avoid dead trees (Table 2). Alternatively, we should consider that many nests found in dead trees were established while the trees were still alive.

The living trees with nests corresponded to 59 individuals and 42 species (Table 3; Fig 1). Only one nest was found in most tree individuals and species and a maximum of 3-4 nests were associated to six tree species. This overall framework suggests that availability of cavities is common

**Table 2**. Density of living and dead trees (trees/ha) in the initial (IS) and advanced (AS) stages of forest regeneration distributed by circumference at breast height (CBH): CBH 1: 21-50 cm; CBH 2: 51-80 cm; CBH 3: 81-110 cm and CBH 4: above 110 cm).

| CBH | Living trees | | Dead trees | |
| --- | --- | --- | --- | --- |
| | AS | IS | AS | IS |
| 1 | 536.5 | 599.7 | 8.50 | 5.00 |
| 2 | 257.1 | 165.1 | 7.25 | 9.25 |
| 3 | 92.9 | 61.2 | 8.25 | 9.75 |
| 4 | 96.3 | 41.1 | 4.25 | 2.50 |

in many tree species and trees selectivity by stingless bees is very low or nonexistent in this forest habitat.

Among the 194 tree species recorded in MER (Rocha-Santos & Talora, 2012) 41 (21%) showed Meliponini nests, therefore apparently the appropriate tree hollows for these bees are well spread in the flora. If we consider the relatively high values of alpha and beta tree diversity in the Atlantic rainforest (e.g, Guedes et al., 2005), the data on MER refute the argument of Hubbell and Johnson (1977) that most of the nests of stingless bees would be found in relatively few species of trees out of the total available in any vegetation.

In a disturbed area of Atlantic rainforest, with depressed richness of trees, Batista et al. (2003) found only 18 tree species with nests of stingless bees; while Eltz et al. (2003) recorded nests in 38 species of trees in lowland dipterocarp forests, Malaysia. In areas of Caatinga, Castro (2001), Martins et al. (2004), Teixeira and Viana (2005) and Souza et al. (2008) found most nests of stingless bees in only five tree species (*Caesalpinia pyramidalis* Tul.; *Commiphora leptophloeos* Mart. J.B. Gillett.; *Schinopsis brasiliensis* Engl.; *Copaifera coriacea* Mart. and *Amburana cearensis* Schwacke & Taub.). In savannah-Cerrado areas, Kerr (1971), Aquino et al. (2007), Antonini and Martins (2003) and Serra et al. (2009) found a predominance of nests in the species *Caryocar brasiliense* Cambess., *Qualea parviflora* Mart. and *Salvertia convallariaeodora* A. St.-Hil.. In mixed forest of Araucaria, Witter and colleagues (2010) reported certain

**Table 1.** Characteristics of stingless bees' nests found in tree hollows in Atlantic Rainforest (Michelin Ecological Reserve): Advanced Stage (or old growth mature; AS) and Initial Stage (IS) of forest regeneration; Circumference at breath height (CBH)= 1.30m.

| | Number of nests AS   IS | Average height | Height Range | CBH Mean | CBH Range | Nesting Trees |
| --- | --- | --- | --- | --- | --- | --- |
| *Melipona scutellaris* Latreille | 6    5 | 8.4m (±4.2m) | 2.5-13m | 125cm (±39.7cm) | 59-197cm | Living and Dead |
| *Plebeia droryana* (Friese) | -    2 | 85.5cm (±62.9cm) | 41-130cm | 55.5cm (±14.8cm) | 45-66cm | Living and Dead |
| *Scaptotrigona bipunctata* (Lepeletier) | 9    5 | 5.1m (±5.1m) | 171cm-15m | 120.1cm (±34.1cm) | 70-180cm | Living and Dead |
| *Scaptotrigona xanthotricha* (Lepeletier) | 13    8 | 3.2m (±4.2m) | 25cm-13m | 127.2cm (±34.8cm) | 76-180cm | Living |
| *Tetragonisca angustula* (Latreille) | 14   20 | 3.1m (±4.7m) | 1cm-11m | 119.3cm (±85.3cm) | 41-232cm | Living and Dead |
| *Trigona fuscipennis* Friese | -    1 | 4.5m | - | 119.3cm | - | Living |

**Table 3**. Tree species most used as nesting substrate by stingless bees in initial and advanced stages of forest regeneration. Ecological Group: T = tolerant to shade; I = Intolerant to shade; NC = Not Classified (Lorenzi, 2002a; 2002b; 2009; Rocha-Santos & Talora, 2012). *S.b* (*Scaptotrigona bipunctata*); *S.x* (*Scaptotrigona xanthotricha*); *T.a* (*Tetragonisca angustula*); *M.s* (*Melipona scutellaris*) and *T.f* (*Trigona fuscipennis*). (*) = sample not identified.

| Family | Tree species | Ecological group and hardness | Number of nests | | Meliponini species |
|---|---|---|---|---|---|
| | | | IS | AS | |
| Anacardiaceae | *Thyrsodium spruceanum* Salzm. Ex Benth. | T (0.41 g/cm) | 2 | 1 | *T.a; S.b* |
| Apocynaceae | *Symphonia globulifera* L.f. | T (0.14 g/cm) | 1 | 0 | *T.a* |
| Araliaceae | *Dendropanax bahiensis* Fiaschi. | NC (0.35 g/cm) | 0 | 1 | *S.x* |
| Burseraceae | *Protium icicariba* (DC.) Marchand. | T (0.53 g/cm) | 0 | 3 | *S.x; M.s* |
| Burseraceae | *Protium* sp. | NC(0.64 g/cm) | 2 | 0 | *M.s; T.a* |
| Chrysobalanaceae | *Licania hypoleuca* Benth. | T (*) | 1 | 0 | *T.f* |
| Clusiaceae | *Aspidosperma c.f. spruceanum* Benth. ex Müll.Arg. | T (0.38 g/cm) | 1 | 0 | *T.a* |
| Clusiaceae | *Clusiacea* sp.2 | NC (0.69 g/cm) | 1 | 0 | *S.b* |
| Clusiaceae | *Clusiacea* sp.1 | NC(0.60 g/cm) | 0 | 1 | *S.x* |
| Clusiaceae | *Garcinia macrophylla* Mart. | T (0.17 g/cm) | 1 | 0 | *S.x* |
| Clusiaceae | *Tabernaemontana flavicans* Willd. Ex Roem. & Schult. | T (*) | 0 | 1 | *M.s* |
| Cunoniaceae | *Lamanonia ternata* Vell. | T (0.66 g/cm) | 1 | 0 | *S.x* |
| Elaeocarpaceae | *Sloanea guianensis* (Aubl.) Benth. | T (0.83 g/cm) | 2 | 1 | *P.d* |
| Elaeocarpaceae | *Sloanea obtusifolia* (Moric.) K. Schum. | I (0.15 g/cm) | 0 | 1 | *T.a* |
| Euphorbiaceae | *Pera glabrata* (Schott) Poepp. ex Baill. | I (0.23 g/cm) | 0 | 1 | *T.a* |
| Fabaceae | *Albizia pedicellaris* (DC.) L.Rico. | I (0.091 g/cm) | 0 | 1 | *S.b* |
| Fabaceae | *Arapatiella psilophylla* (Harms) R.S.Cowan. | T (0.13 g/cm) | 1 | 0 | *S.x* |
| Fabaceae | *Inga capitata* Desv. | T (0.63 g/cm) | 1 | 0 | *T.a* |
| Fabaceae | *Inga edulis* Mart. | T (0.70 g/cm) | 1 | 0 | *T.a* |
| Fabaceae | *Inga thibaudiana* DC. | I (0.86 g/cm) | 1 | 0 | *M.s* |
| Fabaceae | *Inga* sp.1 | NC(0.78 g/cm) | 1 | 0 | *S.b* |
| Fabaceae | *Peltogyne confertiflora* (Mart. Ex Hayne) Benth. | T (0.74 g/cm) | 0 | 1 | *M.s* |
| Fabaceae | *Stryphnodendron pulcherrimum* Willd. Hochr. | I (0.69 g/cm) | 4 | 0 | *S.x* |
| Fabaceae | *Swartzia polita* (R.S.Cowan) Torke | T (0.79 g/cm) | 0 | 1 | *S.b* |
| Fabaceae | *Swartzia* sp.1 | NC(0.89 g/cm) | 0 | 1 | *S.x* |
| Hypericaceae | *Vismia guianensis* (Aubl.) Choisy | I (0.90 g/cm) | 1 | 0 | *T.a* |
| Hypericaceae | *Vismia guianensis* (Aubl.) Choisy | I (0.90 g/cm) | 1 | 0 | *T.a* |
| Icacinaceae | *Emmotum nitens* Miers. | T (0.83 g/cm) | 0 | 1 | *T.a* |
| Lauraceae | *Ocotea* cf. *canaliculata* (Rich.) Mez | I (0.64 g/cm) | 1 | 0 | *S.b* |
| Lauraceae | *Ocotea longifolia* Kunth. | T (0.43 g/cm) | 1 | 0 | *M.s* |
| Melastomataceae | *Henriettea succosa* (Aubl.) DC. | T (0.75 g/cm) | 1 | 1 | *T.a* |
| Melastomataceae | *Tibouchina* sp. | NC(*) | 1 | 0 | *S.x* |
| Meliaceae | *Trichilia lepidota* Mart. | T (0.31 g/cm) | 0 | 3 | *T.a; S.b* |
| Moraceae | *Artocarpus heterophyllus* Lamk. | T (0.92 g/cm) | 1 | 0 | *T.a* |
| Myrsinaceae | *Myrsine* sp.1 | T (0.62 g/cm) | 0 | 1 | *T.a* |
| Myristicaceae | *Virola gardneri* (A.DC.) Warb | T (*) | 0 | 1 | *S.x* |
| Peraceae | *Pogonophora schomburgkiana* Miers ex Benth. | I (*) | 1 | 0 | *T.a* |
| Phyllanthaceae | *Amanoa guianensis* Aubl. | I (0.79 g/cm) | 0 | 1 | *T.a* |
| Rubiaceae | *Psycotria carthagenensis* Jacq. | T (0.86 g/cm) | 0 | 2 | *T.a; S.x* |
| Salicaceae | *Casearia* sp.1 | NC (0.84 g/cm) | 0 | 1 | *S.x* |
| Sapotaceae | *Pouteria venosa* (Mart.) Baehni. | T (0.93 g/cm) | 0 | 1 | *S.x* |
| Urticaceae | *Pourouma velutina* Mart. Ex Miq. | I (0.16 g/cm) | 2 | 0 | *T.a; S.x* |

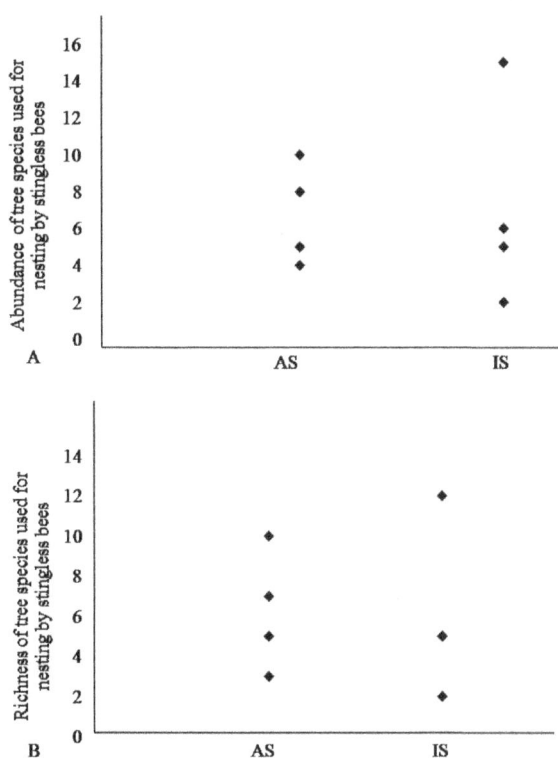

**Fig 1.** Abundance (A) and richness (B) of tree species used for nesting by stingless bees in each replica forest habitat in advanced stage (AS) and initial stage (IS) of regeneration in the Michelin Ecological Reserve (MER).

nesting specificity by *Melipona bicolor schencki* Gribodo in a few species of Lauraceae (59.52%). In urban area in the center of Curitiba city (Paraná State, Brazil), Taura and Laroca (1991) also found a greater number of nests of stingless bees in only two tree species, *Jacaranda mimosaefolia* D. Don. and *Platanus* sp.

A common denominator to these studies with stingless bees in non-forested habitats is the dominance of some tree species and/ or the availability of tree hollows suitable for nesting in a small proportion of arboreal flora. By analogy, the plant family with highest frequency of stingless bee nests in MER was Fabaceae (Table 3) that was also highlighted in the review of nesting trees by Cortopassi-Laurino et al. (2009). In MER, Fabaceae surpasses the others in richness and abundance in all stages of forest regeneration (Rocha-Santos & Talora, 2012), and this predisposes its use as nesting trees by stingless bees: this usage reflects availability and not preference or selectivity. Often, the authors who have studied stingless bee communities in forest habitats just point out the use of "suitable tree hollows for nesting" (mainly with apparent suitable size), regardless of tree species (Hubbell & Johnson, 1977; Oliveira et al., 1995; Eltz et al., 2003). When the availability of tree hollows is widely distributed in the tree flora, as in the MER rainforest, the apparent tree selectivity by Meliponini disappears.

Trees used as nesting sites by Meliponini have wood density (or hardness) ranging from 0.13 to 0.93 g/cm³ and about 70% of these trees have wood hardness above 0.6 g/cm³. The hardness had no significant relationship with nest abundance (p=0.4210) and richness (p=0.2779) of stingless bees (Fig 2). There is no significant variation in the distribution of total nests related to trees' hardness,

however, *Plebeia droryana* (Friese) and *Melipona scutellaris* Latreille often used trees with higher densities (0.86 g/cm³ and 0.75 g/cm³, respectively), while *Tetragonisca angustula* (Latreille) often occupies cavities in trees with low hardness (below 0.55 g/cm³). In these cases, we can rule out the effect of trees size used for nesting, as a confounding variable, mainly in the case of *T. angustula*, a generalist species able to use small hollows (Silva et al., 2013, 2014).

It was detected a difference in hardness of nesting trees between the two stages of forest regeneration (p=0.0002), in a first analysis of the four replicates. However, one of the replicas was detected as an outlier, with very low values of hardness in 'IS' (0.13g/cm³) in comparison with the others replicas of 'IS' stage of forest regeneration (around or above 0.5 g/cm³) and, a posteriori, it was excluded from the analysis. With this procedure, the differences in wood hardness disappeared between the two stages of

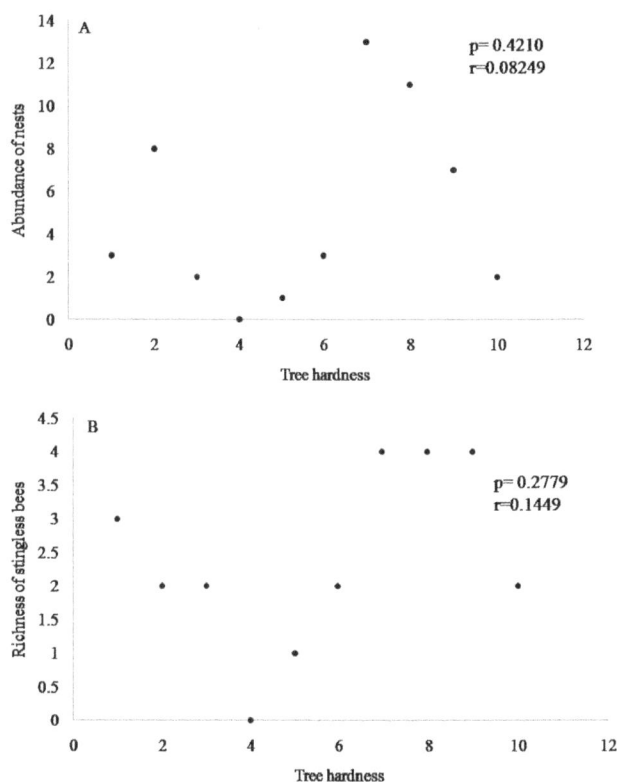

**Fig 2.** Correlation between the abundance of nests (A), species richness of stingless bees (B) regarding the hardness of trees that were found with nest of stingless bees.

forest (p >0.05). This result support the previous argument of similar hollow availability in trees regardless of the wood hardness or ecological group (e.g., tolerant or intolerant to shade; Table 3). Moreover, one nest per tree/family with varying wood hardness was the predominant pattern in both forest stages of regeneration (Table 3), indicating that the hardness *per se* would not be relevant in the process of hollow formation and availability for Meliponini.

In summary, probably the hardness and durability of hollows in the trees have not exerted significant selective pressure on stingless bees in the rainforest, or at least the hardness variation range has had no influence on colonies reproductive success and longevity.

The average CBH of nesting trees in MER was 132.3cm (±66.1cm). The CBH had negative significant relationship (p=0.0224, r=-0.4953) with nest abundance and no significant relationship with stingless bee richness (p=0.1072) (Fig 3). The lower occurrence of nests in larger trees should be a sampling artifact, however, reflecting the pattern of availability of CBHs categories, i.e. the great reduction in abundance of larger trees (Table 2). In fact, the variation in the abundance of nests of stingless bees is not significant (p=0.944) between the largest categories 3 and 4 of CBH.

Samejima et al. (2004) indicated that the low density of stingless bee nests in disturbed areas of forest would be related primarily to the absence of largest trees for nesting (CBH above 150cm or diameter above 24cm). Hubbell and Johnson (1977) argued that the colonization of secondary forests by stingless bees would depend on the tree size, and the initial stage of forest regeneration should be colonized primarily by small bees and later by larger ones. The small nesting trees should be less accessible to Meliponini species with large colonial biomass (Hubbell & Johnson, 1977; Roubik, 1983; Samejima et al., 2004). Sampling with trap nests also supports the argument that different species might respond to different thresholds of minimum size of cavities in the forest (Silva et al., 2014). A positive relationship between stingless bees body size and minimum diameter of nesting trees was also detected (Kleinert 2006), probably because the minimum biomass of colonies would also be lower in smaller species of stingless bees.

However, in MER rainforest, the stingless bees with very different body sizes that would fit the profile of "large colonial biomass" (*S. xanthotricha*, *S. bipunctata* and *M. scutellaris*) show no significant variation (p=0.2666, r=-02540) in the occupancy of trees with different CBHs (Fig 3C). Likewise, the use of tree hollow sizes does not group the stingless bee taxa or sizes; on the contrary, affinities are random (Fig 4). For instance, stingless bee species with more sampled nests also used a higher number of tree hollows and tend to have greater nesting overlap (e.g. *T. angustula* and *S. xanthotricha*). These results support the argument of weak selectivity for different hollow sizes among different species of stingless bees.

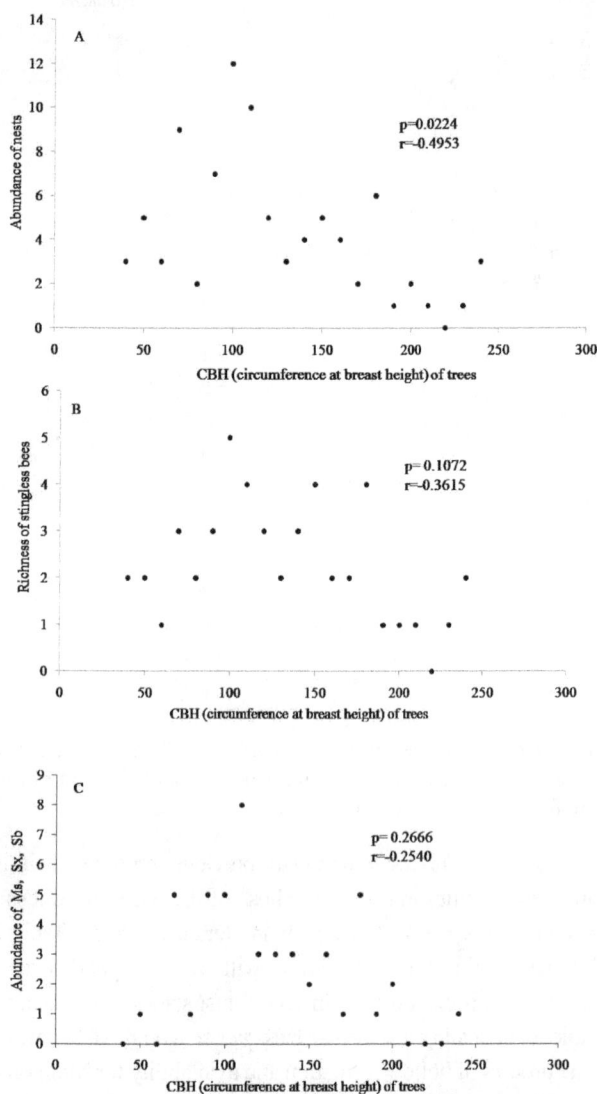

**Fig 3.** Correlation between the abundance of nests (A), species richness of stingless bees (B) and abundance of nests of *Melipona scutellaris* (Ms), *Scaptotrigona xanthotricha* (Sx) and *S. bipunctata* (Sb) (C) regarding the CBH (circumference at breast height) of trees that were found with nest of stingless bees.

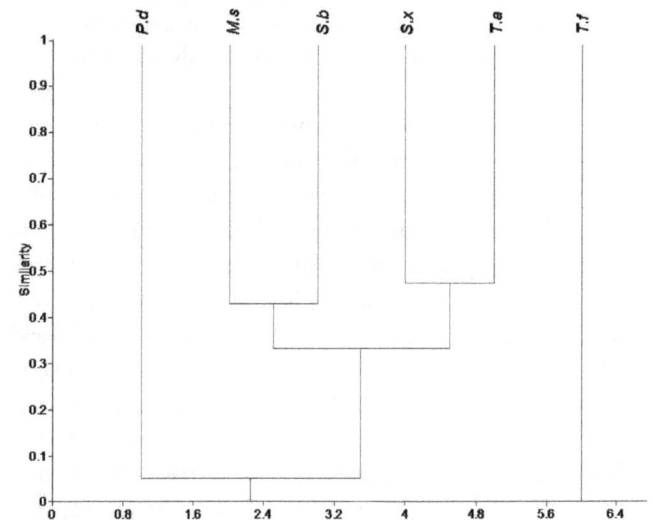

**Fig 4.** Similarity among Meliponini species in use of plant families. Bray-Curtis coefficient. Caption: *Plebeia droryana* (P.d); *Trigona fuscipennis* (T.f); *Scaptotrigona bipunctata* (S.b); *Melipona scutellaris* (M.s); *Scaptotrigona xanthotricha* (S.x); *Tetragonisca angustula* (T.a).

The lack of variation in stingless bee richness with CBH trees (Fig 3) and the lack of differences (species and abundance) between stages of forest regeneration support the argument that trees above a CBH's threshold (or hollow size threshold) are used and shared by most stingless bee species in local communities. For example, trees above 80cm of CBH (Table 2) or trap nests above 3L (Silva et al., 2014) were used by stingless bees with wide range of body sizes and colonial biomasses.

If there is any influence of the diameter of the trees on the stingless bees' community in MER, it is unlikely that this factor is sufficient to explain the absence of spatial dynamics in the community between stages of forest regeneration. Even if the longevity of trees and tendency to develop hollows were influenced by the type of wood (e.g. hardness, chemicals, fiber, etc.) or by characteristics of tree growth (tolerant or intolerant to shade; Table 3), there are not evidences that these attributes have effective expression on nesting site availability and usage by stingless bees of Atlantic rainforest in MER, contrary to suppositions made by several authors.

Of the total number of nesting trees in MER, 51.6% were found in the early forest stage of regeneration and 49.4% in the old growth stage of forest regeneration. The differences were not significant between the two stages regarding the abundance of living trees with nests (p=0.9170) and richness of nesting trees (p=0.8326). According to Silva and colleagues (2013) the abundance of nests and the species richness of stingless bees were also similar in both stage of forest regeneration. Probably, there is no variation in the availability of suitable tree hollows per unit area between the two different stages of forest regeneration.

There are contrasting approaches and mainly conflicting interpretations about the relative importance of pioneer or slow growth trees as nesting substrate for Meliponini in tropical forests (Hubbell & Johnson, 1977; Batista 2003; Eltz et al., 2003; Samejima et al., 2004). Often, a higher wood hardness of slow growth trees (e.g. tolerant to shade) corresponds to a higher mechanical strength and natural wood durability (Florsheim, 1992), which would mean less exposure to fungi and insects and weathering action (Burger & Richter, 1991). However, the similar availability of hollows on pioneer and slow growth trees is likely a key determinant of abundance and spatial distribution of stingless bees nests in forests at different stages of regeneration, and not the physical characteristics of these trees *per se*.

Alternatively, the range of variation in tree hollow availability between the stages of forest is well above the demand by the bees' community. For example, Hubbell and Johnson (1977) estimated that the stingless bees should occupy 34% of tree hollows available in a tropical dry forest in Costa Rica. The low proportion of nests per trees with suitable sizes (1 nest per 100 trees with CBH >60cm), in Atlantic rainforest of MER (Silva et al., 2013) also supports that availability of tree hollows overcomes the stingless bees demands in both stage of forest regeneration.

The MER data support the general argument of oversupply of arboreal substrates for nesting in the forest for stingless bees, suggesting access control mechanisms operating at the community level that would also explain the low nest density (2.8 nests/ha; Silva et al., 2013). On the other hand, selectivity for tree species does not exist or at least is not relevant in spatial structuring (or temporal) of communities of stingless bees in this rainforest.

The extreme disruption of forest habitats, accompanied by extensive deforestation and savannization of landscapes, however, tends to reduce the diversity of trees and change considerably the pattern of tree hollows supply, exposing stingless bees to greater convergence in the use of nesting substrates. In this process, generalist species in using cavities with higher swarm rates should become dominant in the communities, as probably is happening with *T. angustula* (Silva et al., 2013, 2014). In such a scenario, the management of stingless bees in the forest should be closely associated with the management of diversity of trees for regeneration of tropical rainforests.

## Acknowledgements

To the Pollination Ecology Lab staff (ECOPOL-IB/UFBA) for helping with field work. To Dr. Lazaro Benedito da Silva - Laboratory of Plant Morphology and Anatomy (IB/UFBA) - for helping in determining the wood density of the trees. To Michelin for logistical support. To CAPES for the first author's doctoral scholarship. To CNPq (Processes No. 481113/478271/2008 and 474313/2011-5) and FAPESB (APR0114/2006) for research support.

## References

Aguilar-Monge, I. (1999). El potencial de las abejas nativas sin aguijón (Apidae, Meliponini) en los sistemas agroforestales. http://www.fao.org/ag/aga/agap/frg/afris/espanol/document/agrof99/aguilari.htm. (accessed date: 23 April 2014).

Anderson, M.J. (2005). Permanova: Permutational multivariate analysis of variance – A computer program. Departament of Statistics, University of Auckland, New

Zealand. http://www.stat.auckland.ac.nz/~mja/Programs.htm/. (accessed date: 23 December 2013).

Antonini, Y. & Martins, R.P. (2003). The value of a tree species (*Caryocar brasiliense*) for a stingless bee *Melipona quadrifasciata quadrifasciata*. J. Insect. Conserv. 7: 167-174. doi: 10.1023/A:1027378306119

Aquino, F.G., Walter B.M.T. & Ribeiro J.F. (2007). Woody community dynamics in two fragments of "cerrado" stricto sensu over a seven-year period (1995-2002), MA, Brazil. Rev. Bras. Bot. 30: 113-121. doi: 10.1590/S0100-84042007000100011

Araújo, E.D., Costa, M., Chaud-Netto, J. & Fowler, H.G. (2004). Body size and flight distance in stigless bees (Hymenoptera: Meliponini): inference of flight range and possible ecological implications. Braz. J. Biol. 64: 563-568. doi: 10.1590/S1519-69842004000400003

Batista, M.A., Ramalho, M. & Soares, A.E.E. (2003). Nesting sites and abundance of meliponini (Hymenoptera: Apidae) in heterogeneous habitats of the atlantic rain forest, Bahia, Brazil. Lundiana 4: 19-23. doi:10.1590/S1519-566X2007000100005

Burger, L.M. & Richter, H.G. (1991). Anatomia da madeira. São Paulo: Nobel S/A, 154 p.

Castro M.S. (2001). A comunidade de abelhas (Hymenoptera; Apoidea) de uma área de caatinga arbórea entre os Inselbergs de Milagres (12°53'S; 39°51'W), Bahia. Thesis. São Paulo: Universidade de São Paulo.

Cortopassi-Laurino, M., Alves, D.A. & Imperatriz-Fonseca, V.L. (2009).

Árvores neotropicais, recursos importantes para a nidificação de abelhas sem ferrão (Apidae, Meliponini). Mens. Doce 100: 21-28.

Eltz, T., Brühl, C.A., Kaars, S.V. & Linsenmair, K.E. (2002). Determinants of stingless bee nest density in lowland dipterocarp forests of Sabah, Malaysia. Oecologia 131: 27-34. doi: 10.1007/s00442-001-0848-6

Eltz, T., Bruhl, C.A., Imiyabir, Z. & Linsenmair, K.E. (2003). Nesting and nest trees of stingless bees (Apidae:Meliponini) in lowland dipterocarp forests in Sabah, Malaysia, with implications for forest management. Forest Ecol. Manag. 172: 301-13. doi:10.1016/S0378-1127(01)00792-7

Flesher, K.M. (2006). The biogeography of the medium and large mammals in a umandominated landscape in the Atlantic Forest of Bahia, Brazil: evidence for the role of agroforestry systems as wildlife habitat. Thesis. New Jersey: The State University of New Jersey. 624 p.

Florsheim, S.M.B. (1992). Variações da estrutura anatômica e densidade básica da madeira de árvore de aroeira Myracrodruon urundeuva F.F. & M.F. Alemão (Anacardiaceae). Dissertation. São Paulo: Universidade de São Paulo.

Guedes, M.L., Batista, M.A., Ramalho.M., Freitas, H.M.B. & Silva, E.M. (2005). Breve incursão sobre a biodiversidade na Mata Atlântica. In C.R. Franke, P.L.B. Rocha, W. Klein & S.L. Gomes (Orgs.), Mata Atlântica e Biodiversidade (pp. 39-92). Salvador: Editora da UFBA.

GraphPad-Software (1998). [Computer Program] Version 3.05 San Diego (CA): GraphPad Software, Inc.

Hubbell, S.P. & Johnson, L.K. (1977). Competition and nest spacing in a tropical stingless bee community. Ecology 58: 950-963. doi: 10.2307/1936917

Kenney, A.J. & Krebs, C.J. (2000). Programs for Ecological Methodology, 2nd ed. Vancouver: Dept. of Zoology, University of British Columbia.

Kerr, W.E. (1971). Contribuição à ecogenética de algumas espécies de abelhas. Ciênc. Cult. 23(suppl): 89-90.

Kleinert, A.M.P. (2006). Demografia de ninhos de meliponíneos em biomas neotropicais. Thesis. São Paulo: Universidade de São Paulo, 93 p.

Lorenzi, H. (2002a). Árvores Brasileiras - Manual de identificação e cultivo de plantas arbóreas nativas do Brasil - vol. 01, 4th Ed. Nova Odessa: Instituto Plantarum, 384 p.

Lorenzi, H. (2002b). Árvores Brasileiras - Manual de identificação e cultivo de plantas arbóreas nativas do Brasil - vol. 02, 2nd Ed. Nova Odessa: Instituto Plantarum, 384 p.

Lorenzi, H. (2009). Árvores Brasileiras - vol. 03, 2nd Ed. Nova Odessa: Instituto Plantarum, 389 p.

Martins, C.F., Cortopassi-Laurino, M., Koedam, D. & Imperatriz-Fonseca, V.L. (2004). Espécies arbóreas utilizadas para nidificação por abelhas sem ferrão na caatinga (Seridó, PB; João Câmara, RN). Biota Neotrop. 4(2): 1-8. doi: 10.1590/S1676-06032004000200003

Michener, C.D. (2000) The bees of the World. Baltimore: Johns Hopkins University, 913 p.

Moreno, F. & Cardoso, A. (1997). Abundância de abejas sin aguijón (Meliponinie) en especies maderables del Estado Portuguesa, Venezuela. Vida Silvestre Neotrop. 6: 53-56.

Nogueira-Neto, P. (1997). Vida e criação de abelhas indígenas sem ferrão. São Paulo: Nogueirapis, 445 p.

Nogueira-Neto, P., Imperatriz-Fonseca, V.L., Kleinert-Giovannini, A., Viana, B.F. & Castro, M.S. (1986). Biologia e manejo das abelhas sem ferrão. São Paulo: Editora Tecnapis, 54 p.

Oliveira, M.L., Morato, E.F. & Garcia, M.V.B. (1995). Diversidade de espécies e densidade de ninhos de abelhas sociais sem ferrão (Hymenoptera, Apidae, Meliponinie) em floresta de terra firme na Amazônia Central. Rev. Bras. Zool. 12(1): 13-24. doi: 10.1590/S0101-81752005000400041

Roubik D.W. (1983). Nest and colony characteristcs of stingless bees from Panama (Hymenoptera: Apidae). J. Kansas Ent. Soc. 56(3): 327-355.

Roubik, D.W. (1989). Ecology and natural history of tropical bees. Cambridge: University Press, 514 p.

Rocha-Santos, L. & Talora, D.C. (2012). Recovery of Atlantic Rainforest areas altered by distinct land-use histories in northeastern Brazil. Trop. Conserv. Sci. 4: 475-494.

Samejima, H., Marzuki, M., Nagamitsu, T. & Nakasizuka, T. (2004) The effects of human disturbance on a stingless bee community in a tropical rainforest. Biol. Conserv. 120(4): 577-587. doi: 10.1016/j.biocon.2004.03.030

Serra, B.D.V., Drummond, M.S., Lacerda, L.M. & Akatsu, I.P. (2009). Abundância, distribuição espacial de ninhos de abelhas meliponini (Hymenoptera, Apidae, Apini) e espécies vegetais utilizadas para nidificação em áreas de cerrado do Maranhão. Iheringia 99(1): 12-17. doi: 10.1590/S0073-47212009000100002

Silva, M.D., Ramalho, M. & Monteiro, D. (2013). Diversity and habitat use by stingless bees (Apidae) in the Brazilian Atlantic Forest. Apidologie 44: 699-707. doi: 10.1007/s13592-013-0218-5

Silva, M.D., Ramalho, M. & Monteiro, D. (2014). Communities of social bees (Apidae: Meliponini) in trap-nests: the spatial dynamics of reproduction in an area of Atlantic Forest. Neotrop. Entomol. 43: 307-313. doi: 10.1007/s13744-014-0219-8

Souza, B.A., Carvalho, C.A.L. & Alves, R.M.O. (2008). Notas sobre a bionomia de Melipona asilvai (Apidae: Meliponini) como subsídio à sua criação racional. Arch. Zootec. 57(217): 53-62.

Sutherland, W.J. (Ed.) (2006). Ecological census techiniques. 2nd Ed. Cambridge: Cambridge Univ. Press, 432 p.

Taura, H.M. & Laroca, S. (1991). Abelhas altamente sociais (Apidae) de uma área restrita em Curitiba (Brasil): Distribuição dos ninhos e abundância relativa. Acta Biol. Parana. 20(1,2,3,4): 85-101.

Teixeira, A.F.R. & Viana, B.F. (2005). Distribuição e densidade dos sítios nidificados pelos meliponíneos (Hymenoptera Apidae) das dunas do médio São Francisco. Rev. Nordes. Zool. 2: 5-20. doi: 10.1590/S0085-56262002000400012

Vital, B.R. (1984). Métodos de determinação da densidade da madeira. Boletim técnico 1. Viçosa: Sociedade de Investigações Florestais, 21 p.

Witter, S., Lopes L.A., Lisboa, B.B., Blochtein, B., Mondin, C.A. & Imperatriz-Fonseca, V.L. (2010). Ninhos da abelha guaraipo (Melipona bicolor schencki), espécie ameaçada, em remanescente de Mata com Araucária no Rio Grande do Sul. Série Técnica Fepagro 5, 37 p.

# 22

# Does ant community richness and composition respond to phytophysiognomical complexity and seasonality in xeric environments?

EM Silva[1]; AM Medina[1]; IC Nascimento[2]; PP Lopes[1]; KS Carvalho[2]; GMM Santos[1]

1 - Universidade Estadual de Feira de Santana, Feira de Santana, Bahia, Brazil
2 - Universidade Estadual do Sudoeste da Bahia, Campus Jequié, Bahia, Brazil.

**Keywords**
Environmental Heterogeneity, Species Composition, Formicidae, Seasonality

**Corresponding author**
Ivan C. Nascimento
Univ. Estadual do Sudoeste da Bahia
Departamento de Ciências Biológicas
Av. Jose Moreira Sobrinho
Jequiezinho
Jequié, BA, Brazil
45200-000
E-mail: icardoso@hotmail.com

**Abstract**

This study aimed to analyze how the vegetation structure (physiognomy) and seasonal changes between seasons (wet and dry) influence richness, diversity and composition of ant species of arboreal and shrubby savanah (Caatinga) environments. The vegetation structure was significantly different among the three strata for all parameters (mean diameter of vegetation, level of herbaceous cover, degree of coverage and depth of litter and percentage of canopy cover). We collected 127 ant species. The mean number of species was approximately two times higher in the rainy season than in the dry season. There was no difference in species richness between the arboreal and shrubby Caatinga physiognomies nor interaction between season and physiognomy. Despite the similarity in richness, species composition differed between physiognomies, however we found no difference in composition between seasons. The seasonal differentiation may be mainly related to the variation in the overall numbers of individuals circulating in the environment, since the enhancement of resource availability during rainy season allows the colony to grow or expand foraging activities, which increases local diversity. Water restriction explains the limited diversity in both environments, while the occurrence of species with greater resource specificity may determine differences in ant composition. Differences in composition of each of Caatinga's physiognomy enhance beta diversity, therefore, raising the overall diversity in the Caatinga Domain.

## Introduction

It is common that communities suffer changes in species composition and richness in seasonal environments, such as tropical dry forests (Murphy & Lugo, 1986). A remarkable feature of these vegetation types is the loss of leaves by trees during the dry season (Veloso et al., 1991), which interspersed with wetter periods and higher productivity, determine changes in the amount and quality of resources and, consequently, the structure of local communities (Sánchez-Azofeifa et al., 2005). This explains why certain species may specialize in the use of resources under more severe conditions, resulting in a temporal partition of the same, reflecting a temporal variation of community composition (Pianka, 1980). Apart from climate change, another factor that can influence the animal community is the structural complexity of vegetation. The ecological prediction states that the occurrence of more species in a community is seen as a response to the greater complexity of vegetation structure (Pacheco et al., 2009; Corrêa et al., 2006), which provides a greater amount of realizable niches by species of animals and therefore a greater number of species in a given community (Tews et al., 2004). To test this hypothesis, several studies involving the comparison of areas with distinct physiognomies have been done throughout the world, using ant communities, such as Armbrecht and Ulloa-Chacón (1999) in Colombia; Fisher and Robertson (2002) in Madagascar; Wilkie et al. (2009) in peruvian Amazonic Forest; Lindsey and Shinner (2001) in South Africa. In Brazil we highlight the studies of Fowler et al. (2000) comparing forests from Bahia and Pará States; Corrêa et al. (2006) in forest patches from Mato Grosso do Sul and Delabie et al. (2007) comparing shaded cocoa agroecosystem developed under Atlantic Forest vegetation or other native vegetation in Bahia State.

The Caatinga, a native dryland registered in the Northeastern Brazil, shows a marked variation in its vegetation structure (Andrade-Lima, 1981), particularly with regard to the density and size of the plants. These differences can be perceived at the local level, where even within a few dozen meters we recognize differences, usually related to a clearly identifiable environmental change as rock extrusions ('lajedos' formations), that determine shallow soils and lower water availability (Amorim et al., 2005). However, studies with animal communities are still incipient, particularly in relation to communities of ants (Neves et al., 2006; Leal, 2003).

In this context, the present study attempted to answer the following question: Does the type of vegetation (physiognomy) that makes up the Caatinga and seasonal variations between the dry and rainy seasons in this environment, influence the richness, diversity and species composition of ants? To answer this question, the following hypotheses were formulated: The richness, diversity and composition of ant species respond positively to the increased structural complexity of the environment and negatively to conditions of the dry season due to the scarcity of food and nesting resources.

## Material and Methods

### Study Area

In this study, we compared two physiognomies of Caatinga, an arboreal and a shrubby Caatinga, both located in the city of Milagres, Bahia, Brazil (12°52'36S 39°51'22W). The Arboreal Caatinga is characterized by having tall trees reaching up to 20 meters, straight stems and understory consisting of smaller trees and ephemeral subshrubs (Ferreira, 1997). The shrubby Caatinga, in turn, is marked by more sparse trees and greater representation of Cactaceae and Euphorbiaceae with a formation that resembles the vegetation of fields (Ross, 2001).

The region has a semi-arid tropical climate with an average temperature of 24.3°C and average rainfall of 551 mm/year, although large variations between years may occur (142 to 1206 mm/year). The rainy season generally extends from December to February, although there are annual variations, with at least five dry months during the year (Bahia, 1994).

In order to characterize each vegetation type in each sampled area we evaluated vegetation variables which were compared using two-sample independent tests (t test or Mann-Whitney): CBH (Circumference at Breast Height), herbaceous cover, litter cover, litter depth and percent cover of vegetation canopy, measured at each sample point of fauna. CBH of trees was measured at 1.30m above ground, within a 5m radius circular plot (78,5m²) marked from the sampling points; in plants smaller than 1.50m in height, CBH was replaced by the circumference of the trunk below the first branch (Soares, 1999). The herbaceous cover is given by counting the herbaceous plants in a radius of 1.50m from the sample point. Coverage of litter was measured according to the scale of Fornier (1974): 1 (0-25%=small); 2 (26-50%=medium 1); 3 (51-75%=medium 2); 4 (76-100%=large). The depth of the litter was classified according to Pacheco et al. (2009) comprising four classes of arbitrary amplitude: very shallow (0-2cm), shallow (2-4cm), deep (4-6cm) and very deep (>6cm). The percentage cover of the vegetation canopy was evaluated through a modification of the methodology for indirect estimation of canopy proposed by Monte et al (2007). We used a Sony Cybershot camera (7.2 MP) to capture the canopy image, and through the Photoshop 7.0 software we created a binary image (black-white) in order to quantify the amount of black pixels, estimating canopy coverage.

### Ant sampling

Between May 2009 and January 2010 we carried out four field incursions, two during the dry season and two during the rainy season. In each incursion we sampled the ant fauna associated with three areas of arboreal Caatinga and three areas of shrubby Caatinga.

Samples were taken on a transect of 350 meters in each area, each transect containing 15 pitfall traps 25m distant from one another. The traps were kept active for 48 hours in the field. Additionally, we installed 15 traps with attractive bait sardines in vegetable oil (1cm³), at each transect, exposed for a period of 30min. To avoid interference, the baits were installed only after the removal of the pitfalls. This collection protocol had three replicates at each physiognomy and repeated in all four incursions in the field totaling 12 transects in arboreal Caatinga and 12 transects in the shrubby Caatinga.

Collected ants were identified following the classification proposed by Bolton et al. (2006) and witness individuals were deposited in the Prof. Johann Becker Entomological Collection from Zoology Museum of the Universidade Estadual de Feira de Santana (MZFS) and in the entomological collection from the Myrmecology Laboratory from the Comissão Executiva de Pesquisa da Lavoura Cacaueira (CEPEC/CEPLAC), in Itabuna, Bahia.

### Data Analysis

We tested our hypothesis with analyzes based on the components of community structure. The first analysis was based on species richness and used the sample points as local units. For this, we used a generalized linear mixed model with Poisson distribution and log link function to assess the influence of vegetation type and season (explanatory variables) on species richness (response variable). Furthermore, we use the sampling points, and collection areas incursions as random factors to control the temporal pseudoreplication. We conducted this analysis in R software (R Development Core Team, 2013) using the lme4 package (Bates et al., 2013).

The second analysis was based on species composition and areas used as sampling points. For this purpose, we build a similarity matrix using the Jaccard index and from this we performed a non-metric multidimensional scaling (NMDS). This technique is an ordering method more robust to non-linear situations and often summarizes more information in fewer axes than other indirect ordination techniques (Legendre & Legendre, 1998). We tested changes in species composition between seasons (dry and wet) and between vegetation types (arboreal and shrubby Caatinga) using a two level similarity analysis (Two-way ANOSIM). We conducted this analysis using Primer® 6.0 software (Clarke & Gorley, 2006).

## Results

### Vegetation structure

The vegetation structure of arboreal and shrubby Caatinga were different in all traits, with the arboreal physiognomy presenting a CBH 1.4 times greater than in shrubby Caatinga (arboreal = 30.0±21.4mm; shrubby = 42.2±22.03mm; t = 2.652; n = 45; $P \leq 0.001$), herbaceous cover twice as large (arboreal = 1.9±0.9, shrubby = 0.7±0.4; t = 7.0; n = 45; $P$ <0.001), larger leaf litter coverage (U = 122.5; n = 90; $P$ <0.001), larger leaf litter depth (U = 701; n = 90; $P$ <0.001), and canopy coverage was two times larger (arboreal = 70.9±9.2%, shrubby = 33.6±22.2%, t = 14.74; d.f. = 118.57; $P$ < 0.001).

### Mirmecofauna

We collected 127 ant species (Apendix 1), with the most frequent species being *Dinoponera quadriceps* (68%), *Camponotus* sp6 (56%) and *Ectatomma muticum* (41%). A total of 32 ant species was recorded exclusively on arboreal Caatinga and 29 species exclusively in the area of shrubby Caatinga. Between stations, 18 species of ants were collected only in the dry season and 20 species of ants were collected exclusively in the wet season.

The mean number of species by point sampled was approximately two times higher in the wet season than in the dry season ($\chi^2$ = 5.45; d.f. = 1; $P$ <0.05). Moreover, there was no difference in species richness among the Caatinga physiognomies ($\chi^2$ = 0.4796; d.f.=1; $P$ = 0.49) nor interaction between physiognomy and season ($\chi^2$ = 6.95; d.f. = 3; $P$ = 0.07) (Fig. 1).

Despite the similarity in richness, composition of ant species differed between the physiognomies (R = 0.849; $P$ <0.01), however we found no difference in species composition of ants between the dry and wet seasons (R = 0.118; $P$ = 0.10) (Fig. 2).

## Discussion

The fact that there were differences in species composition between the physiognomies, despite the similarity of species richness, indicates that there are two distinct communities in the Caatinga - a specialist in arboreal Caatinga

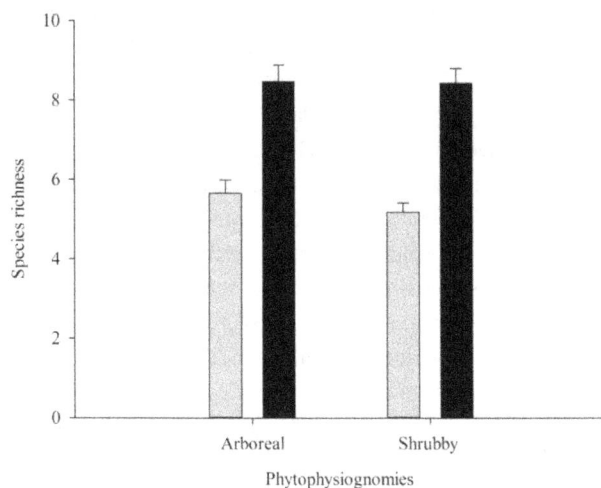

Figure 1. Mean of ant species richness for point sampled in two Caatinga phytophysiognomies (arboreal and shrubby) ($\chi^2$=0.4796; d.f.=1; $P$ = 0.49) in dry (gray columns) and rainy (black columns) seasons. The mean number of species for point sampled was approximately two times higher in the wet season than in the dry season ($\chi^2$=5.45; d.f.=1; $P$ < 0.05). Bars represent standard errors.

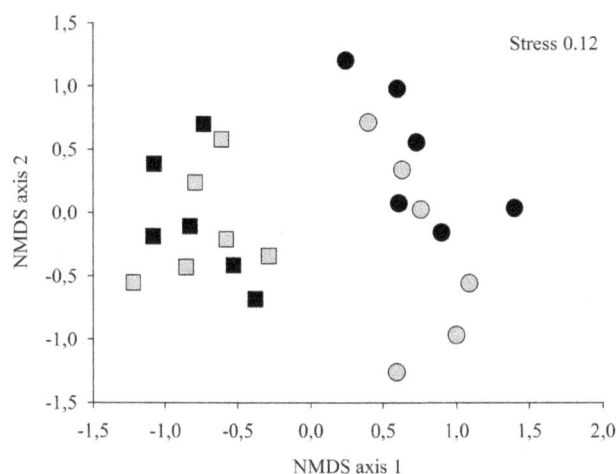

Fig 2. Non-metric dimensional scaling (NMDS) based on Jaccard Similarity Index, comparing ant species composition in Milagres, Bahia, Brazil. Circles represent arboreal Caatinga and squares represent shrubby Caatinga. Dry season in gray and rainy season in black.

and another in shrubby Caatinga. Increasing environmental complexity can change the types of resources and their availability. Once the resources are different, the environment may become less advantageous to the dominant species and completely change the structure and composition of the community (Perfecto & Vandermeer, 1996). An example of how the heterogeneity of vegetation may determine the occurrence of specialist species is the presence of *Gnamptogenys concinna* in the area of arboreal Caatinga. It was believed that this ant species was restricted to wet forest environments, and recently recorded for the state of Bahia in cacao shaded by large trees (Delabie et al., 2010). *G. concinna* is the only one belonging to the arboreal specialist genus (Lattke, 1990; Longino, 1998), with strong links with canopies with high numbers of epiphytes of the families Bromeliaceae and Orchidaceae (Delabie et al., 2010). A link to these epiphytes may explain its occurrence in the area of arboreal Caatinga, considering that the area where it was found has a canopy with lots of large epiphytic Bromeliaceae.

Studies focusing on the influence of grazing on the ant community in semi-arid regions also found effects restricted to species composition (Bestelmeyer & Wiens, 2001). However, in the present study species richness was similar because once the two areas are subject to stress caused by lack of rain, the food resources must be scarce in both environments.

We found no effect of the interaction between environmental complexity and seasonality on the ant diversity, but we found a difference in the richness of ant species between seasons, but without a difference in species composition. Due to the unpredictability in the acquisition of resources caused by the shortage of rainfall, most species in Caatinga should be able to pull through in xeric conditions (like ants in american desertlands e.g: Whitford et al., 1999), and the species should only reduce the number of active individuals in the colony during the dry season and invest in foraging activities when the environment has more resources (Bernstein, 1979). If there are less active individuals, the probability of finding more species will be smaller. It is worthwhile to point out that, although fluctuations in abundance may occur between seasons, we work only with occurrences, missing evidence to support this hypothesis. A study addressing ants that use floral resources in a tropical dry forest demonstrated that dietary overlap is higher in the dry season (Brito et al., 2012) indicating a decrease in food resources.

Since there are large differences in species composition in the two physiognomies, this leads to an increase in beta diversity of Caatinga on a regional scale, similar to what happened with other groups such as bees (Martins, 2002). Given that there is more than one type of physiognomy in the Caatinga, the importance of environmental heterogeneity for increasing beta diversity in the region shown in the present study may be underestimated.

Two factors influence the effect of the plant structure on ants and are characteristics that distinguish arboreal and shrubby Caatinga. The first is the increase in shading in the environment caused by the presence of trees (Reyes-López et al., 2003). The second is the deposition of litter which is a factor that positively influence the activity of ant species (Perfecto & Vandermeer, 1996). Similar to that reported in this study, other studies have also found no influence of environmental complexity in species richness of ants (Corrêa et al., 2006; Neves & Braga, 2010). However, the use of only one feature as a surrogate of environmental complexity may fail to find an effect even if it exists (see discussion on the review by Tews et al., 2004), so an indirect analysis can more

easily detect the environmental heterogeneity even without setting a key structure, provided focusing on the species composition.

Our study shows the following pattern: while the variation in rainfall is responsible for the increase in the number of species that may be present in the environment on a local scale, the difference of environmental complexity on both physiognomies is detectable in species composition and is responsible for increasing the beta diversity through differences in species composition.

## Acknowledgments

Many colleagues contributed to this study in different ways, especially J.J. Resende who helped during sampling. We thank two anonymous reviewers for valuable comments. The Brazilian National Council for Scientific and Technological Development (CNPq) supported this study. G.M.M. Santos received productivity fellowship from CNPq.

## References

Amorim, I.L., Sampaio, E.V.S.B. & Araújo, E.L. (2005). Flora e estrutura da vegetação arbustiva-arbórea de uma área de Caatinga do Seridó, RN, Brasil. Acta Botanica Brasilica, 19: 615-623. doi.org/10.1590/S0102-33062005000300023.

Andrade-Lima, D. (1981). The caatingas dominium. Revista Brasileira de Botânica, 4: 149-153.

Armbrecht, I. & Ulloa-Chacón, P. (1999). Rareza y diversidad de hormigas de bosques secos colombianos y sus matrices. Biotropica, 31: 646 - 653. doi 10.1111/j.1744-7429.1999. tb00413.x

Bahia, Centro de Estatística e Informação (1994). Informações básicas dos municípios baianos. Recôncavo Sul 8. 279-299, Salvador, 761p.

Bates, D., Maechler, M. & Bolker, B. (2013). lme4: Linear mixed-effects models using S4 classes. Vienna, Austria: The Comprehensive R Archive Network (CRAN).

Bernstein, R.A. (1979). Schedules of foraging activity in species of Ants. Journal of Animal Ecology, 48: 921-930.

Bestelmeyer, B. & Wiens, J. (2001). Ant biodiversity in semi-arid landscape mosaics: the consequences of grazing vs. natural heterogeneity. Ecological Applications, 11: 1123-1140. doi: 10.1890/1051-0761(2001)011[1123:ABISLM]2.0.CO;2

Bolton, B., Alpert, G., Ward, P.S. & Nasrecki, P. (2006). Bolton's Catalogue of ants of the world. Harvard University Press, Cambridge, Massachusetts, CD-ROM.

Brito, A.F., Presley, S.J. & Santos, G.M.M. (2012). Temporal and trophic niche overlap in a guild of flower-visiting ants in a seasonal semi-arid tropical environment. Journal of Arid Environments, 87: 161-167. doi: .org/10.1016/j.jaridenv.2012.07.001

Clarke, K.R. & Gorley, R.N. (2006). Primer v.6: Computer Program and User Manual/tutorial. PRIMER-E Ltd, Plymouth, United Kingdom.

Corrêa, M., Fernandes, W. & Leal, I. (2006). Ant diversity (Hymenoptera: Formicidae) from capões in Brazilian Pantanal: relationship between species richness and structutural complexity. Neotropical Entomology, 35: 724-730. doi. org/10.1590/S1519-566X2006000600002.

Delabie, J.H.C., Alves, H.S.R., França, V.C., Martins, P.T.A. & Nascimento, I.C. (2007). Biogeografia das formigas predadoras do gênero *Ectatomma* (Hymenoptera: Formicidae: Ectatomminae) no leste da Bahia e regiões vizinhas. Centro de Pesquisas do Cacau, Ilhéus, Bahia, Brasil. Agrotrópica, 19: 13-20.

Delabie, J.H.C., Rocha, W.D., Feitosa, R.M., Devienne, P., Fresneau, D. (2010). *Gnamptogenys concinna* (F. Smith, 1858): nouvelles données sur sa distribution et commentaires sur ce cas de gigantisme dans le genre *Gnamptogenys* (Hymenoptera, Formicidae, Ectatomminae). Bulletin de la Société Entomologique de France, 115: 269-277.

Ferreira, M.I.M.M. (1997). Análise temporal do uso das terras em área do município de Viçosa do Ceará - Ce. (Tese de Mestrado) Fortaleza, Universidade Federal do Ceará, 51p.

Fisher, B.L. & Robertson, H.G. (2002). Comparison and origin of forest and grassland ant assemblages in the high plateau of Madagascar (Hymenoptera: Formicidae). Biotropica, 34: 155-167. doi: 10.1111/j.1744-7429.2002.tb00251.x

Fowler, H.G., Delabie, J.H.C. & Moutinho, P.R.S. (2000). Hypogaeic and epigaeic ant (Hymenoptera: Formicidae) assemblages of atlantic costal rainforest and dry mature and secondary Amazon forest in Brazil: Continuums or communities. Tropical Ecology, 41: 73-80.

Fournier, LA. (1974). Un método cuantitativo para la medición de características fenológicas en árboles.Turrialba, 24:422-423.

Lattke, J. E. (1990). Revisión del género *Gnamptogenys* Mayr para Venezuela. Acta Terramaris, 2: 1-47.

Leal, I. R. (2003). Diversidade de formigas em diferentes unidades de paisagem Caatinga. In: I.R. Leal, M. Tabarelli & J.M.C. Silva (Org.), Ecologia e Conservação da Caatinga, 1 ed. Recife: Editora Universitária UFPE, v.1, pp.435-461

Legendre, P. & Legendre, L. (1998). Numerical ecology, 2nd English ed. - Elsevier, 852 p.

Lindsey, P.A. & Skinner, J.D. (2001). Ant composition and activity patterns as determined by pitfall trapping and other methods in three habitats in the semi-arid Karoo. Journal of Arid Environments, 48: 551-568.

Longino, J.T. 1998. http://academic.evergreen.edu/projects/ants/Genera/Gnamptogenys/species/concinna/concinna.html, 1998.

Martins, C.F. (2002). Diversity of the bee fauna of the Brazilian Caatinga. In P. Kevan & V. Imperatriz Fonseca (eds.), Pollinating bees - The conservation link between agriculture and nature (pp. 131-134). Brasília: Ministério do Meio Ambiente.

Monte, M. A., Reis, M. das G. F., Reis G. G., Leite H. G. & Stocks J. J. (2007). Métodos indiretos de estimação da cobertura de dossel em povoamentos de clone de eucalipto. Pesquisa Agropecuária Brasileira, 42: 769-775.

Murphy, P.G., & Lugo, A.E. (1986). Ecology of Tropical Dry Forest. Annual Review of Ecology and Systematic, 17: 67-88.

Neves, F.S., Braga, R.F. & Madeira, B.G. (2006). Diversidade de formigas arborícolas em três estágios sucessionais de uma floresta estacional decidual no norte de Minas Gerais - Dossiê Florestas Estacionais Deciduais: Uma abordagem multidisciplinar. Unifontes Científica, Montes Claros, 8: 59-68.

Neves, F. & Braga, R. (2010). Diversity of arboreal ants in a Brazilian tropical dry forest: effects of seasonality and successional stage. Sociobiology, 56: 1-18.

Pacheco, R., Silva, R.R., Morini, M.S.C. & Brandão, C.R.F. (2009). A comparison of the leaf-litter ant fauna in a secondary Atlantic Forest with an adjacent Pine plantation in Southeastern Brazil. Neotropical Entomology, 38: 55-65. doi: org/10.1590/S1519-66X2009000100005.

Perfecto, I. & Vandermeer, J. (1996). Microclimatic changes and the indirect loss of ant diversity in a tropical agroecosystem. Oecologia, 108: 577-582.

Pianka, E.R. (1980). Guild structure in desert lizards. Oikos, 35: 194-201.

R Development Core Team (2013). R: A language and environment for statistical computing. R Foundation for Statistical Computing, Vienna, Austria.URL http://www.R-project.org/.

Reyes-López, J., Ruiz, N. & Fernández-Haeger, J. (2003). Community structure of ground-ants: the role of single trees in a Mediterranean pastureland. Acta Oecologica, 24: 195-202.

Ross, J.L.S. (2001). Geografia do Brasil. ed.2. EDUSP.

Sánchez-Azofeifa, G.A., Quesada, M., Rodríguez, J.P., Nassar, J.M., Stoner, K.E., Castillo., A., Garvin, T., Zente, E.L., Calvo-Alvarado, J.C. , Kalacska, M.E.R., Fajardo, L., Gamon, J.A. & Cuevas-Reyes, P. (2005). Research priorities for neotropical dry forests. Biotropica, 37: 477-485. doi: 10.1046/j.0950-091x.2001.00153.x-i1

Soares, M.L. (1999). Estrutura vegetal e grau de perturbação dos manguezais da Lagoa da Tijuca, Rio de Janeiro, RJ, Brasil. Revista Brasileira de Biologia, 59: 503-515.

Tews, J., Brose, U., Grimm, V., Tielbörger, K., Wichmann, M.C., Schwager, M. & Jeltsch, F. (2004). Animal species diversity driven by habitat heterogeneity/diversity: the importance of keystone structures. Journal of Biogeography, 31:

79-92. doi: 10.1046/j.0305-0270.2003.00994.x

Veloso, H.P., Rangel-Filho, A.L.R., LIMA, J.C.A. (1991). Classificação da vegetação brasileira adaptada a um sistema universal. Rio de Janeiro: IBGE, Departamento de Recursos Naturais e Estudos Ambientais, 124p.

Whitford, W., Zee, J. Van & Nash, M. (1999). Ants as indicators of exposure to environmental stressors in North American desert grasslands. Environmental Monitoring and Assessment, 54: 143-171.

Wilkie, K.T.R., Mertl, A.L. & Traniello, J.F.A. (2009). Diversity of ground-dwelling ants (Hymenoptera: Formicidae) in primary and secondary forests in Amazonian Ecuador. Myrmecological News, 12: 139-147.

**Appendix 1**. List of ant species collected in areas of Arboreal Caatinga and Shrubby Caatinga, during dry and rainy seasons in Milagres, Bahia, Brazil.

| Species | Phytophisiognomy | | | |
| --- | --- | --- | --- | --- |
| | Arboreal Caatinga | | Shrubby Caatinga | |
| | Dry Season | Rainy Season | Dry Season | Rainy Season |
| *Acanthoponera* sp01 | X | | | X |
| *Acanthostichus* sp01 | X | | | |
| *Acromyrmex* sp01 | | | X | X |
| *Anochetus* sp02 | X | | | |
| *Apterostigma* sp01 | X | | | |
| *Atta sexdens rubropilosa* | X | X | X | X |
| *Azteca* sp01 | | X | | |
| *Azteca* sp02 | | X | | |
| *Azteca* sp03 | | X | | |
| *Brachymyrmex* sp01 | X | X | | |
| *Brachymyrmex* sp02 | | X | | |
| *Brachymyrmex* sp03 | | | X | X |
| *Brachymyrmex* sp04 | | | X | X |
| *Camponotus* sp01 | X | X | X | X |
| *Camponotus* sp02 | X | X | X | X |
| *Camponotus* sp03 | X | X | X | X |
| *Camponotus* sp04 | X | X | X | X |
| *Camponotus* sp05 | X | X | X | |
| *Camponotus* sp06 | X | X | X | X |
| *Camponotus* sp07 | X | | X | X |
| *Camponotus* sp08 | X | X | X | X |
| *Cephalotes* prox. *goeldi* | X | X | | |
| *Cephalotes atratus* | X | X | | |
| *Cephalotes clypeatus* | | X | X | X |
| *Cephalotes depressus* | X | X | | X |
| *Cephalotes grandinosus* | | X | | X |
| *Cephalotes minutus* | X | X | X | X |
| *Cephalotes pussilus* | X | X | X | X |
| *Cephalotes* sp01 | X | | | |
| *Cephalotes ustus* | | X | | |
| *Crematogaster* sp01 | X | | X | X |
| *Crematogaster* sp02 | | X | X | X |
| *Crematogaster* sp03 | | | X | X |
| *Crematogaster* sp04 | | | | X |
| *Crematogaster* sp05 | | | X | X |
| *Crematogaster* sp06 | X | X | X | X |
| *Crematogaster* sp07 | X | X | | X |
| *Cyphomyrmex* sp01 | | | X | X |
| *Cyphomyrmex* sp02 | | | X | X |
| *Dinoponera quadriceps* | X | X | X | X |
| *Dorymyrmex* sp01 | | | X | X |
| *Dorymyrmex thoracicus* | | | X | X |

| | | | | |
|---|---|---|---|---|
| *Ectatomma edentatum* | X | X | X | X |
| *Ectatomma muticum* | X | X | X | X |
| *Ectatomma* sp01 | | | X | X |
| *Ectatomma* sp02 | | | X | X |
| *Ectatomma suzanae* | X | X | X | X |
| *Forelius brasiliensis* | | | X | X |
| *Gnamptogenys concinna* | X | | | |
| *Gnamptogenys* sp01 | | | X | X |
| *Gnamptogenys* sp02 | | | X | |
| *Hylomyrma balzani* | X | X | | X |
| *Hylomyrma* sp01 | X | | | |
| *Labidus coecus* | X | X | | |
| *Labidus mars* | X | | | |
| *Labidus praedator* | X | X | | |
| *Linepithema humile* | | X | | |
| *Linepithema* sp01 | | X | X | X |
| *Linepithema* sp02 | | X | X | X |
| *Linepithema* sp03 | | | | X |
| *Linepithema* sp04 | | | X | |
| *Mycetophylax* sp01 | | | | X |
| *Neivamyrmex* sp01 | X | | X | |
| *Nylanderia* sp01 | X | X | | |
| *Ochetomyrmex* sp01 | | X | | |
| *Octostruma* sp03 | | X | | |
| *Odontomachus chelifer* | X | X | | |
| *Odontomachus haematodus* | X | X | X | X |
| *Oxyepoecus* sp02 | | | | X |
| *Pachycondyla bucki* | X | X | | |
| *Pachycondyla* prox. *magnifica* | X | | | |
| *Pachycondyla* prox. *venusta* | | X | X | X |
| *Pachycondyla striata* | | X | | |
| *Pheidole* sp01 | X | X | X | X |
| *Pheidole* sp02 | X | X | X | X |
| *Pheidole* sp03 | X | X | X | X |
| *Pheidole* sp04 | X | X | | X |
| *Pheidole* sp05 | X | X | X | |
| *Pheidole* sp06 | | | | X |
| *Pheidole* sp07 | X | X | X | X |
| *Pheidole* sp08 | X | X | | X |
| *Pheidole* sp09 | X | X | | X |
| *Pheidole* sp10 | X | X | X | X |
| *Pheidole* sp11 | | | X | X |
| *Pheidole* sp12 | X | X | X | X |
| *Pheidole* sp13 | X | X | X | X |
| *Pheidole* sp14 | X | X | X | X |
| *Pheidole* sp15 | X | X | X | |
| *Pheidole* sp16 | X | | X | X |

| | | | | |
|---|---|---|---|---|
| *Pheidole* sp17 | X | X | | |
| *Pheidole* sp18 | | | X | |
| *Pheidole* sp19 | X | X | X | X |
| *Pheidole* sp20 | X | | | X |
| *Pheidole* sp21 | X | X | X | X |
| *Pheidole* sp22 | X | X | X | X |
| *Pheidole* sp23 | X | X | X | X |
| *Pheidole* sp24 | X | X | X | X |
| *Pheidole* sp25 | X | | X | |
| *Pogonomyrmex (E.) naogeli* | X | | X | X |
| *Procryptocerus* sp01 | X | | | |
| *Pseudomyrmex elongatus* | | X | | X |
| *Pseudomyrmex gracilis* | | | | X |
| *Pseudomyrmex* sp01 | X | X | X | X |
| *Pseudomyrmex* sp02 | | | X | X |
| *Pseudomyrmex* sp03 | X | | | |
| *Pseudomyrmex* sp04 | | X | | X |
| *Pseudomyrmex* sp05 | X | | | |
| *Pseudomyrmex tenuis* | X | X | | X |
| *Pseudomyrmex termitarius* | | | X | X |
| *Rogeria* sp01 | X | | | |
| *Solenopsis* sp02 | X | X | X | X |
| *Solenopsis* sp03 | X | X | X | X |
| *Solenopsis* sp04 | X | X | X | X |
| *Solenopsis* sp05 | | X | X | X |
| *Solenopsis* sp06 | X | | X | X |
| *Solenopsis* sp07 | X | X | X | X |
| *Solenopsis* sp08 | | | | X |
| *Solenopsis* sp09 | | X | X | |
| *Strumigenys* sp02 | | | | X |
| *Tapinoma* sp01 | | | | X |
| *Tapinoma* sp02 | X | | | X |
| *Tapinoma* sp03 | | | X | X |
| *Trachymyrmex* sp01 | X | X | X | X |
| *Trachymyrmex* sp02 | X | | X | |
| *Wasmannia* sp01 | | X | X | X |
| *Wasmannia* sp02 | | X | X | X |
| *Wasmannia* sp03 | | X | X | |

# Pattern of the daily flight activity of *Nannotrigona testaceicornis* (Lepeletier) (Hymenoptera: Apidae) in the Brazilian semiarid region

WP Silva, M Gimenes

Universidade Estadual de Feira de Santana (UEFS), Feira de Santana, BA, Brazil

**Keywords**
Stingless bees, Daily flight activity, Meliponini, Biological rhythm.

**Corresponding author**
Miriam Gimenes
Departamento de Ciências Biológicas,
Universidade Estadual de Feira de Santana
Av. Transnordestina, s/nº, Novo Horizonte,
44036-900, Feira de Santana, BA, Brazil
E-Mail: mgimenes@uefs.br

**Abstract**

The flight activities of Meliponini (stingless bees) are associated with a series of particular behaviors as collection of pollen, nectar, resin, and clay during the day, which can be influenced by extrinsic (e.g. abiotic factors) or intrinsic factors (e.g. internal conditions of the colony, morphology, physiology of the individuals, and others). This study aims to analyze the pattern of the daily flight activity of *Nannotrigona testaceicornis* (Lepeletier) and the influence of climatic factors (temperature, relative humidity, and light intensity) on these activities in Chapada Diamantina, Bahia, Brazil. The study was conducted every two months between October 2010 and October 2011 during three consecutive days in two colonies (one managed and the other unmanaged). The managed colony was more active than the unmanaged one in all the months. Nevertheless, both colonies showed regularity of times for their first activities and for the preferential time for the most part of the activities analyzed, which occurred generally in the morning, until noon. This daily pattern of activity of both colonies was mainly influenced by light intensity. In this sense, these activities began with the sunrise and become more intense with the increase of the light intensity in the environment.

## Introduction

The daily activity patterns of eusocial stingless bees are mainly associated with activities as collection of pollen, nectar, resin, and clay. These activities occur during the day, but in higher intensity at specific moments, influenced by climatic factors (Hilário et al., 2001; Souza et al., 2006; Rodrigues et al., 2007; Ferreira-Junior et al., 2010). According to Corbet et al. (1993) there would be an ideal temperature range in which the eusocial bees are active, able to carry out activities that are external to the nest. The light intensity is another climatic factor that influences the flight activity of the bees (Kleinert-Giovannini, 1982; Corbet et al., 1993), acting as an indicator for the onset of the flight activities (Lutz, 1931). Heard and Hendrikz (1993) studying *Tetragonula carbonaria* (Smith) verified a diurnal pattern of activity with the influence of temperature and radiation. For these authors, there was no good correlation between activity and temperature, but it is not conclusive.

Factors such as rainfall, for *Melipona asilvai* Moure (Nascimento & Nascimento, 2012), and relative air humidity, for *Plebeia remota* (Holmberg) (Hilário et al., 2007), can also influence the daily activities of the stingless bees.

The time to supply with floral resources also exerts great influence on the foraging of eusocial bees. Since most plants show another dehiscence in the morning, the presence of a large quantity of bees in the flowers is common during this period for pollen collection (Roubik, 1989).

Many factors, such as those mentioned above are important for interpreting the patterns of daily activity, mainly related to the behavior in the collection of resources. However, the patterns of daily activity can also be manifestations of the circadian timing system, which can be related to many of the aspects of rhythmic daily activities of Meliponini (Bloch, 2008). Endogenous rhythmicity in the activity of Meliponini has already been observed by Bellusci and Marques (2001). According to Moore (2001), eusocial bees present a daily rhythm in the collection of floral resources such as nectar and pollen and these rhythms are generally synchronized with environmental (light/dark) or climatic (temperature) cycles. In this sense, there are very few studies in literature that analyze the daily activities based on the biological rhythm. Moreover, the pattern of daily activity can occur in a different way in the different months of the year, depending on the climatic characteristics of every month, as noted by Nunes-Silva et al. (2010) when they observed the flight activity of *P. remota*.

In order to verify the pattern of daily activity of Meliponini and the influence of environmental and climatic factors, an analysis was performed on two nests of *Nannotrigona testaceicornis* (Lepeletier), a Neotropical bee that occurs in several regions of Brazil (Moure et al., 2012).

## Material and Methods

### Study area

The work was developed in the Valley of Capão, municipality of Palmeiras, Chapada Diamantina (Bahia State, Brazil) (12° 31'S, 41° 33'W, at 1000 m of altitude). This area has low, fairly homogenous, and dense vegetation. The canopy is high, reaching 10 m height with trees emerging over 15 m mainly in the lower altitudes, which can be characterized as dry forest (semideciduous) (Queiroz et al., 2005). The climate of Palmeiras varies between sub-humid and dry with a yearly average temperature of 24.3 °C (Köppen-Geiger: BSh). The rainy period occurs between October and July and the annual rainfall index is 1,361.7 mm (SEI, 2011).

The climatic data were collected at intervals of one hour, from 5:00 to 18:00 during the observations. The data of temperature and relative air humidity were collected with a digital thermo hygrometer fixed above the soil (ca. 1.5 m) in a shadowy area and the data of light intensity (illuminance) were measured with a digital luximeter (Lutron LX-107) at a distance of approximately 1 m from the soil. Sunrise and sunset times were obtained from the almanac of the National Observatory (http://euler.on.br/ephemeris/index.php).

### Flight activity

For this study were used two colonies of *N. testaceicornis*, one managed and one unmanaged. The managed colony of *N. testaceicornis* was captured in the Valley of Capão (Chapada Diamantina), and had been maintained in a wooden box, under a roof, for eight years. The unmanaged colony was located in the trunk of a tree which was transferred from Caatinga environment (semiarid region, located around Palmeiras, Bahia, Brazil) to Valley of Capão, approximately four years ago and had been kept in an open area.

Individuals of *N. testaceicornis* were collected and deposited in the Entomological Collection "Johann Becker" at the Museum of Zoology of the State University of Feira de Santana.

The observations of the managed colony were made every two months from October 2010 to October 2011 while the unmanaged colony was observed in February, June, and October 2011. The observations in each colony were conducted between 5:00 and 19:00 (intervals of 15 min/hour) over three consecutive days. Quantification of the number of bees was made by counting the bees that entered and left the nest (observation of the individuals at the entrance of the colonies). The bees were separated in categories: a) did not carry apparent material; b) carrying pollen, and c) carrying resin on the corbiculae.

### Statistical analyses of data

The analyses of the daily activities of *N. testaceicornis* colonies were performed using the Rayleigh test of the Circular Statistics Method (Batschelet, 1980; Zar, 2010). Preferential times (or acrophases) of activities were considered for values with the significant vector (r) above of 0.7, which can range from 0 to 1, according to dispersion of data (p<0.05). The analyses were only applied when the number of each activity observed reached 10 or more in the month. The daily activities of both colonies were compared using the Watson-Williams test (p<0.05) (Batschelet, 1980; Zar, 2010).

Pearson's correlation coefficient (r) was applied for verifying the correlation between the abiotic variables (temperature, relative humidity, and light intensity in open area and light intensity in hive entrance) and biotic variables (entrance, entrance with pollen, entrance with resin and exit). Correlations were considered statistically significant when p<0.01, by Student's t test. Correlations were made using the program PAST version 1.85 (Hammer et al., 2001). They were considered moderate correlations when the values of r (Pearson's correlation coefficient) were between 0.39 and 0.69 and strong correlations when the r values were above 0.7 (Dancey & Reidy, 2005 in Figueiredo-Filho et al., 2009).

## Results

During the observations of the daily flight activities of *N. testaceicornis* in the study area, the temperature varied from 17 °C to 28 °C with an average temperature variation of 25 °C (October 2010 and February 2011) to 21°C (June 2011) (Table 1). The lowest values of temperature and light luminosity were registered close to sunrise and sunset and the highest values were registered in the afternoon between 12:00 and 17:00, generally higher than 25 °C. The relative humidity varied from 41% to 90% with the average varying from 55% (August) to 83% (December) (Table 1). Sunrise times oscillated between 5:10 (October) and 6:00 (June) and sunset occurred between 17:27 (June) and 18:19 (February).

**Table 1.** Time of the first activities of the unmanaged and managed colonies of *Nannotrigona testaceicornis* and the meteorological factors in these times (light intensity at the entrance of the colony, LI, sunrise, temperature, Temp. at the onset of the activity and monthly average, relative humidity, RH at the onset and monthly average) from October 2010 to October 2011 in Valley of Capão (Chapada Diamantina, Bahia, Brazil).

| Months | Hour | LI (lux) | Sunrise | Temp. (°C) | Temp. Average | RH (%) | RH Average |
|---|---|---|---|---|---|---|---|
| **Managed Colony** | | | | | | | |
| Oct/10 | 7:00 | 504 – 1110 | 5:15 | 19 | 25 | 82 | 59 |
| Dec/10 | 6:00 | 28 – 126 | 5:11 | 20 | 24 | 90 | 83 |
| Feb/11 | 6:00 | 1 – 95 | 5:39 | 20 | 25 | 77 | 61 |
| Abr/11 | 6:00 | 78 – 85 | 5:50 | 20 | 23 | 76 | 70 |
| Jun/11 | 8:00 | 410 – 786 | 6:03 | 18 | 21 | 80 | 69 |
| Aug/11 | 8:00 | 12500 – 19400 | 6:00 | 19 | 23 | 70 | 55 |
| Out/11 | 6:00 | 125 – 157 | 5:14 | 20 | 24 | 78 | 59 |
| **Unmanaged Colony** | | | | | | | |
| Feb/11 | 6:00 | 3 – 230 | 5:39 | 20 | 25 | 77 | 61 |
| Jun/11 | 8:00 | 2100 – 3850 | 6:03 | 18 | 21 | 80 | 69 |
| Oct/11 | 6:00 | 798 – 1333 | 5:14 | 20 | 24 | 78 | 59 |

## Monthly activities

The managed colony of *N. testaceicornis* showed daily activity in all the seven months of observation. The numbers of total daily flight activities (entrance plus exit), entrance with pollen, and entrance with resin were always higher for the managed colony than for the unmanaged colony in all the months when both colonies were observed (Figs 1 and 2). In October 2011 (Temperature average: 24°C), a higher number for total activity and entrance with pollen in both colonies could be observed and also in August 2011 (Temperature average: 23°C) for the managed colony (Fig 1).

## Daily Activities

The opening and closing of the cerumen tube at the nest entrance by the bees were observed daily and these movements marked respectively the starting and the ending of the activities for both colonies. The first activities of both colonies of *N. testaceicornis* occurred at the same time of the day. However, these times varied in different months of the year (Table 1; Fig 1). The first total activities and the first entrance activities with pollen occurred earlier (interval between 6:00 and 7:00) in December 2010, February 2011, April 2011, and October 2011 for the managed colony and in February 2011 and October 2011 for the unmanaged one. The first activities of the bees occurred later (between 8:00 and 9:00) in June 2011 (managed and unmanaged colonies) and August 2011 (managed colonies). During

these activities, the temperature usually ranged from 18°C (June 2011) to 20°C (in the other months), the relative humidity ranged from 76% to 90% and the values of light intensity were always higher than 1 lux.

The last total activities and collection of pollen by the bees of the two colonies occurred between 17:00 and 18:00, generally a little before sunset while the light intensity still above 1 lux, with the exception of February 2011 in which the activities ended later (between 18:00 and 19:00), month that values of temperature were higher (media 24.5°C) (Fig 1) and sunset occurred later (18:19).

Total activities (Fig 1) and the entrance with pollen (Fig 2) in the managed and unmanaged colonies occurred at a higher intensity in the morning until 12:00, declining after this time. Through the analysis of Circular Statistics was observed the presence of preferential times (or acrophases) for most of the flight activities of the bees for both colonies (Table 2).

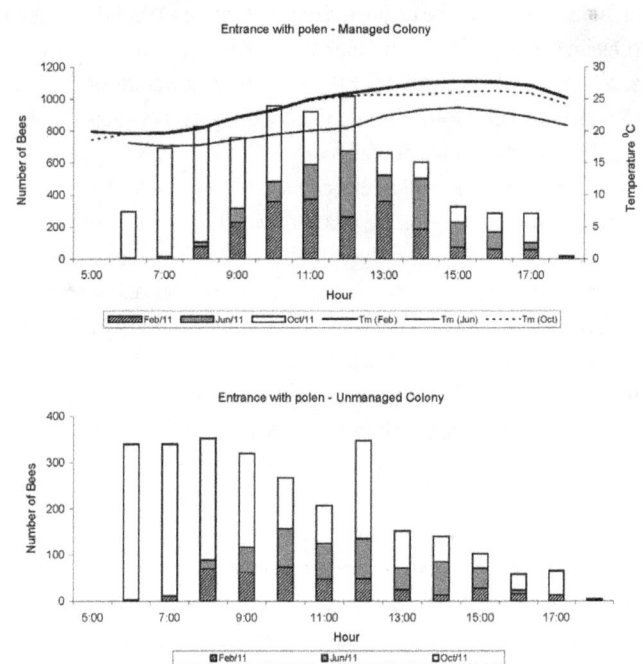

**Fig 2.** Entrance activity with pollen of *Nannotrigona testaceicornis* and temperature values in colony unmanaged and managed, in February, June and October/11 in the Valley of Capão (Chapada Diamantina) Bahia, Brazil. Tm: Temperature

Although no difference was found between the times of the first and last activities of the managed and unmanaged colonies, there was a difference in the acrophases of the colonies. In the unmanaged colony, located in an open area, the acrophases of all activities occurred generally earlier than in the managed colony (Table 2). The acrophases values for the unmanaged colony varied from 8:57 (October 2011) for the activity of entrance with pollen to 12:00 (June 2011) for the activity of exit. In the managed colony, the times of the acrophases varied from 9:31 (October 2011) for the entrance with pollen to 12:46 (June 2011) for the entrance into the nest (Table 2).

Furthermore, the Watson-Williams test showed significant difference between the acrophases of the two colonies in all

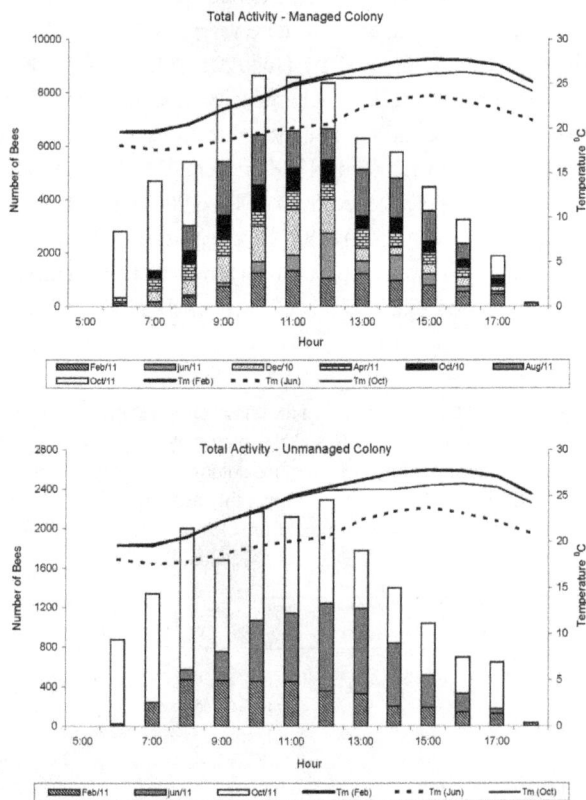

**Fig 1.** Total activity (entrance plus exit, without materials apparent) of *Nannotrigona testaceicornis* and temperature values in colony unmanaged (in February, June and October/11) and managed (from October 2010 to October 2011) in the Valley of Capão (Chapada Diamantina), Bahia, Brazil. Tm: Temperature.

**Table 2**. Acrophases times of exit and entrance activity; entrance with pollen and entrance with resin for *Nannotrigona testaceicornis* from October 2010 to October 2011 in the Valley of Capão (Chapada Diamantina, Bahia, Brazil).

| Bee/activity | Oct/10 | Dec/10 | Feb/11 | Apr/11 | Jun/11 | Aug/11 | Oct/11 |
|---|---|---|---|---|---|---|---|
| **Unmanaged colony** | | | | | | | |
| Entrance | | | 11:01 | | 12:21 | | * |
| Exit | | | 10:49 | | 12:00 | | 10:23 |
| Pollen | | | 10:27 | | 11:42 | | 08:57 |
| Resin | | | | | | | * |
| **Managed colony** | | | | | | | |
| Entrance | 11:25 | 11:03 | 12:18 | 11:12 | 12:46 | 11:43 | 09:45 |
| Exit | 11:20 | 10:48 | 12:02 | 11:06 | 12:22 | 11:32 | 09:51 |
| Pollen | 09:51 | 10:35 | 11:34 | 10:56 | 12:38 | 11:49 | 09:31 |
| Resin | 10:04 | 11:06 | 11:41 | * | 13:26 | 11:11 | |

*value non-significant ($r < 0.7$)

activities that were compared (entrance, exit, and entrance with pollen) in February 2011, June 2011, and October 2011 (Table 3).

All bees' acrophases were later in October 2010 than in October 2011. Although the value of mean temperature was the same in October 2010 and October 2011 (24°C), the light intensity was different. The mean of light intensity in open area was 17,160 lux in 2010 and 39,314 lux in 2011 and in the entrance of the managed colony, a light intensity reached 807 lux in 2010 and 2,446 lux in 2011.

The acrophases for the entrance with resin were detected only for the managed colony; this was generally later than for the other activities, varying from 10:04 (October 2010) to 13:26 (June 2011).

There were significant positive correlations (moderate) between the light intensity and entrance activities ($r = 0.48$, $p<0.01$) and exit ($r = 0.50$, $p<0.01$) in the managed colony at the open area. Also, there were significant positive correlations (moderate) between the light intensity at the entrance of the colony and the activities of entrance ($r = 0.40$, $p<0.01$) and exit ($r = 0.44$; $p<0.01$) in unmanaged colony. There were no significant correlations between the two colonies with temperature and also with the relative humidity (Table 4).

**Table 3**. Comparison of the daily flight activities of the managed and unmanaged colonies of *Nannotrigona testaceicornis* in February, June and October 2011 in the Valley of Capão (Chapada Diamantina, Bahia, Brazil). Significance determined by Watson-Williams test (F) ($p < 0.05$).

| Month/Activity | F | Critical value of F | Statistical significance |
|---|---|---|---|
| February/11 | | | S |
| Entrance | 316.93 | 0.05(1).1.7582 = 3.86 | S |
| Exit | 218.51 | 0.05(1).1.4968 = 3.86 | S |
| Pollen | 83.91 | 0.05(1).1.3719 = 3.86 | S |
| June/11 | | | S |
| Entrance | 62.56 | 0.05(1).1.5517 = 3.86 | S |
| Exit | 41.71 | 0.05(1).1.4268 = 3.86 | S |
| Pollen | 78.06 | 0.05(1).1.2148 = 3.86 | S |
| October/11 | | | |
| Entrance | 61.57 | 0.05(1).1.17794 = 3.86 | S |
| Exit | 90.95 | 0.05(1).1.13387 = 3.86 | S |
| Pollen | 45.45 | 0.05(1).1.5725 = 3.86 | S |

**Table 4**. Correlation analysis between abiotic variables (temperature, relative humidity, RH and light intensity in open area, LI–AO, and in hive entrance, LI-EC) and biotic variables (bees activities), of the managed (from October 2010 to October 2011) and the unmanaged colonies (in February, June and October 2011) in the Valley of Capão (Chapada Diamantina, Bahia).

| Managed colony | Temperature | LI – EC | LI – AO | RH |
|---|---|---|---|---|
| Entrance | 0.15 | 0.32 | 0.48* | -0.18 |
| Exit | 0.09 | 0.31 | 0.50* | -0.15 |
| Polen | -0.02 | 0.26 | 0.39 | -0.03 |
| Resin | 0.05 | 0.14 | 0.06 | 0.17 |
| **Unmanaged colony** | | | | |
| Entrance | 0.08 | 0.40* | 0.31 | -0.05 |
| Exit | -0.01 | 0.44* | 0.32 | 0.05 |
| Polen | -0.03 | 0.27 | 0.26 | 0.09 |
| Resin | -0.01 | 0.19 | 0.05 | -0.02 |

*moderate significance

## Discussion

The highest values of the activities in the managed colony of *N. testaceicornis* observed in all months of the study may be an indication that this colony is stronger than the unmanaged one, since the number of flights of the colony characterized it as weak, medium, or strong (Hilário et al., 2000). In general, the two colonies were more active in October 2011 and the managed colony also in August 2011.

The onset of the daily flight activities of *N. testaceicornis* showed regularity in the months investigated for both colonies. The beginning of activities occurred generally in the early morning and the end of the activities in the late afternoon for both colonies. This regularity was also observed by the occurrence of a preferential time for most part of the flight activities of the bees, which was statistically significant and occurred generally in the morning. Based on the first and last activities and on the acrophases observed in *N. testaceicornis,* it can be suggested that the two colonies showed a daily biological rhythm for most of the activities that were observed in the study. In literature, there is a great number of studies showing that daily flight activities occur at specific times through the day (Iwama, 1977; Kleinert-Giovannini, 1982; Guibu & Imperatriz-Fonseca, 1984; Kleinert-Giovannini & Imperatriz-Fonseca, 1986; Souza et al., 2006). However, few studies relate these activities with rhythmic aspects of the foraging behavior of the bees. Heard and Hendrikz (1993), working with *T. carbonaria* in Australia, noted the presence of activity peaks of the workers, indicating the presence of a daily periodicity independently from the meteorological variables considered (temperature, radiation, and relative humidity). Some studies in Brazil consider the pattern of the daily activity of Meliponini as rhythmic manifestations that may have an endogenous origin. Hilário et al. (2003) studying *Melipona bicolor* Lepeletier in the Southeast, and Gouw and Gimenes (2013) studying *Melipona scutellaris,* Latreille and *Frieseomelitta doederleini* (Friese) in the Northeast of Brazil observed the occurrence of the daily activity rhythm in these species.

The daily activities rhythm observed for the two colonies of *N. testaceicornis* may be synchronized by environmental meteorological factors or even by the time that the floral resources are available to the flower visitors. The influence of light/dark and photoperiodic cycles as synchronizers of the activities of *N. testaceicornis* can be observed in the first and last flight activities of both colonies since these activities occur very close to the time of sunrise and sunset, respectively. Moreover, the acrophases occur earlier in October and December (end of Spring and beginning of Summer, when the sun rises earlier) compared to June (Winter, when the sun rises later) in the area of the study, mainly for the managed colony. Besides this, in February 2011, the activities finished later than in the other months in both colonies, and in this month the sunset occurred later. The entrainment of the daily activities by the light/dark cycle was observed by Bellusci and Marques (2001) for the workers of *Scaptotrigona* aff. *depilis* (Moure), suggesting an endogenous character of this synchronization.

The entrainment of the daily rhythm of insect activities by the daily light/dark cycles and the annual photoperiodic cycle has already been discussed in the chronobiological literature (Saunders, 2002; Dunlap et al., 2004). This entrainment would lead to the adjustment of the time activities of the insects (such as bees) with the most favorable moments of the day, when the abiotic factors are favorable for the flight, allowing the organisms to anticipate the favorable or unfavorable cyclical environmental conditions (Dunlap et al., 2004).

The synchronization of the bees with the flowers may also be related to the time of the presentation of resource, as pollen or nectar. In this study, the workers from both colonies of *N. testaceicornis* presented their acrophases for the entrance with pollen in the morning, perhaps related to the greater availability of pollen by the plants at this time of the day, as also considered by Roubik (1989). Moreover, the light/dark and photoperiodic cycles can act in the synchronization of daily rhythm of foraging bees with the flower opening, the pollen exposure, and nectar production as were observed by Gimenes et al. (1993; 1996) in the interaction between flowers of *Ludwigia elegans* (Cambess.) H.Hara (Onagraceae) and flower-visiting bees.

The daily flight activities of both colonies of *N. testaceicornis* were probably influenced by the light intensity because it showed positive correlations between the activities of entrance and exit in both colonies. Moreover, the bees were never active when the light intensity was under 1 lux. The effects of light intensity can be better observed when the activities of both colonies of *N. testaceicornis* are compared. The fact that the acrophases of the unmanaged colony had occurred earlier may be related to the positive influence of light intensity. This influence is due to the higher values of light intensity that acts on the entrance of the unmanaged colony whose nest was located in an open area favoring the earlier exit of the bees.

Another point to be considered is the fact that in October 2011 all acrophases of the managed colony occurred earlier than in the same month in 2010. In both months, the temperature average did not vary, but the intensity light was lowest in 2010. Also, the light intensity seems influencing the end of the flight activity of *N. testaceicornis* in both colonies. According to Corbet et al. (1993) the flight activities of social bees can be more related to radiation than to temperature, especially at the onset of daily activity, next to sunrise. Lutz (1931) also observed the influence of light intensity at the beginning of flight activity of one species of Meliponini, *Trigona mosquito* (Smith). Such effect of light intensity could be modulator of the daily flight activities of bees. In this regard, Bellusci and Marques (2001) observed the endogenous circadian rhythm having light intensity as a modulator factor, in what clearer days influenced positively the increase of the external activities of *S.* aff. *depilis*.

Statistical analysis showed no significant effect of temperature on the activities of *N. testaceicornis*. However, the onset of the activities always occurred when the temperature was over 18°C. In other studies in the literature, it was observed a correlation between temperature and flight activities of the Meliponini, mainly for *Melipona* spp. (Souza et al., 2006; Ferreira-Junior et al., 2010), *Plebeia pugnax* Moure (in litt.) (Hilário et al., 2001), and *Tetragona clavipes* (Fabricius) (Rodrigues et al., 2007). The effect of the temperature on the activities of the bees may be related to the body size; species considered small (such as *N. testaceicornis*) start their activity at a higher temperature than those that are of a larger size. According to Teixeira and Campos (2005) the onset of the external activity of bees of a small body size such as *Plebeia droryana* (Friese), *Frieseomelitta varia* (Lepeletier), *Friesella schrottkyi* (Friese), and *N. testaceicornis* occurred at temperature values ranging from 18°C to 21°C, while larger Meliponini such as *Melipona quadrifasciata anthidioides* Lepeletier and *M. bicolor* started their activities at lower temperatures (ca. 12°C).

The fact of no significant correlation between the daily flight activities of *N. testaceicornis* and the temperature may be related to the fact that bees living in tropical regions, where the average temperature ranges from 20°C to 30°C, so, there are not great thermal stresses (Heinrich, 1993). However, Heard and Hendrikz (1993) correlated daily activity with both climatic factors, temperature and light intensity in flight activities of *T. carbonaria*.

The results from this study showed the influence of light/dark cycle in different months of the year and also the light intensity in the daily flight activities of *N. testaceicornis* in an area of the Brazilian semiarid, where temperatures are relatively favorable to the flight of bees. Therefore, the synchronization of the flight activities with the times of day, and favorable climatic factors can ensure bee survival, especially for Meliponini of small size, as *N. testaceicornis*.

## Acknowledgments

The authors would like to thank Dr. Favízia F. de Oliveira (Universidade Federal da Bahia - UFBA), for identifying the bee species; Mr. Lars Erich Rellstab for their permission to work in the area with their bees; and CAPES for the Masters study grant.

# References

Batschelet, E. (1980). Circular Statistics in Biology. London: Academic Press, 371 p.

Bellusci, S. & Marques, M.D. (2001). Circadian activity rhythm of the foragers of a eusocial bee (*Scaptotrigona* aff. *depilis*, Hymenoptera, Apidae, Meliponinae) outside the nest. Biol. Rhythm Res. 32: 117-124.

Bloch, G. (2008). Socially mediated plasticity in the circadian clock of social insects. In: J. Gadau & J. Fewell (Eds.), Organization of Insect Societies-From Genomes to Socio-Complexity. Cambridge: Harvard University Press.

Corbet, S.A., Fussell, M., Ake, R., Fraser, A., Gunson, C., Savage, A. & Smith, K. (1993). Temperature and pollination activity of social bees. Ecol. Entomol. 18: 17-30.

Dunlap, J.C., Loros, J.J. & Decoursey, P.J. (2004). Chronobiology: biological timekeeping. Massachusetts: Inc. Publishers, Sinauer Associates, 382 p.

Ferreira-Junior, N.T., Blochtein, B. & Moraes, J.F. de. (2010). Seasonal flight and resource collection patterns of colonies of the stingless bee *Melipona bicolor schencki* Gribodo (Apidae, Meliponini) in an Araucaria Forest area in southern Brazil. Rev. Bras. Entomol. 54: 630-636.

Figueiredo-Filho, D.B., Silva Jr., J.A. (2009). Desvendando os Mistérios do Coeficiente de Correlação de Pearson (r). Rev. Política Hoje 18: 115-146.

Gimenes, M., Benedito-Silva, A.A. & Marques, M.D. (1993). Chronobiologic aspects of a coadaptive process: the interaction of *Ludwigia elegans* flowers and its more frequent bee visitors. Chronobiol. Int. 10: 20-30.

Gimenes, M., Benedito-Silva, A.A. & Marques, M.D. (1996). Circadian rhythms of pollen and nectar collection by bees on the flowers of *Ludwigia elegans* (Onagraceae). Biol. Rhythm Res. 27: 281-290.

Gouw, M.S. & Gimenes, M. (2013). Differences of the Daily Flight Activity Rhythm in two Neotropical Stingless Bees (Hymenoptera, Apidae). Sociobiology 60: 183-189.

Guibu, L.S. & Imperatriz-Fonseca, V.L. (1984). Atividade externa de *Melipona quadrifasciata* Lepeletier (Hymenoptera, Apidae, Meliponinae). Ciênc. Cult. 36: 623.

Hammer, O., Harper, D.A.T. & Ryan, P.D. (2001). Past: Paleontological Statistics software package for education and analysis. http://palaeo-electronica.org/2001_1/past/issue1_01.htm> (accessed date: December, 2011).

Heard, T.A. & Hendrikz, J.K. (1993). Factors influencing flight activity of colonies of the stingless bee *Trigona carbonaria* (Hymenoptera: Apidae). Austral J. Zool. 41: 343-353.

Heinrich, B. (1993). The hot-blooded insects. Strategies and mechanisms of thermoregulation. Massachusetts: Harvard University Press, 600 p.

Hilário, S.D., Imperatriz-Fonseca, V.L. & Kleinert, A.M.P. (2000). Flight activity and colony strength in the stingless bee *Melipona bicolor bicolor* (Apidae, Meliponinae). Rev. Brasil. Biol., 60: 299-306.

Hilário, S.D., Imperatriz-Fonseca, V.L. & Kleinert, A.M.P. (2001). Responses to climatic factors by foragers of *Plebeia pugnax* Moure (In Litt.) (Apidae, Meliponinae). Rev. Brasil. Biol. 61: 191-196.

Hilário, S.D., Gimenes, M. & Imperatriz-Fonseca, V.L. (2003). The influence of colony size in diel rhythms on flight activity of *Melipona bicolor* Lepeletier, 1836. In G.A.R. Melo & I. Alves dos Santos (Eds.), Apoidea Neotropica: Homenagem aos 90 anos de Jesus Santiago Moure (pp. 191-197). Criciúma: UNESC.

Hilário, S.D., Ribeiro, M.F. & Imperatriz-Fonseca, V.L. (2007). Impacto da precipitação pluviométrica sobre a atividade de voo de *Plebeia remota* (Holmberg, 1903) (Apidae, Meliponini). Biota Neotrop. 7: 135-143.

Iwama, S. (1977). A influência dos fatores climáticos na atividade externa de *Tetragonisca angustula* (Apidae, Meliponinae). Bol. Zool. Univ. S. Paulo 2: 189-201.

Kleinert-Giovannini, A. (1982). The influence of climatic factors on flight activity of *Plebeia emerina* Friese (Hymenoptera, Apidae, Meliponinae) in winter. Rev. Bras. Entomol. 26:1-13.

Kleinert-Giovannini, A. & Imperatriz-Fonseca, V.L. (1986). Flight activity and responses to climatic conditions of two subspecies of *Melipona marginata* Lepeletier (Apidae, Meliponinae). J. Apic. Res. 25: 3-8.

Lutz, F.E. (1931). Light as a factor in controlling the start of daily activity of a wren and stingless bees. Am. Mus. Nat. Hist. 468: 1-4.

Moore, D. (2001). Honey bee circadian clocks: behavioral control from individual workers to whole-colony rythms. J. Insect Physiol. 47: 843-857.

Moure J.S., Urban D. & Melo G.A.R. (2012). Catalogue of Bees (Hymenoptera, Apoidea) in the Neotropical Region - online version. http://www.moure.cria.org.br/catalogue (accessed date: 25 June, 2012).

Nunes-Silva, P., Hilário, S.D., Santos Filho, P.S. & Imperatriz-Fonseca, V.L. (2010). Foraging activity in *Plebeia remota*, a stingless bees specie, is influenced by the reproductive state of a colony. Psyche, 2010, Article ID 241204, doi:10.1155/2010/2412041-16.

Queiroz, L.P., França, F., Giulietti, A.M., Melo, E., Gonçalves, C.N., Funch, L.S., Harley, R.M., Funch, R.R. & Silva, T.R.S. (2005). Caatinga. In F.A. Juncá, L.S. Funch & W. Rocha. (Eds.), Biodiversidade e conservação da Chapada Diamantina (pp. 95-120). Brasília: Ministério do Meio Ambiente.

Rodrigues, M., Santana, W.C., Freitas, G.S. & Soares A.E.E. (2007). Flight activity of *Tetragona clavipes* (Fabricius,

1804) (Hymenoptera, Apidae, Meliponini) at the São Paulo University campus in Ribeirão Preto. Biosci. J. 23: 118-124.

Roubik, D.W. (1989). Ecology and natural history of tropical bees. Cambridge: Cambridge University Press, 514 p.

Saunders, D.S. (2002). Insect Clocks. Amsterdam: Elsevier Science, 576 p.

SEI. (2011). Superintendência de Estudos Econômicos e Sociais da Bahia. Estatística dos municípios baianos. Salvador, 434 p.

Souza, B.A., Carvalho, C.A.L. & Alves, R.M.O. (2006). Flight activity of *Melipona asilvai* Moure (Hymenoptera: Apidae). Braz. J. Biol. 66: 731-737.

Teixeira, L.V. & Campos, F.N.M. (2005). Início da atividade de vôo em abelhas sem ferrão (Hymenoptera, Apidae): influência do tamanho da abelha e da temperatura ambiente. Rev. Bras. Zoociênc. 7: 195-202.

Zar, J.H. (2010).Biostatistical Analysis. 5th Ed. New Jersey: Pearson Prentice-Hall, Upper Sandller River, 944 p.

# 24

# Response of Ants to the Leafhopper *Dalbulus quinquenotatus* DeLong & Nault (Hemiptera: Cicadellidae) and Extrafloral Nectaries Following Fire

G. Moya-Raygoza[1], K.J. Larsen[2]

*1 - Universidad de Guadalajara, CUCBA, Jalisco, Mexico.*

*2 - Department of Biology, Luther College, Decorah, USA.*

## Keywords

*Acacia pennatula,* five-spotted gamagrass leafhopper, *Tripsacum dactyloides,* mutualism

## Corresponding author

Gustavo Moya-Raygoza, Ph.D.
Departamento de Botánica y Zoología
CUCBA, Universidad de Guadalajara
Km 15.5 carretera Guadalajara-Nogales
Las Agujas, Zapopan, C.P. 45110
Apdo. Postal 139, Jalisco, México
E-mail: moyaraygoza@gmail.com

## Abstract

Previous investigations of mutualistic associations between ants and plants bearing extrafloral nectaries (EFNs) or between ants and trophobiont leafhoppers have studied these relationships separately, but nothing is known on how ant abundance responds to these two food resources occurring in the same habitat when that habitat is disturbed by fire. The objectives of this study are to document ant abundance with the trophobiont five-spotted gamagrass leafhopper, *Dalbulus quinquenotatus* DeLong & Nault, and with EFNs on trees of *Acacia pennatula* (Schlecht & Cham.) Benth. (Fabaceae) that occur in the same habitat, and how ant abundance in both of these mutualisms is affected after disturbance by fire. This study was performed at several sites in central Mexico where the perennial gamagrass *Tripsacum dactyloides* L. (Gramminae) and *A. pennatula* both occur. More ants were collected in association with the leafhopper *D. quinquenotatus* than with EFNs of *A. pennatula*. At sites where dry season fire occurred, new green leaves were produced by both *T. dactyloides* and *A. pennatula* after the burn. On these new leaves after fire, significantly more ants tended *D. quinquenotatus* leafhoppers on *T. dactyloides* than visited EFNs on *A. pennatula*. In burned sites the ants *Anoplolepis gracilipes* Smith, *Brachymyrmex obscurior* Forel and *Pheidole* sp. live in association with the leafhoppers, whereas EFNs on *A. pennatula* were associated with the ants *A. gracilipes, B. obscurior, Camponotus* sp., *Crematogaster* sp. and *Solenopsis* sp.

## Introduction

Ants (Hymenoptera: Formicidae) often live in mutualistic relationships with trophobiont insects that excrete honeydew, or with plants bearing extrafloral nectaries (EFNs) that produce nectar (Hölldobler & Wilson, 1990). Some species of aphids, whiteflies, scale insects, mealybugs, treehoppers, and leafhoppers (Hemiptera) live in facultative or obligatory mutualistic relationships with ants (Way, 1963; Buckley, 1987; Blüthgen et al., 2006; Gibb & Cunningham, 2009; Fagundes et al., 2013). In these associations, the insect provides honeydew, a sugary excretion of carbohydrates, amino acids, and water for the ants, whereas ants protect the hemipterans from natural enemies (Delabie, 2001; Heil & Mckey, 2003; Zhang et al., 2012; Zhang et al., 2013). Plants in over one hundred families bear EFNs that produce secretions rich in sugars, amino acids, and lipids that attract ants, and in return these ants protect those plants from herbivores (González-Teuber & Heil, 2009; Byk & Del-Claro, 2011; Marazzi et al., 2013; Weber & Keeler, 2013).

Ants are attracted to high quality sugar resources as food (Heil & McKey, 2003). Previous studies have shown that when both honeydew and extrafloral nectar are offered to ants, ants are more abundant at the honeydew rather than at exudates of EFNs (Fiala, 1990; Rashbrook et al., 1992; Del-Claro & Oliveira, 1993; Blüthgen et al., 2000; Katayama et al., 2013). Ants were more abundant tending the hemipterans, particularly when greater numbers of hemipterans are present because of the larger quantities of honeydew produced (Katayama & Suzuki, 2010). Blüthgen et al. (2000) found greater numbers of ants at honeydew resources as opposed to EFN resources because honeydew is apparently a higher quality food resource, rich in amino acids. Moreover, Katayama and Suzuki (2003) demonstrated that if an aphid colony increases in size, ants stop using EFNs and strengthen their mutualistic association with aphids.

Fire affects the growth of plants because some perennial species such as grasses and plants bearing EFNs quickly re-grow after disturbance occurs. New leaves formed after the plants burn are ready to be colonized directly or indirectly by ants, often attracted to food resources such as the honeydew produced by leafhoppers that feed on young grasses (Moya-Raygoza, 1995) or from nectar produced by EFNs (Alves-Silva & Del-Claro, 2013). Ants respond to burned plants with EFNs or hemipterans in the same way. The abundance of ants increased on the shrub *Banisteriopsis campestris* (A. Juss.) which bears EFNs after fire, mainly because of concentrated extrafloral nectar (Alves-Silva & Del Claro, 2013). Alves-Silva (2011) and Koptur et al. (2010) also found a richer ant community guarding plants from herbivory after fire because of the production of extrafloral nectar. Similarly, higher numbers of ants were found tending the honeydew-producing fivespotted gamagrass leafhopper, *Dalbulus quinquenotatus* DeLong & Nault, after its host plant, the perennial gamagrass *Tripsacum dactyloides* L. (Gramminae), was burned (Moya-Raygoza, 1995).

Mutualisms between ants and EFNs-bearing plants and ants and trophobiont hemipterans have been investigated separately after disturbance by fire, but little is known how ant abundance responds to these two food resources when present in the same habitat. This study was performed in central Mexico, where the perennial gamagrass *T. dactyloides* hosts *D. quinquenotatus* leafhoppers and trees of *Acacia pennatula* (Schlecht & Cham.) Benth. (Fabaceae) with EFNs occur together in the same habitats (Fig. 1a). These sites often are accidentally burned, and the fire often kills or drives away insects living on those plants. *Dalbulus quinquenotatus* lives on the basal leaves of *T. dactyloides* in an obligatory mutualism with tending ants (Larsen et al., 1991). Ants tending *D. quinquenotatus* receive honeydew and protect this leafhopper from natural enemies (Moya-Raygoza & Nault, 2000). In contrast, *A. pennatula* have EFNs and live in a mutualistic relationship with ants (Moya-Raygoza, 2005), providing nectar for the ants in return for protection from herbivory.

Fire is an important abiotic factor in mutualisms because it affects plant re-growth and the abundance of ants that depend on exudates produced indirectly by trophobiont insects and directly by EFNs. When fire consumes the foliage of both *T. dactyloides* and *A. pennatula*, the mutualisms involving ants with both species are temporarily disrupted. However, only a few days after being burned, new leaves of both plant species begin to re-grow (Fig. 1b) and are soon colonized by *D. quinquenotatus* and ants in the case of *T. dactyloides*, or by ants visiting EFNs in the case of *A. pennatula*. Measuring the total abundance of ants collecting honeydew from *D. quinquenotatus* and visiting EFNs resources before and after

the host plants are burned helps us understand the ecological importance of mutualisms that can be strong driving forces for community organization (Wimp & Whitham, 2001). The objectives of this study are to document ant abundance with *D. quinquenotatus* leafhoppers and EFNs in the same habitat, and how ant abundance in both of these mutualisms is affected after disturbance to their habitat by fire.

**Materials and Methods**

*Study system*

Nine field sites containing both *T. dactyloides* and *A. pennatula* were chosen for this study. Each site had both species of plant present and covered an area of 0.05-0.25 ha. All sites were in the state of Jalisco in Central Mexico. The sites were: 1) El Arenal: 1,501 m elev, 20°46.032′N, 103°40.766′W; 2) Los Chorros: 1,371 m elev, 20°41.211′N, 103°41. 558′W; 3) San Isidro: 1,266 m elev, 20°49.014′N, 103°20. 262′W; 4) Agua Caliente: 1,385 m elev, 20°25.770′N, 103°41.485′W; 5) Cocula: 1,273 m elev, 20°25.595′N, 103°44.601′W; 6) San Agustin: 1,638 m elev, 20°30.682′N, 103°28.796′W; 7) La Mimila: 1,649 m elev, 20°44.411′N, 103°37.686′W; 8) El Molino: 1,608 m elev, 20°23.938′N, 103°32.760′W; and 9) Zapopan: 1,631 m elev, 20°44.283′N, 103°30.805′W (Fig. 1c). The closest sites were 5.45 km apart (Agua Caliente and Cocula) while the most distant sites (San Isidro and Cocula) were 60.44 km apart. All sites had similar habitat characteristics and belong to pine-oak ecosystem (Rodríguez-Trejo & Myers, 2010). Plants of both species live on steep slopes or beside roadways and grow on limestone soils (Wilkes, 1972). The sites had similar vegetation consisting of a plant community containing *T. dactyloides* interspersed with *A. pennatula* trees and few other plants such *Lysilona* sp. Each *T. dactyloides* population was composed of scattered clumps consisting of clusters of stems. *Tripsacum dactyloides* can use rhizomes to spread across the landscape and does not possess extrafloral nectaries. Moreover, ants are present on *T. dactyloides* only when the plants are hosts for *D. quinquenotatus* leafhoppers as compared with plants without *D. quinquenotatus* (Larsen, et al. 1991).

All sites were sampled to determine the numbers of ants when leafhoppers and EFNs were available. *Acacia pennatula* has actively secreting extrafloral nectaries on young leaves primarily from April to June (McVaugh, 1987; Moya-Raygoza, 2005), whereas leafhoppers are present on *T. dactyloides* primarily during the wet season from June to September (Moya-Raygoza, 1995) when these habitats are not burned. We observed that when the habitats were burned, both plant species started to produce new green leaves within several days, and this altered the food resources available for visiting ants. Fires generally occur from March to May towards the end of each dry season. The dry season in Jalisco generally occurs from October to May and is characterized by lower rainfall, lower temperatures and shorter days as compared with the wet season which typically lasts from June to September (Mosino-Aleman & Garcia, 1974). After burning, both honeydew and EFN nectar food resources for ants are found in May and June within the same plant community. The highest nectar secretion rates have been documented from EFNs on young leaves of damaged plants (Heil et al., 2004), while high numbers of *D. quinquenotatus* leafhoppers have been found on *T. dactyloides*

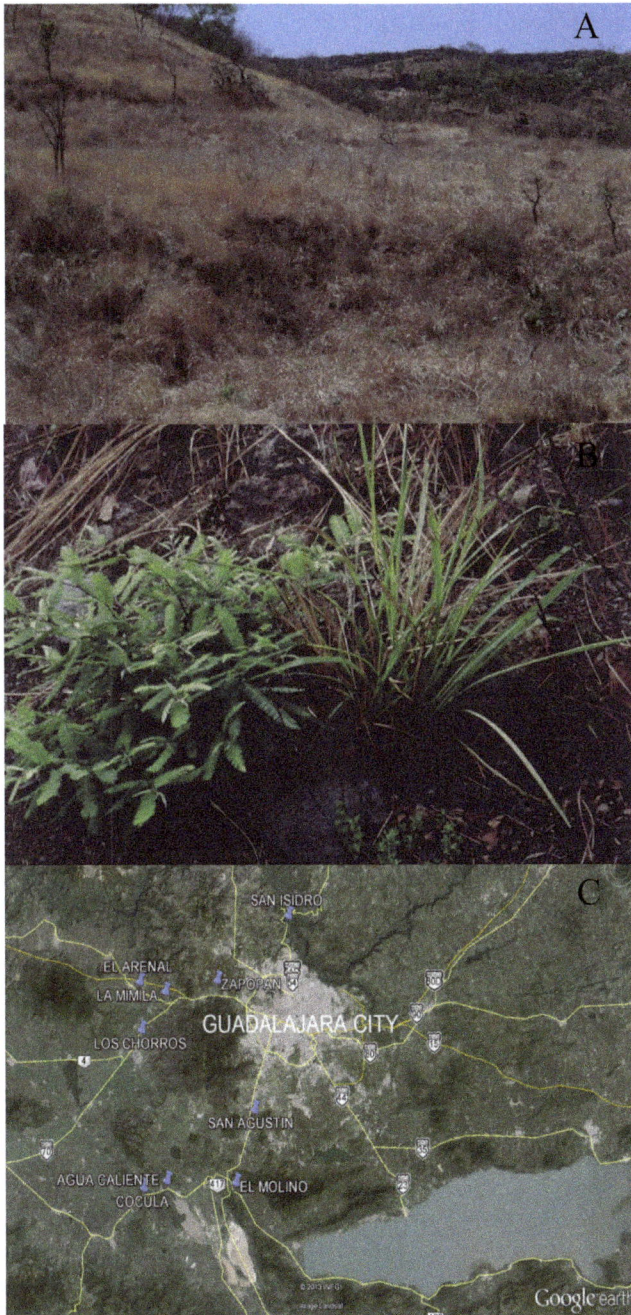

**Fig 1**. Tritrophic interaction trophobiont five-spotted gamagrass leafhopper-Ants-*Acacia pennatula*. A) Hillside in Jalisco, Mexico late in the dry season covered with *T. dactyloides* hosting *D. quinquenotatus* leafhoppers, interspersed with trees of *A. pennatula* bearing EFNs in unburned site. B) Young green leaves growing on *A. pennatula* (left) and *T. dactyloides* (right) several days after being burned by fire. C) Location of field sites containing both *T. dactyloides* and *A. pennatula* from the state of Jalisco in central Mexico.

after fire (Moya-Raygoza, 1995). No data were collected between October and April because ants do not visit either of these food resources during that time. The wet season begins in June, and no fires occur once the rains begin to fall.

*Sampling*

We confirmed the presence of ants associated with EFNs of *A. pennatula* and *D. quinquenotatus* at each site. Once these fires took place, we sampled ants on burned and unburned sites. We selected *A. pennatula* trees at each site and neighboring clumps of *T. dactyloides*. Ten terminal branches on each selected *A. pennatula* tree and one basal leaf from each of ten different *T. dactyloides* clumps were randomly selected. Terminal branches of *A. pennatula* were selected because the highest concentration of EFNs occurs on these branches, whereas basal leaves of *T. dactyloides* were selected because this is where the highest numbers of *D. quinquenotatus* are found. We collected all nymphs and adults of *D. quinquenotatus* leafhoppers and all tending ants from the basal 10 cm of each selected *T. dactyloides* stem.

All EFNs were counted and ants collected from the terminal 10 cm of each selected *A. pennatula* branch. Therefore ant abundance at each resource was quantified on one stem or branch for each of 10 separated plants of *T. dactyloides* and *A. pennatula* by site. We selected the same 10 cm surface on both plant species to have approximately the same area of food resource available for the ants. Sampling at all sites was performed between 09:00 and 14:00 h, one site per day during the last week of May 2007, first week of June 2012, and the second week of September 2012. The Arenal and Los Chorros sites were burned in May 2007, while the Zapopan and Los Chorros sites were burned in June 2012. *Dalbulus quinquenotatus*, EFNs and ants were sampled approximately one month after each fire. Ants were sampled at these times because both extrafloral nectar produced by *A. pennatula* and honeydew produced by *D. quinquenotatus* was present. All collected insects were stored in 70% ethanol and returned to the lab for identification.

Analysis of Deviance, using R.3.1.0 for Windows (R Project), was performed to evaluate the interaction (resource for ants, honeydew-extrafloral nectar × disturbance, fire-without fire) on the number of ants. This comparison included the ant abundance obtained on the three sampling dates. Furthermore, the total number of ants tending *D. quinquenotatus* on *T. dactyloides* was compared vs the total number of ants on *A. pennatula* bearing EFNs with a Wilcoxon test using SPSS 12 for Windows. Therefore a comparison of ant abundance at leafhoppers vs EFNs was conducted when combining both burned and unburned resources in the three sample dates. Average and standard error were determined for the number of *D. quinquenotatus* nymphs and adults, tending ants, and EFNs for each burned and unburned site.

## Results

Ant species collected from burned *T. dactyloides* associated with the leafhopper *D. quinquenotatus* were *Anoplolepis gracilipes* Smith, *Pheidole* sp., and *Brachymyrmex obscurior* Forel. Ants found on unburned *T. dactyloides* tending *D. quinquenotatus* included *A. gracilipes* and *B. obscurior*. Greater ant species richness was associated with *A. pennatula*. *Anoplolepis gracilipes*, *B. obscurior*, *Camponotus* sp., *Crematogaste* sp. and *Solenopsis* sp. were found at EFNs when *A. pennatula* was burned. Ant taxa found visiting the EFNs of unburned *A. pennatula* included *A. gracilipes*, *B. obscurior*, *Crematogaster* sp., *Dorymyrmex* sp. and *Pheidole* sp. We observed these ants differed in body size and likely collect and store honeydew or extrafloral nectar differently.

New green leaves were produced by both *T. dactyloides* and *A. pennatula* after they were burned. Disturbance by fire does not have the same effect on the numbers of ants tending *D. quinquenotatus* and visiting EFNs. We found an interaction between fire and plant species, and significantly more ants were found tending *D. quinquenotatus* leafhoppers on *T. dactyloides* than visiting EFNs ($Z = 7.63$; $P = 0.001$). Rapid colonization of new growth on *T. dactyloides* by ants and leafhoppers was observed after burning in the last week of May 2007. At this time only adult leafhoppers were observed in the two burned sites tended by a great number *B. obscurior* ants, while EFNs were visited by few ants of *Solenopsis* sp. at the two burned sites (Table 1 and Fig. 2). In June 2012, leafhoppers were tended by *Pheidole* sp. and a great number of nymphs were tended by great numbers of *B. obscurior* ants at the two 2012 burned sites (Table 2 and Fig. 3). Near the end of the wet season in September 2012, four months after the June fire, a large number of leafhopper nymphs were tended by larger numbers of *B. obscurior* ants, while low numbers of *A. gracilipes*, *Camponotus* sp.,

*Crematogaster* sp. and *B. obscurior* ants visited the EFNs at the two burned sites (Table 3 and Fig. 4).

The number of ants tending leafhoppers was significantly higher than the number of ants found visiting EFNs of *A. pennatula* when combining both burned and unburned resources in the three sample dates (Wilcoxon = 299.50; $Z = 3.04$; $P = 0.002$). Leafhoppers and ants were found together at the end of the dry season in May 2007 on the six unburned sites, while only in two of the six unburned sites ants visited the EFNs of *A. pennatula* (Table 1 and Fig. 2). In June 2012, at the end of the dry season, no ants or leafhoppers were found on the leaves of unburned *T. dactyloides* plants that were dried out (Table 2 and Fig. 3). In September 2012, at the end of the wet season, only in one of the four unburned sites ants visited the EFNs of *A. pennatula*, whereas in these four unburned sites ants tended the leafhoppers (Table 3 and Fig. 4).

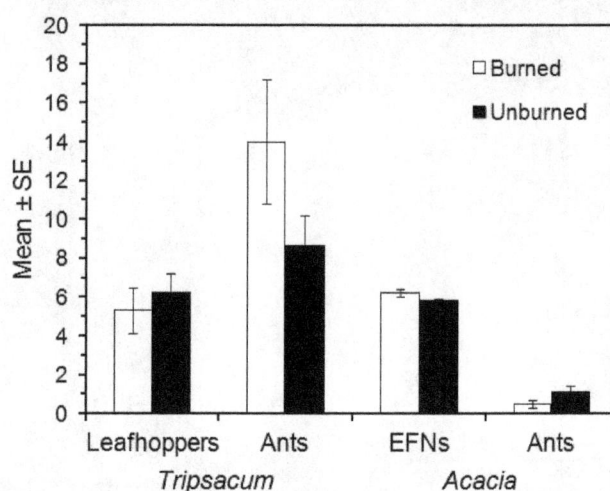

**Fig 2.** Average number (± standard error) of leafhoppers, ants tending *D. quinquenotatus* leafhoppers on *T. dactyloides*, EFNs, and ants visiting EFNs on *A. pennatula* from burned and unburned sites in Jalisco, Mexico in the last week of May 2007.

**Table 1.** Average number (± standard error) of *Dalbulus quinquenotatus* nymphs, adults, and tending ants (and species of tending ant), *Acacia pennatula* EFNs, and ants on 10 stems and 10 branches of *T. dactyloides* and *A. pennatula* respectively in burned (in May 2007) and unburned sites at locations in Jalisco, Mexico at the end of the dry season in May 2007.

| Site | Ant/Leafhopper interaction on *Tripsacum dactyloides* | | | | Ant/*Acacia* interaction | | |
|---|---|---|---|---|---|---|---|
| | *Dalbulus quinquenotatus* | | Ants | Ant species | *A. pennatula* EFNs | Ants | Ant Species |
| | Nymphs | Adults | | | | | |
| Both resources burned | | | | | | | |
| 1. Arenal | 0 | 9.2 ± 1.7 | 7.7 ± 1.4 | *B. obscurior* | 5.6 ± 0.1 | 1.0 ± 0.4 | *Solenopsis* sp. |
| 2. Los Chorros | 0 | 1.4 ± 0.3 | 20.4 ± 5.7 | *B. obscurior* | 6.9 ± 0.3 | 0 | - |
| Both resources unburned | | | | | | | |
| 3. San Isidro | 3.9 ± 2.2 | 1.7 ± 0.7 | 15.1 ± 5.7 | *B. obscurior* | 5.8 ± 0.2 | 0 | - |
| 4. Agua Caliente | 0.7 ± 0.5 | 0.5 ± 0.4 | 1.7 ± 1.3 | *B. obscurior* | 5.4 ± 0.1 | 1.0 ± 0.2 | *Pheidole sp.* |
| 5. Cocula | 10.2 ± 1.2 | 1.2 ± 0.4 | 4.3 ± 0.8 | *B. obscurior* | 5.4 ± 0.1 | 0 | - |
| 6. San Agustin | 6.5 ± 3.3 | 1.9 ± 1.1 | 13.0 ± 5.1 | *B. obscurior* | 6.6 ± 0.1 | 0 | - |
| 7. La Mimila | 3.7 ± 1.9 | 1.1 ± 0.5 | 5.7 ± 4.3 | *B. obscurior* | 5.5 ± 0.2 | 0 | - |
| 8. El Molino | 2.0 ± 0.6 | 3.9 ± 1.2 | 12.1 ± 2.9 | *B. obscurior* | 6.6 ± 0.1 | 5.8 ± 1.4 | *B. obscurior* |

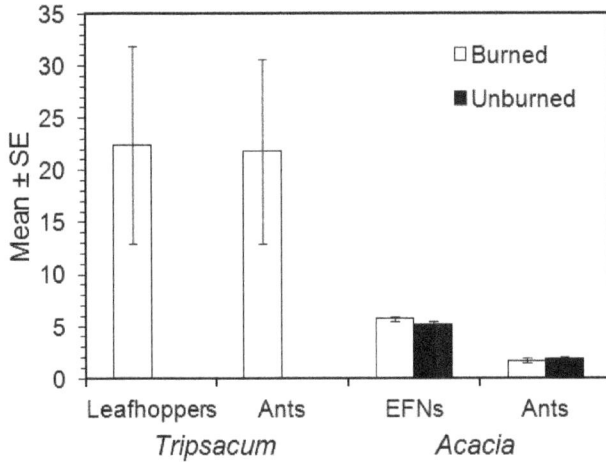

**Fig 3.** Average number (± standard error) of leafhoppers, ants tending *D. quinquenotatus* leafhoppers on *T. dactyloides*, EFNs, and ants visiting EFNs on *A. pennatula* from burned and unburned sites in Jalisco, Mexico in the first week of June 2012.

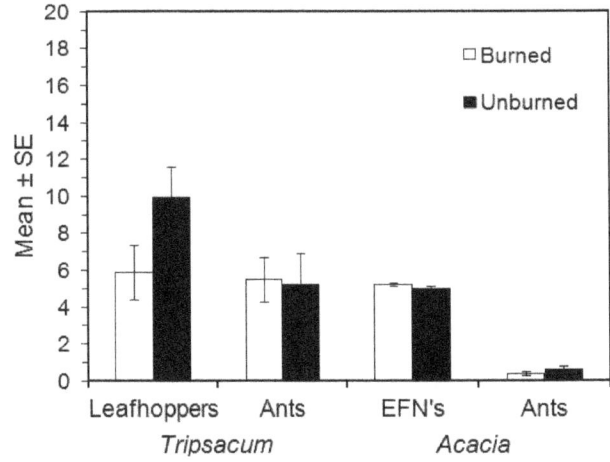

**Fig 4.** Average number (± standard error) of leafhoppers, ants tending *D. quinquenotatus* leafhoppers on *T. dactyloides*, EFNs, and ants visiting EFNs on *A. pennatula* from burned and unburned sites in Jalisco, Mexico in the second week of September 2012.

## Discussion

The exudates honeydew and extrafloral nectar are key factors determining the abundance of ants when both food resources for ants are present (Buckley, 1983; Fiala, 1990; Rashbrook et al., 1992; Del-Claro & Oliveira, 1993; Blüthgen et al., 2006; Katayama et al., 2013). Considering the abundance of ants tending the leafhopper *D. quinquenotatus* compared with the abundance of ants visiting EFNs, more ants were collected in association with *D. quinquenotatus* than with EFNs on *A. pennatula*. This finding is similar to the results of other studies (Fiala, 1990; Rashbrook et al., 1992; Del-Claro & Oliveira, 1993; Blüthgen et al., 2000; Katayama & Suzuki, 2003; Katayama & Suzuki, 2010; Katayama et al., 2013) comparing ant abundance at honeydew-producing insects with plants with EFNs in non-disturbed conditions. In the

rainforest canopy, ants are usually more abundant at honeydew than extrafloral nectar, as honeydew is apparently a more valuable resource to ants than nectar from EFNs (Blüthgen et al., 2000). Ants (*Camponotus* sp.) also did not stop tending the honeydew-producing membracids (*Guayaquila xiphias* Fabricius) when an alternative EFN sugar source was available on *Didymopanax vinosum* (Cham. & Schltdl.), their host plant (Del-Claro & Oliveira, 1993). Recently Katayama et al. (2013) demonstrated that the ant *Lasius japonicus* Santsci switches from visiting EFNs on the bean plant *Vicia faba* L. to the aphid *Aphis craccivora* Koch, because the density and total food reward to ants from the aphids exceed that from EFNs.

We ascribe the difference in abundance between ants visiting the leafhopper *D. quinquenotatus* and EFNs on *A. pennatula* to several factors. First, *D. quinquenotatus* leaf-

**Table 2.** Average number (± standard error) of *Dalbulus quinquenotatus* nymphs, adults, and tending ants (and species of tending ant), *Acacia pennatula* EFNs, and ants on 10 stems and 10 branches of *T. dactyloides* and *A. pennatula* respectively in burned (in June 2012) and unburned sites at locations in Jalisco, Mexico in June 2012.

| Site | Ant/Leafhopper interaction on *Tripsacum dactyloides* | | | | Ant/*Acacia* interaction | | |
|------|----------------|--------|------|-------------|----------------------|------|-------------|
| | *Dalbulus quinquenotatus* | | Ants | Ant species | *A. pennatula* EFNs | Ants | Ant Species |
| | Nymphs | Adults | | | | | |
| Both resources burned | | | | | | | |
| 1. Zapopan | 4.4 ± 2.3 | 2.4 ± 0.7 | 4.8 ± 1.8 | *Pheidole* sp. | 5.6 ± 0.1 | 1.9 ± 0.2 | *A. gracilipes* Camponotus sp. |
| 2. Los Chorros | 37.0 ±17.9 | 1.1 ± 0.5 | 38.9 ± 16.2 | *B. obscurior* | 5.8 ± 0.3 | 1.5 ± 0.4 | *Crematogaster* sp. *B. obscurior* |
| Both resources unburned | | | | | | | |
| 3. San Isidro | 0 | 0 | 0 | - | 6.5 ± 0.4 | 2.5 ± 0.5 | *Dorymyrmex* sp. *Crematogaster* sp. |
| 4. San Agustin | 0 | 0 | 0 | - | 6.2 ± 0.4 | 0.3 ± 0.1 | *B. obscurior* |
| 5. La Mimila | 0 | 0 | 0 | - | 1.4 ± 0.7 | 1.1 ± 0.5 | *A. gracilipes* |
| 6. El Arenal | 0 | 0 | 0 | - | 6.3 ± 0.4 | 3.3 ± 0.9 | *A. gracilipes* |

**Table 3.** Average number (± standard error) of *Dalbulus quinquenotatus* nymphs, adults, and tending ants (and species of tending ant), *Acacia pennatula* EFNs, and ants on 10 stems and 10 branches of *T. dactyloides* and *A. pennatula* respectively in burned (in June 2012) and unburned sites at locations in Jalisco, Mexico at the end of the wet season in September 2012.

| Site | Ant/Leafhopper interaction on *Tripsacum dactyloides* | | | | Ant/*Acacia* interaction | | |
| | *Dalbulus quinquenotatus* | | Ants | Ant species | *A. pennatula* EFNs | Ants | Ant Species |
| | Nymphs | Adults | | | | | |
|---|---|---|---|---|---|---|---|
| Both resources burned | | | | | | | |
| 1. Zapopan | 0.9 ± 0.7 | 0.5 ± 0.2 | 1.1 ± 0.5 | *A.gracilipes* | 5.5 ± 0.1 | 0.9 ± 0.2 | *A. gracilipes* *Camponotus* sp. |
| 2. Los Chorros | 9.5.0 ±2.1 | 1.0 ± 0.2 | 10.0 ± 1.3 | *B. obscurior* | 5.0 ± 0.2 | 0 | - |
| Both resources unburned | | | | | | | |
| 3. San Isidro | 15.1± 5.4 | 1.5± 0.5 | 14.3± 6.1 | *B. obscurior* | 5.6 ± 0.1 | 0 | - |
| 4. San Agustín | 8.9± 1.3 | 1.9± 0.3 | 1.4± 0.4 | *B. obscurior* | 5.4± 0.1 | 0 | - |
| 5. La Mimila | 1.2± 0.3 | 0.9± 0.3 | 1.4± 0.3 | *A. gracilipes* | 4.4 ± 0.2 | 0 | - |
| 6. El Arenal | 6.9± 2.8 | 3.9± 1.3 | 3.9± 1.4 | *A. gracilipes* | 4.9 ± 0.1 | 2.8 ± 0.6 | *A. gracilipes* |

hoppers produce honeydew at a consistent rate (Larsen et al.,1992), whereas EFNs are highly variable in nectar production over the course of a day, resulting in a less predictable resource for the ants. For example, nectar production is highly variable in the plant *Macaranga tanarious* (L.) Muell. Arg. (Heil et al., 2000). Second, *D. quinquenotatus* is sedentary and gregarious (Heady & Nault 1985), resulting in a higher density of both nymphs and adult leafhoppers on the basal leaves of *T. dactyloides*. At higher leafhopper densities, more honeydew is produced in a concentrated area allowing easy collection by the ants. Third, *D. quinquenotatus* responds to the stroking of their abdomen by antennae of tending ants by excreting and holding honeydew droplets until droplets are removed by ants (Larsen et al., 1992). Ant-tended *Dalbulus quinquenotatus* leafhoppers secrete three to six times the volume of honeydew compared with other species of non-myrmecophilous *Dalbulus* leafhoppers (Larsen et al., 1992), increasing the availability of honeydew for tending ants. In contrast, EFNs of *A. pennatula* do not respond to antennation by ants by increasing extrafloral nectar secretions. However, this is not universal as *Inga* plants have been shown to increase nectar production in response to tending ants (Bixenmann et al., 2011). Fourth, *D. quinquenotatus* leafhoppers and their tending ants often live together in mud shelters made by tending ants on the basal leaves of the gamagrass. Within these shelters, high densities of ants and leafhoppers occur and parasitism is reduced (Moya-Raygoza & Larsen, 2008). These shelters help to increase the quantity of honeydew for tending ants by concentrating the leafhoppers, whereas *A. pennatula* does not provide shelters for ants in the form of big thorns as is found on other *Acacia* species. Providing shelter for members of the mutualism is important in establishing obligatory relationships (Speight et al., 1999). Fifth, the honeydew of myrmecophilous hemipterans contains melezitose that provide nitrogen and is a higher quality nectar than nectar from EFNs (Cook & Davidson, 2006). Sixth, excess *D. quinquenotatus* leafhoppers are sometimes eaten by tending ants (Moya-Raygoza & Nault,

2000), making the leafhoppers a high quality source of protein. Ant colony growth and reproduction requires substantial quantities of protein (Davidson et al., 2003).

Moreover, this *D. quinquenotatus* leafhopper-ant association is an obligate and highly specialized mutualism as compared with the more general and facultative ant-*A. pennatula* mutualism. Moya-Raygoza (2005) found that the ant *B. obscurior* visits active EFNs of *A. pennatula* but does not protect this species of *Acacia* from herbivores. Lack of protection by ants against herbivores is common among plants with EFNs (Buckley, 1983; Heads, 1986; Oliveira et al., 1999; Ruhren, 2003). In contrast, both Moya-Raygoza and Nault (2000) and Larsen et al. (2001) have shown that tending ants protect both nymph and adult *D. quinquenotatus* from predators. Thus, this mutualism between *D. quinquenotatus* and ants is obligatory, as these leafhoppers apparently cannot live without tending ants.

*Post-fire response*

Both *T. dactyloides* and *A. pennatula* respond quickly to a fire event with new growth, producing young leaves ready to be colonized by herbivorous insects. Previous studies conducted in the tropics have found that some species of plants respond to fire with vigorous growth, which can be colonized rapidly by herbivores (Prada et al., 1995; Vieira et al., 1996). We found that ants are adapted to colonize plants quickly after fire, taking advantage of new resources such as honeydew offered by *D. quinquenotatus* feeding on *T. dactyloides* and extrafloral nectar produced by EFNs of *A. pennatula*, resulting in the reestablishment of these mutualistic interactions only a few days after fire.

We found more ants tending leafhoppers than visiting EFNs at burned sites where both *T. dactyloides* and *A. pennatula* were found. Fire does not kill *T. dactyloides*, but instead stimulates the growth of new stems from *T. dactyloides* rhizomes. These new stems are the first food resources that appear within the community and are quickly recolonized by *D. quinquenotatus*. These leafhoppers may come from contiguous unburned sites. These immigrant leafhoppers start to feed and produce large quantities of honeydew that attract large numbers of ants. The numbers of ants revealed this fast recolonization by leafhoppers and ants at sites where fire occurred in either May 2007 or June 2012. In contrast, in June 2012 no leafhoppers or ants were found on *T. dactyloides* leaves at unburned sites because those leaves were dried out. Although EFNs at unburned sites were actively producing extrafloral nectar at that time, few ants were present.

No previous studies have compared the ant abundance at leafhoppers and EFNs on fire-disturbed habitats when both resources are available at the same time. Schowalter (2006), reported that ants and sap-sucking insects such as leafhoppers dominate early-successional tropical forests as they contain an abundance of young, succulent leaf tissue that favor sap-sucking hemipterans and tending ants. In North American grasslands, populations of some leafhopper species are significantly greater following fire due to immigration from unburned areas into rapidly growing burned areas (Warren et al., 1987). Previously, Moya-Raygoza (1995) found that *D. quinquenotatus* leafhoppers were found in larger numbers and tended by a greater number of ants in burned than unburned *T. dactyloides* colonies, because recently burned plants produce new young leaves with higher concentrations of nitrogen.

Similar results have been found in the interaction between ants and EFN-bearing plants in other systems after disturbance. For example, pruned plants (*Conocarpus erectus* L.) grew faster and produced higher numbers of extrafloral nectaries and attracted a higher density of ants (Piovia-Scott, 2011). Leaf damage also increases the production of extrafloral nectar in different plants (Heil et al., 2001). In another case, higher abundance of ants was found in the shrub *B. campestris* after fire because of a high concentration of extrafloral nectar (Alves-Silva & Del-Claro, 2013). Similarly Alves-Silva (2011) and Koptur et al. (2010) found a more diverse ant fauna guarding plants from herbivory after fire occurred due to the high production of extrafloral nectar. This is not surprising as ants are attracted to high quality sugar resources produced by plants with EFNs (Heil & McKey, 2003).

Therefore, the availability of honeydew and extrafloral nectar to ants after fire is important because it can regulate ecological dominance, affecting the ant trophobiont and plant communities. Greater numbers of ants tending leafhoppers may result in better protection of these honeydew producers by ants compared with the ant protection of plants with EFNs that can also occur in these fire-prone sites. Moreover, colonization by ants after fire is important to initiate these mutualisms with both hemipterans and EFNs. Our results highlight the importance of investigating mutualisms not only in paired species, but also among multiple mutualisms involving ants when a system is disturbed.

## Acknowledgments

We are grateful to Miguel Vasquez Bolaños for the identification of some of the ant taxa. We also appreciate the comments and suggestions of two anonymous reviewers.

## References

Alves-Sila, E. (2011). Post fire resprouting of *Banisteriopsis mallifolia* (Malpighiacea) and the role of extrafloral nectaries on the associated ant fauna in a Brazilian Savanna. Sociobiology, 58: 327-340.

Alves-Silva, E. & Del-Claro, K. (2013). Effect of post-fire resprouting on leaf fluctuating asymmetry, extrafloral nectar quality, and ant-plant-herbivore interactions. Naturwissenschaftlen, 100: 525-532. doi: 10.1007/s00114-013-1048-z

Bixenmann, R.J., Coley, P.D. & Kursar, T.A. (2011). Is extrafloral nectar production induced by herbivores or ants in a tropical facultative ant-plant mutualism? Oecología 165: 417-425. doi: 10.1007/S00442-010-1787-x

Blüthgen, N., Verhaagh, M., Goitía, W., Jaffé, K., Morawetz, W. & Barhlott, W. (2000). How plants shape the ant community in the Amazonian rainforest canopy: the key role of extrafloral nectaries and homopteran honeydew. Oecologia, 125: 229-240. doi: 10.1007/s004420000449

Blüthgen, N., Mezger, D. & Linsenmair, K.E. (2006). Ant-hemiptera trophobiosis in a Bornean rainforest - diversity, specificity and monopolization. Insectes Sociaux, 53: 194-203.doi: 10.1007/s00040-005-0858-1

Buckley, R.C. (1983). Interactions between ants and membracid bugs decreases growth and seed set of host plant bearing extrafloral nectaries. Oecologia, 58: 132-136.

Buckley, R.C. (1987). Interactions involving plants, Homoptera, and ants. Annual Review of Ecology and Systemtics, 18: 111-135.doi:0066/4162/87/1120-0111

Byk, J. & Del-Claro, K. (2011). Ant-plant interaction in the Neotropical savanna: direct benefical effects of extrafloral nectaries on an ant colony fitness. Population Ecology, 53: 327–332. doi: 10.1007/s10144-010-0240-7

Cook, S.C. & Davidson, D.W. (2006). Nutritional and functional biology of exudate-feeding ants. Entomologia Experimentalis et Applicata, 118: 1-10.

Davidson, D.W., Cook, S.C., Snelling, R.R. & Chua, T.H. (2003). Explaining the abundance of ants in lowland tropical rainforest canopies. Science, 300: 969-972.

Delabie, J. (2001). Trophobiosis between Formicidae and Hemiptera (Sternorrhyncha and Auchenorrhyncha): an overview. Neotropical Entomology, 30: 501-516. doi: 10.1590/51519-566x2001000400001.

Del-Claro, K. & Oliveira, P.S. (1993). Ant-homoptera interactions: do alternative sugar sources distract tending ants? Oikos, 68: 202-206.

Fagundes, R., Riveiro, S.P. & Del-Claro, K. (2013). Tending-ants increase survivorship and reproductive success of *Calloconophora pugionata* Dietrich (Hemiptera, Membracidae), trophobiont herbivore of *Myrcia obovata* O. Berg (Myrtales, Myrtaceae). Sociobiology, 60: 11-19.

Fiala, B. (1990). Extrafloral nectaries vs ant-Homoptera mutualism: a comment on Becerra and Venable. Oikos, 59: 281-282.doi: 10.2307/3545545

Gibb, H. & Cunningham, S.A. (2009). Does the availability of arboreal honeydew determine the prevalence of ecologically dominant ants in restored habitats? Insectes Sociaux, 56: 405-412. doi: 10.1007/s00040-009-0038-9

González-Teuber, M. & Heil, M. (2009). Nectar chemistry is tailored for both attraction of mutualists and protection from exploiters. Plant Signaling and Behavior, 4: 809-813.

Heads, P.A. (1986). Bracken, ants and extrafloral nectaries. IV. Do wood ants (*Formica lugubris*) protect the plant against insect herbivores? Journal of Animal Ecology, 55: 795-809.

Heady, S.E. & Nault, L.R. (1985). Escape behavior of *Dalbulus* and *Baldulus* leafhoppers (Homoptera: Cicadellidae). Environmental Entomology, 14: 154-158.

Heil, M. & McKey, D. (2003). Protective ant-plant interactions as model systems in ecological and evolutionary research. Annual Review of Ecology, Evolution and Systematics, 34: 425-453. doi: 10.1146/annurev.ecolsys.34.011802.132410

Heil, M., Fiala, B., Baumann, B. & Linsenmair K.L. (2000). Temporal, spatial and biotic variation in extrafloral nectar secretions by *Macaranga tanarius*. Functional Ecology, 14: 749-757.

Heil, M., Koch, T., Hilpert, A., Fiala, B., Boland, W. & Linsenmair K.L. (2001). Extrafloral nectar production of the ant-associated plant, *Macaranga tanarius*, is an induced, indirect, defensive response elicited by jasmonic acid. PNAS, 98: 1083-1088.doi: 10.1073/pnas.031563398

Heil, M., Greiner, S., Meimberg, H., Krüger, R., Noyer, Jean-Louis,, Heubl, G., Linsenmair, K.E. & Boland W. (2004). Evolutionary change from induced to constitutive expression of an indirect plant resistance. Nature, 430: 205–208.

Hölldobler, B. & Wilson, E.O. (1990). The Ants. Cambridge: Harvard University Press, 732 p.

Katayama, N. & Suzuki, N. (2003). Changes in the use of extrafloral nectaries of *Vicia faba* (Leguminosae) and honeydew of aphids by ants with increasing aphid density. Annals of the Entomological Society of America, 96: 579-584. doi: 0013-8746/03/0579-0584

Katayama, N. & Suzuki, N. (2010). Extrafloral nectaries indirectly protect small aphid colonies via ant-mediated interactions. Applied Entomology and Zoology, 45: 505-511.

Katayama, N., Hembry, D.H., Hojo, M.K. & Suzuki, N. (2013). Why do ants shift their foraging from extrafloral nectar to aphid honeydew? Ecological Research, 28: 919-926. doi: 10.1007/s11284-013-1074-5

Koptur, S., William, P. & Olive, Z. (2010). Ants and plants with extrafloral nectaries in fire successional habitats on Andros (Bahamas). Florida Entomologist, 93: 89-99.

Larsen, K.J., Vega, F.E., Moya-Raygoza, G. & Nault, L.R. (1991). Ants (Hymenoptera: Formicidae) associated with the leafhopper *Dalbulus quinquenotatus* (Homoptera: Cicadellidae) on gamagrasses in Mexico. Annals of the Entomological Society of America, 84: 498-501. doi: 0013- 8746/91/0498-0501

Larsen, K.J., Heady, S.E. & Nault, L.R. (1992). Influence of ants (Hymenoptera: Formicidae) on honeydew excretion and escape behaviors in a myrmecophile, *Dalbulus quinquenotatus* (Homoptera: Cicadellidae), and its congeners. Journal of Insect Behavior, 5: 109-122. doi: 0892-7553/92/0100-0109

Larsen, K.J., Staehle, L.M. & Dotseth, E.J. (2001). Tending ants (Hymenoptera: Formicidae) regulate *Dalbulus quinquenotatus* (Homoptera: Cicadellidae) population dynamics. Environmental Entomology, 30: 757–762. doi: 0046-225x/01/0757-0762

Marazzi, B., Bronstain, J.L. & Koptur, S. (2013). The diversity, ecology and evolution of extrafloral nectaries: current perspectives and future challenges. Annals of Botany, 111: 1243-1250. doi: 001.10.1093/aob/mct109

McVaugh, R. (1987). Flora Nova-Galiciana (Leguminosae), Vol 5. University of Michigan Press, Ann Arbor, 786 p.

Mosino-Aleman, P.A. & Garcia, E. (1974). The climate of Mexico. In R.A. Bryson & F.K. Hare (Eds). Climate of North America, vol. 11. World Survey of Climatology (pp 345-404). Elsevier Scientific, New York.

Moya-Raygoza, G. (1995). Fire effects on insects associated with the gamagrass *Tripsacum dactyloides* in Mexico. Annals of the Entomological Society of America, 88: 434-440. doi: 0013-8746/95/0434-0440

Moya-Raygoza, G. (2005). Relationships between the ant *Brachymyrmex obscurior* (Hymenoptera, Formicidae) and *Acacia pennatula* (Fabaceae). Insectes Sociaux, 52: 105-107. doi:10.1007/s00040-004-0777-6

Moya-Raygoza, G. & Nault, L.R. (2000). Obligatory mutualism between *Dalbulus quinquenotatus* (Homoptera: Cicadellidae) and attendant ants. Annals of the Entomological Society of America, 93: 929–940. doi: 0013-8746/00/0929-0940

Moya-Raygoza, G. & Larsen, K.J. (2008). Positive effects of shade and shelter construction by ants on a leafhopper-ant mutualism. Environmental Entomology, 37: 1471-1476. doi: 0046- 225x/08/1471-1476

Oliveira, P.S., Rico-Gray, V., Diaz-Castelazo, C. & Castillo-Guevara, C. (1999). Interactions between ants, extrafloral nectaries and insect herbivores in Neotropical coastal sand dunes: herbivore deterrence by visiting ants increases fruit set in *Opuntia stricta* (Cactaceae). Functional Ecology, 13: 623–631. doi:10.1046/j.1365-2435.1999.00360.x

Piovia-Scott, J. (2011). The effect of disturbance on ant-plant mutualism. Oecologia, 166: 411-420. doi:10.1007/s00442-010-1851-6

Prada, M., Marini-Filho, O.J. & Price, P.W. (1995). Insects in flower heads of *Aspilia foliacea* (Asteraceae) after a fire in a Central Brazilian Savanna: Evidence for the plant vigor hypothesis. Biotropica, 27: 513-518.

Rashbrook, V.K., Compton, S.G. & Lawton, J.H. (1992). Ant-herbivore interactions: reasons for the absence of benefits to a fern with foliar nectaries. Ecology, 73: 2167-2174.

Rodríguez-Trejo, D.A. & Myers, R.L. (2010). Using oak characteristics to guide fire regime restoration in Mexican pine-oak forests. Ecological Research, 28: 304–323. doi: 10.3368/er.28.3.304

Ruhren, S. (2003). Seed predators are undeterred by nectar-feeding ants on *Chamaescrista nictitans* (Caesalpineaceae). Plant Ecology, 166: 189-198.

Schowalter, T.D. (2006). Insect Ecology: An Ecosystem Approach. Academic Press, Elseiver, New York, 350 p.

Speight, M.R., Hunter, M.D. & Watt, A.D. (1999). Ecology of Insects Concepts and Applications. Blackwell Science, Oxford, UK, 572 p.

Vieira, E.M., Andrade, I. & Price, P.W. (1996). Fire effects on a *Palicaurea rigida* (Rubiaceae) gall midge: A test of the plant vigor hypothesis. Biotropica, 28: 210-217.

Warren, S.D., Scifres, C.J. & Teel, P.D. (1987). Response of grassland arthropods to burning: a review. Agriculture, Ecosystems and Environment, 19: 105-130. doi:10.10167-8809-(87)90012-0

Way, M.J. (1963). Mutualism between ants and honeydew-producing Homoptera. Annual Review of Entomology, 8: 307-344.

Weber, M.G. & Keeler, K.H. (2013). The phylogenetic distribution of extrafloral nectaries in plants. Annals of Botany, 111: 1251-1261. doi: 10.1093/aob/mcs225

Wilkes, H.G. (1972). Maize and its wild relatives. Science, 117: 1071-1077.

Wimp, G. M. & Whitham, T.G. (2001). Biodiversity consequences of predation and host plant hybridization on an aphid-ant mutualism. Ecology, 82: 440-452.

Zhang, S., Zhang, Y. & Ma, K. (2012). Distribution of ant-aphid mutualism in canopy enhances the abundance of beetles on the forest floor. PLoS ONE, 7 (4): e35468. doi: 10.1371/journal.pone.0035468.

Zhang, S., Zhang, Y. & Ma, K. (2013). The ecological effects of ant-aphid mutualism on plants at a large spatial scale. Sociobiology, 60: 236-241. doi: 10.13102/sociobiology.v60i3.236-241.

# 25

# *Oxytrigona tataira* (Smith) (Hymenoptera: Apidae: Meliponini) as a collector of honeydew from *Erechtia carinata* (Funkhouser) (Hemiptera: Membracidae) on *Caryocar brasiliense* Cambessèdes (Malpighiales: Caryocaraceae) in the Brazilian Savanna

FH Oda[1], AF Oliveira[2], C Aoki[3]

1 Universidade Estadual de Maringá (UEM), Maringá, PR, Brazil
2 Universidade Federal de Mato Grosso do Sul (UFMS), Campo Grande, MS, Brazil
3 Universidade Federal de Mato Grosso do Sul (UFMS), Aquidauana, MS, Brazil

**Keywords**
Hemipterans, Exudate, Feeding behavior, "Pequi", Stingless bee

**Corresponding author**
Fabrício Hiroiuki Oda
Laboratório de Ictioparasitologia
Núcleo de Pesquisas em Limnologia
Ictiologia e Aqüicultura
Universidade Estadual de Maringá
Bloco G-90 Avenida Colombo, 5.790
87020-900, Maringá, PR, Brazil
E-Mail: fabricio_oda@hotmail.com

**Abstract**
Trophobiont insects are of general interest to behavioral ecology due to the fact that the outcomes of their interactions with hymenopterans can result in strong facultative mutualisms. In this paper we present observations of *Oxytrigona tataira* (Smith) collecting honeydew from *Erechtia carinata* (Funkhouser) on *Caryocar brasiliense* Cambessèdes in a cerrado *sensu stricto* fragment from Brazilian Savanna. Our observations showed that these bees stimulate *E. carinata* individuals touching them with the antennas in the gaster to collect honeydew. *O. tataira* has not been previously recorded collecting honeydew on *E. carinata*. Thus, we present these observations as a novel bee-treehopper interaction.

Mutualisms are interspecific interactions where individuals of two species experience higher fitness when they occur together than when they occur alone (Bronstein, 1998). This interaction is an important process for the structure and composition of communities (Bascompte & Jordano, 2007; Lange & Del-Claro, 2014). Many species of phytophagous hemipterans (e.g. aphids, coccids and treehoppers) are mutualistically associated with ants (Flatt & Weisser, 2000; Offenberg, 2001; Fagundes et al., 2013). In these relationships, ants protect their hemipteran partners against predators and parasites (Schultz & McGlynn, 2000; Fagundes et al., 2013) and often rely on honeydew as one of their most important food source (Davidson et al., 2003), besides extrafloral nectar (Del-Claro et al., 2013) and lepidopteran secretions (Alves-Silva et al., 2013).

Although ants are the most common hymenopteran interacting with treehoppers, bees and wasps can also be associated with them (Lin, 2006). Mutualistic interactions have been reported involving the aetalionid treehopper *Aetalion reticulatum* (Linnaeus) with wasps or ants (Letourneau & Choe, 1987; Ramoni-Perazzi et al., 2006), and even *Trigona* Jurine, a stingless bee, which take advantage of treehoppers sugary excretions, the honeydew (Castro, 1975; Vieira et al., 2007; Oda et al., 2009; Barônio et al., 2012). Godoy et al. (2006) reported eight associations between membracids and bees, including the interaction between *Potnia* Stål and *Oxytrigona tataira* (Smith). Azevedo et al. (2007) reported the associations between *Trigona spinipes* (Fabricius) and five membracids species, and observed that the bees repelled some insects that approached the membracids, although with low aggressiveness.

This kind of interaction has not been reported between the stingless bee *O. tataira* (Apidae: Meliponini) and the treehopper *Erechtia carinata* (Funkhouser) (Auchenorrhyncha: Membracidae). This membracid belongs to the tribe Talipedini and it was recently reassessed, and *Trinarea* Goding is considered a junior synonym of *Erechtia* Walker (Sakakibara, 2012).

*Oxytrigona tataira* has a wide distribution, occurring in eight Brazilian States (Silveira et al., 2002; Lima et al., 2013). *Oxytrigona* Cockerell is commonly known in Brazil as "tataíra" or "cospe-fogo" (fire spitting), due to caustic substance (formic acid) produced by worker bees for defense. These bees are highly aggressive, and also cleptobiotic, being colony's robbers of other meliponine species (Roubik et al., 1987; Roubik, 1992).

*Caryocar brasiliense* Cambessèdes has a wide distribution, occurring in 15 Brazilian States and Paraguay (Lopes et al., 2006). It is commonly known as "pequi", found from shrubs to trees of 1.5 to 11 meters high, with semi-deciduous behavior of foliar change (Lopes et al., 2006). Although the use of secretions from sucking insects (Hemiptera) by species of *Trigona* and *Oxytrigona* has been reported before (Cortopassi-Laurino, 1977; Letourneau & Choe, 1987), herein we present first observations of interaction between *O. tataira* and *E. carinata* on *C. brasiliense* in the Brazilian Savanna.

During three non-consecutive days in January 2012, individuals of *O. tataira* were observed collecting honeydew from *E. carinata* in a cerrado *sensu stricto* fragment (17° 21' S, 53° 18' W) at Alto Araguaia municipality, State of Mato Grosso, mid-western Brazil. The observations were carried out once per day, between 7:00 and 10:00 h. Each observation lasted 15 to 20 minutes.

Voucher specimens were collected, processed and given to specialists for identification and deposition in the Entomological Collection Padre Jesus S. Moure (DZUP) of the Department of Zoology at the Universidade Federal do Paraná (UFPR).

We observed variable numbers of individuals of *O. tataira* (mean = 55 individuals ± 42.7 SD) collecting honeydew of approximately 130 nymphs and one adult of *E. carinata*. The number of *O. tataira* was lower during the second day of observations after heavy rain.

The bees stimulated *E. carinata* individuals by primarily touching the treehoppers at the proximal upper abdomen with their front legs and antennas. The bees repeated the stimulus toward the distal part of the abdomen where the exudate droplet was collected with aid of the first pair of legs and quickly sucked. This behavior of the bees was recorded mainly in the treehoppers nymphs because only one adult was with the nymphs on the branch (Fig 1).

As a result of our presence the bees defended the food source by beating their wings as a warning sign (Fig 2), and some individuals started a short flight over the *E. carinata* colony. In the first day of observations,

**Fig 1**. *Oxytrigona tataira* (Smith) (Hymenoptera: Apidae: Meliponini) collecting honeydew from *Erechtia carinata* (Funkhouser) (Hemiptera: Membracidae) on branch of *Caryocar brasiliense* Cambessèdes (Malpighiales: Caryocaraceae).

inattentive researchers were attacked by the bees. We did not record the presence of ants patrolling the treehoppers.

The behavior of *E. carinata* stimulation by *O. tataira* resembles that observed in species of *Trigona* in association with *A. reticulatum*: the bees touch their antennas to the head of the treehoppers, and then the first two pairs of legs on the back of the abdomen, after that touching the antennas on the distal part of the abdomen and quickly sucking the honeydew droplet released after stimulation (Vieira et al., 2007; Oda et al., 2009; Barônio et al., 2012). This behavior is also similar to that reported between ants and honeydew-producing hemipterans, including membracids, in which the ants actively touch the body of the insects with their antennae to stimulate the releasing of honeydew droplets (Stefani et al., 2000; Pfeiffer & Linsenmair, 2007; Guerra et al., 2011; Gjonov & Gjonova-Lapeva, 2013).

Previous studies suggest that this possibly facultative mutualistic association is beneficial for both species (Vieira et al., 2007; Oda et al., 2009; Barônio et al., 2012), promoting a protection against natural enemies of the treehoppers, and supplying part of the bees diet with the honeydew rich in carbohydrates (Way, 1963; Fagundes et al., 2013). Additionally, the association with the bees can reduce fungus attack because of the removal of the contaminant honeydew (Way, 1963; Buckley, 1987), decreasing the chances of local extinction of these hemipterans (Buckley, 1987).

On ant-tending, the mortality risks of the tended insects can be considerably reduced due to protection against predators and parasitoids (Flatt & Weisser, 2000; Fagundes et al., 2013). However, we did not test it. Thus, it is important to emphasize the necessity of future studies, which are essential to know the true nature of this interaction.

**Fig 2**. Agonistic behavior displayed by *Oxytrigona tataira* (Smith) (Hymenoptera: Apidae: Meliponini) as a result of feeling threatened by the presence of the researchers.

## Acknowledgements

We are grateful to Albino M. Sakakibara and Sebastião Laroca for the identification of treehopper and bee species, respectively. To editor and two anonymous reviewers for their constructive comments, that helped us to improve the manuscript. Guilherme H. Barbosa, Frederico Pereira, Vinícius Cerqueira and Patrícia G. Ferreira for field assistance. Mariana F. Felismino reviewed the English language. The Biocev provided logistical support during fieldwork.

## References

Alves-Silva, E., Bächtold, A., Barônio, G. J. & Del-Claro, K. (2013). Influence of *Camponotus blandus* (Formicinae) and flower buds on the occurrence of *Parrhasius polibetes* (Lepidoptera: Lycaenidae) in *Banisteriopsis malifolia* (Malpighiaceae). Sociobiology, 60: 30-34.

Azevedo, R. L., Carvalho, C. A. L., Bomfim, Z. V. & Vicente, M. A. A. (2007). Interações entre auquenorrincos (Aethalionidae e Membracidae), abelhas (Apidae) e formigas (Formicidae) em plantas de feijão guandu. Rev. Ecossistema, 32: 81-86.

Barônio, G. J., Pires, A. C. V. & Aoki, C. (2012). *Trigona branneri* (Hymenoptera: Apidae) as a collector of honeydew from *Aethalion reticulatum* (Hemiptera: Aethalionidae) on *Bauhinia forficata* (Fabaceae: Caesalpinoideae) in a Brazilian Savanna. Sociobiology, 59: 1-8.

Bascompte, J. & Jordano, P. (2007). Plant-animal mutualistic networks: The architecture of biodiversity. Annu. Rev. Ecol. Evol. Syst., 38: 567-593. doi: 10.1146/annurev. ecolsys.38.091206.095818

Bronstein, J. L. (1998). The contribution of ant-plant protection studies to our understanding of mutualism. Biotropica, 30: 150-161.

Buckley, R. C. (1987). Interactions involving plants, Homoptera and ants. Ann. Rev. Ecol. Syst., 18: 111-135.

Castro, P. R. C. (1975). Mutualismo entre *Trigona spinipes* (Fabricius 1793) e *Aethalion reticulatum* (L. 1767) em *Cajanus indicus* Spreng. na presença de *Camponotus* spp. Cienc. Cult., 27: 537-539.

Cortopassi-Laurino, M. (1977). Notas sobre associações de *Trigona* (*Oxytrigona*) *tataira* (Apidae, Meliponinae). Bol. Zool., 2: 183-187.

Davidson, D. W., Cook, S. C., Snelling, R. R. & Chua, T. C. (2003). Explaining the abundance of ants in lowland tropical rainforest canopies. Science, 300: 969-972. doi:10.1126/science.1082074

Del-Claro, K., Guillermo-Ferreira, R., Almeida, E. M., Zardini, H. & Torezan-Silingardi, H. M. (2013). Ants visiting the post-floral secretions of pericarpial nectaries in *Palicourea rigida* (Rubiaceae) provide protection against leaf herbivores but not against seed parasites. Sociobiology, 60: 217-221. doi:10.13102/sociobiology.v60i3.217-221.

Fagundes, R., Ribeiro, S. P. & Del-Claro, K. (2013). Tending-ants increase survivorship and reproductive success of *Calloconophora pugionata* Drietch (Hemiptera, Membracidae), a trophobiont herbivore of *Myrcia obovata* O.Berg (Myrtales, Myrtaceae). Sociobiology, 60: 11-19.

Flatt, T. & Weisser, W. W. (2000). The effects of mutualistic ants on aphid life history traits. Ecology, 81: 3522-3529. doi: 10.1890/0012-9658(2000)081[3522:TEOMAO]2.0.CO;2

Godoy, C., Miranda, X. & Nishida, K. (2006). Treehoppers of America Tropical. Costa Rica: INBio, 352 p.

Gjonov, I. & Gjonova-Lapeva, A. (2013). New data on ant-attendance in leafhoppers (Hemiptera: Cicadellidae). North-West J. Zool., 9: 433-437.

Guerra, T. J., Camarota, F., Castro, F. S., Schwertner, C. F. & Grazia, J. (2011). Trophobiosis between ants and *Eurystethus microlobatus* Ruckes (Hemiptera: Heteroptera: Pentatomidae) a cryptic, gregarious and subsocial stinkbug. J. Nat. Hist., 45: 1101-1117. doi: 10.1080/00222933.2011.552800

Lange, D. & Del-Claro, K. (2014). Ant-plant interaction in a Tropical Savanna: May the network structure vary over time and influence on the outcomes of associations? PLoS ONE, 9(8): 1-10. doi:10.1371/journal.pone.0105574

Letourneau, D. K. & Choe, J. C. (1987). Homopteran attendance by wasps and ants: the stochastic nature of interactions. Psyche, 94: 81-91. doi: 10.1155/1987/12726

Lima, F. V. O., Silvestre, R. & Balestieri, J. B. P. (2013). Nest entrance types of stingless bees (Hymenoptera: Apidae sensu lato) in a Tropical Dry Forest of mid-Western Brazil. Sociobiology, 60: 421-428. doi: 10.13102/sociobiology.v60i4.421-428

Lin, C-P. (2006). Social behaviour and life history of membracine treehoppers. J. Nat. Hist., 40: 1887-1907. doi: 10.1080/00222930601046618

Lopes, P. S. N., Pereira, A. V., Pereira, E. B. C., Martins, E. R. & Fernandes, R. C. (2006). Pequi. In R. F. Vieira, T. S. A. Costa, D. B. Silva, F. R. Ferreira & S. M.Sano (Eds.), Frutas nativas da região Centro-Oeste do Brasil (pp. 248-287). Brasília: Embrapa Recursos Genéticos e Biotecnologia.

Oda, F. H., Aoki, C., Oda, T. M., Silva, R. A. & Felismino, M. F. (2009). Interação entre abelha *Trigona hyalinata* (Lepeletier, 1836) (Hymenoptera: Apidae) e *Aethalion reticulatum* Linnaeus 1767 (Hemiptera: Aethalionidae) em *Clitoria fairchildiana* Howard (Papilionoideae). Entomobrasilis, 2: 58-60.

Offenberg, J. (2001). Balancing between mutualism and exploitation: the symbiotic interaction between *Lasius* ants and aphids. Behav. Ecol. Sociobiol., 49: 304-310. doi: 10.1007/s002650000303

Pfeiffer, M. & Linsenmair, K. E. (2007). Trophobiosis in a tropical rainforest on Borneo: giant ants *Camponotus gigas* (Hymenoptera: Formicidae) herd wax cicadas *Bythopsyrna circulata* (Auchenorrhyncha: Flatidae). Asian Myrmecol., 1: 105-119.

Ramoni-Perazzi, P., Bianchi-Pérez, G. & Bianchi-Ballestereos, G. (2006). Primer registro de asociación entre *Aetalion reticulatum* (Linné) (Hemiptera: Aetalionidae) y *Synoeca septentrionalis* Richards (Hymenoptera: Vespidae). Entomotropica, 21: 129-132.

Roubik, D. W., Smith, B. H. & Carlson, R. L. (1987). Formic acid in caustic cephalic secretions of stingless bee *Oxytrigona* (Hymenoptera, Apidae). J. Chem. Ecol., 13: 1079-1086. doi: 10.1007/BF01020539

Roubik, D. W. (1992). Stingless bees: A guide to Panamanian and Mesoamerican species and their nests (Hymenoptera, Apidae, Meliponinae). In D. Quintero & A. Aiello (Eds.), Insects of Panamá and Mesoamerica (pp. 495-524). Oxford: Oxford University Press.

Sakakibara, A. M. (2012). Taxonomic reassessment of the treehopper tribe Talipedini with nomenclatural changes and descriptions of new taxa (Hemiptera: Membracidae: Membracinae). Zoologia, 29: 563-576. doi: 10.1590/S1984-46702012000600008

Schultz, T. R. & McGlynn, T. P. (2000). The interactions of ants with other organisms. In D. Agosti, J. D. Majer, L. E. Alonso & T. R. Schultz, (Eds) Ants. Standard methods for measuring and monitoring biodiversity (pp. 35-44). Washington and London: Smithsonian Institution Press.

Silveira, F. A., Melo, G. A. R. & Almeida E. A. B. (2002). Abelhas Brasileiras, Sistemática e Identificação. Belo Horizonte: Fundação Araucária, 253 p.

Stefani, V., Sebaio, F. & Del-Claro, K. (2000). Desenvolvimento de *Enchenopa brasiliensis* Strumpel (Homoptera, Membracidae) em plantas de *Solanum lycocarpum* St.Hill. (Solanaceae) no cerrado e as formigas associadas. Rev. Bras.. Zoociênc., 2(1): 21-30.

Vieira, C. U., Rodovalho, C. M., Almeida, L. O., Siqueroli, A. C. S. & Bonetti, A. M. (2007). Interação entre *Trigona spinipes* Fabricius 1793 (Hymenoptera: Apidae) e *Aethalion reticulatum* Linnaeus 1767 (Hemiptera: Aethalionidae) em *Mangifera indica* (Anacardiaceae). Biosci. J., 23: 10-13.

Way, M. J. (1963). Mutualism between ants and honeydew-producing Homoptera. Annu. Rev. Entomot., 8: 307-344. doi: 10.1146/annurev.en.08.010163.001515

# 26

# The effect of toxic nectar and pollen from *Spathodea campanulata* on the worker survival of *Melipona fasciculata* Smith and *Melipona seminigra* Friese, two Amazonian stingless bees (Hymenoptera: Apidae: Meliponini)

ACM Queiroz[1], FAL Contrera[2], GC Venturieri[1]

1 - Embrapa Amazônia Oriental, Belém, PA, Brazil
2 - Universidade Federal do Pará (UFPA), Belém, PA, Brazil

**Keywords**
meliponine bees, plant toxicity, survival, meliponiculture.

**Corresponding author**
Ana Carolina Martins de Queiroz
Embrapa Amazônia Oriental - Laboratório de Botânica
Trav. Dr. Enéas Pinheiro S/N, Caixa Postal 48, CEP 66095-100, Belém/PA, Brazil
E-Mail: carolina.queiroz@embrapa.br

**Abstract**

*Spathodea campanulata* is an African plant introduced into South America and other tropical and subtropical areas for ornamental purposes. This plant has been linked to insect mortality, bees included. However, its effects on the Neotropical *Melipona* are as yet unknown. Thus, the aim of this study was to evaluate the effect of *S. campanulata* nectar and pollen on the survival of *Melipona fasciculata* Smith and *Melipona seminigra* Friese workers. A total of 120 newly emerged workers of each species were divided into groups of 10 individuals and confined in boxes. They were submitted to the following diet treatments: *S. campanulata* nectar or 11% sucrose solution (nectar control); 11% sucrose solution and *S. campanulata* pollen or 11% sucrose solution and the species' original pollen (pollen control). A higher mortality of workers was detected in the groups fed with toxics nectar and pollen (*M. fasciculata*, p<0.01; *M. seminigra*, p<0.01) than on the respective controls. Our results demonstrate that nectar and pollen from *S. campanulata* affected the survival of *M. fasciculata* and *M. seminigra* worker bees. We thus recommend that *S. campanulata* should not be provided as food source for stingless bees.

## Introduction

Nectar and pollen collected from plants are the main food resources of social bees. Pollen is used mainly as a protein source and nectar a carbohydrate source (Roubik, 1989). Some plants, however, are reported as toxic for bees, and the potentially toxic compounds may be present in the pollen or nectar (Roubik, 1989). Many plant species may poison bees due to the toxicity of pollen or nectar, extrafloral nectaries, tree sap or honeydew (Barker, 1990). Pollen or nectar toxicity for bees is a widespread phenomenon, although it is poorly understood. Thus, many hypotheses have been proposed to explain this phenomenon.

According to Adler (2000), toxic nectar would promote pollinator specialization, prevent nectar theft and degradation, and corrupt pollination behaviors. Johnson et al. (2006) demonstrated that secondary compounds in nectar are effective visitor filters, which lead to a specialization in the pollination system. Toxicity

for several animals is normally due to the secondary compounds found in all plant parts, especially on those most important for survival and reproduction (Levin, 1976). These secondary compounds associated with resistance to herbivory have been frequently observed in floral nectar. Adler (2000) detected nectar that was toxic or had secondary compounds in at least 21 different plant families. According to Ott (1988), at least three psychoactive phytotoxin categories occurred in toxic honeys, and, consequently, in the nectar from which it was produced.

Toxins found amongst some plants are nicotine, rotenones, pyrethrins and tannins (Bueno et al., 1990). *Dimorphandra mollis* Benth. (Fabaceae; popularly called in Brazil "fake barbatimão") and *Stryphnodendron adstringens* (Martius) Coville (Fabaceae; the "real barbatimão"), are rich in tannins and may cause serious losses to beekeepers due to larvae mortality and a reduction in adult *Apis mellifera* Linnaeus longevity. The toxicity of barbatimão is attributed to the pollen and nectar, and the pollen is considered more harmful (Carvalho & Message, 2004; Santoro et al., 2004; Cintra et al., 2005).

*Spathodea campanulata* Beauv. (an exotic species of African origin introduced for ornamental purposes [Nogueira-Neto, 1997]) has been reported as toxic to stingless bees (Tribe Meliponini, *sensu* Michener, 2007). Portugal-Araújo (1963) was one of the pioneers in reporting these effects. He recorded dead stingless bees on *S. campanulata* flowers in Gabon, along with Nogueira-Neto (1997) and Oliveira et al. (1991) in Brazil. Trigo and Santos (2000) monitored dead insects in *S. campanulata* flowers for to up to five days after anthesis and stated that meliponine bees represented 97% of the dead insects. Calligaris (2001) confirmed its nectar toxicity on *Scaptotrigona postica* (Latreille) and *A. mellifera* worker bees in laboratory bioassays, although pollen toxicity was not verified.

Thus, in this study we evaluated the effects of *S. campanulata* nectar and pollen consumption on the survival of two species of *Melipona* Illiger worker bees from the Brazilian Amazon used in meliponiculture, *Melipona fasciculata* Smith and *Melipona seminigra* Friese.

## Material and Methods

### Studied species and study site

We used *M. seminigra* and *M. fasciculata* workers to study the effect of *S. campanulata* nectar and pollen on the worker survival. *M. seminigra* occurs in the Brazilian Amazon States of Acre, Amazonas, Maranhão, Mato Grosso, Pará, Rondônia, Roraima, Tocantins, and *M. fasciculata* in the Brazilian States of Maranhão, Mato Grosso, Pará, Piauí and Tocantins (Camargo & Pedro, 2012). All experiments were carried out from January to May 2012 in the meliponary of Embrapa Amazônia Oriental (1°26'11.52''S, 48°26'35.50''W). The area is composed of secondary forests patches of native plants (several hundred species, including trees and shrubs) and several agricultural crops such as assai trees (*Euterpe oleraceae*, Arecaceae) and other species.

*Spathodea campanulata* is a large tree (up to 20m) with numerous large flowers, externally red and internally yellow (Francis, 1990). In its region of origin (Africa), *S. campanulata* is pollinated by birds and bats, but is also visited by bees attracted by the abundant nectar and colorful flowers (Ayensu, 1974; Rangaiah et al., 2004; Corlett, 2005). In the study site, there was only one tree of *S. campanulata*, situated 20 meters from the nests.

### Experimental design
### Effect of nectar and pollen on the survival of workers

To analyze the effect of *S. campanulata* nectar and pollen on the survival of *M. fasciculata* and *M. seminigra* workers, we used newly emerged workers, obtained from eight different nests, four for each bee species. A total of 120 bees of each species were used (30 from each colony), from which 60 were destined for two control groups (30 for each species) and 60 for each experimental group: *S. campanulata* nectar and pollen. The workers were divided into groups of 10 individuals and confined in polyethylene boxes (8 x 8 x 4 cm) without the queen, and. The boxes were kept in a BOD incubator (model DL-SEDT 02) at 28 ± 1°C. Every day the number of live bees was checked, any dead individuals were removed, the plastic box's rubbish dump area was cleared and water was added to maintain humidity (method adapted from Costa & Venturieri, 2009).

Inflorescences with flower buds and newly opened flowers were gathered from trees located at the research campus – Embrapa Amazônia Oriental – to prepare the *S. campanulata* nectar that was offered to *M. fasciculata* and *M. seminigra* workers. The nectar was removed with an automatic micropipette and the percentage of total sugars was measured with a field refractometer adapted to small volumes (Bellinghan-Stanley™). For pollen sampling, anthers of the flower buds were removed and kept in 2 ml microtubes. A total of 1ml of water with 11% sucrose was added to the tubes with the anthers (the same concentration of sugar found in *S. campanulata* nectar) to wash and assist pollen extraction. The material was centrifuged for five minutes at 2000 rpm, the liquid part of the microtube was drained and the accumulated pollen on the bottom was collected and offered to the bees as protein source. A botanical sample of the plant's reproductive structures was identified and stored at the IAN Herbarium (Embrapa) under the number 187659.

Workers were submitted to the following diet, according to the treatment: NSc - bees fed on *S. campanulata* nectar; NeC - bees fed on 11% sucrose solution; PSc - bees fed on an 11% sucrose solution and *S. campanulata* pollen; PoC - bees fed on 11% sucrose solution (nectar control) and its own pollen (*M. fasciculata* or *M. seminigra* pollen; pollen control). Workers were daily fed with: (1) NSc: 240 µL of nectar and (2) NeC: 240 µL of 11% sucrose solution. The workers from the PSc treatments were offered 240 µL of 11% sucrose solution and 0.1 g of *S. campanulata* pollen daily. For the PoC treatment, 240 µL of 11% sucrose solution and 0.1g of the species' pollen was offered. The food was weighed daily in an analytical balance with a 10⁻³ g precision to check the consumption of each item. The carbohydrate (11% sucrose solution or *S. campanulata* nectar) and protein (pollen of *S. campanulata* and *M. fasciculata M. seminigra* colonies) foods were renewed whenever completely eaten (pollen), or daily (sucrose).

### Analyses

Kaplan-Meier survival curves were made for each species' different treatments. The survival of control workers (pollen and nectar) was monitored until half or more of the treatment individuals died. A Cox-Mantel test was carried out, using the software STATISTICA® 8.0, to compare the survival curves of the treatments used for each species (5% significance level). The data regarding workers that were alive at the end of the experiment were treated as censured and from the workers monitored until death as complete data for the survival curve setting (Crawley, 2007).

## Results

### *Effect of* **S. campanulata** *nectar and pollen on the survival of workers*

There was food consumption in all the studied groups (Table 1). The death rate was high for the 30 *M. fasciculata* bees that received *S. campanulata* nectar as a carbohydrate source (NSc). Only nine bees from the group NSc remained alive at the end of the experiment. Furthermore, from the 30 bees from group NeC, only three died on the second day, so that 27 bees remained alive at the end of the experiment (Fig 1a). A similar pattern of mortality due to *S. campanulata* nectar was detected for *M. seminigra* (Fig 1b). Mortality of *M. fasciculata* and *M. seminigra* workers was significantly higher in the experimental group (*S. campanulata* nectar) than in the control group (Cox-Mantel: *M. fasciculata*, p<0.01; *M. seminigra*, p<0.01).

Only nine of the 30 *M. fasciculata* workers submitted on the treatment with *S. campanulata* pollen (PSc) remained alive at the end of the experiment. There was high mortality on the four experiment days. Mortality was lower on the pollen control group (PoC), with 24 live workers at the end of the experiment (Fig 2a). Again, the pattern was similar for *M. seminigra* (Fig 2b). For both species the survival of the control group was significantly higher than for the experimental group (*S. campanulata* pollen) (Cox-Mantel: *M. fasciculata*, p<0.01; *M. seminigra*, p<0.01).

There was no difference on *M. fasciculata* survival between *S. campanulata* nectar (NSc) and pollen (PSc) treatments (Cox-Mantel, p=0.55). However, the intake of *S. campanulata* nectar by *M. seminigra* had a stronger impact on the survival of workers than pollen (Cox-Mantel, p<0.01). There were no significant differences between the longevity of bees that consumed 11% sucrose solution and pollen from their own boxes for the control groups (NSc and PSc) (Cox-Mantel: *M. fasciculata*: p=0.29; *M. seminigra*: p=0.45).

**Table 1**. Daily consumption rate per worker (mean±S.D.) of *Melipona fasciculata* and *Melipona seminigra*, confined in groups of 10 individuals and submitted to dietary treatments. NSc- bees fed on *Spathodea campanulata* nectar; NeC- bees fed on 11% sucrose solution; PSc - bees fed on an 11% sucrose solution and S. campanulata pollen; PoC - bees fed on 11% sucrose solution (nectar control) and its own pollen (pollen control).

| Species/ Treatment | NSc (µl) | NeC (µl) | PSc (mg) | PoC mg) |
|---|---|---|---|---|
| *M. fasciculata* | 19.62 ± 1.91 | 29.73 ± 8.45 | 9.11 ± 11.40 | 5.20 ± 3.70 |
| *M. seminigra* | 27.01 ± 3.41 | 30.24 ± 6.81 | 5.00 ± 3.70 | 1.40 ± 0.70 |

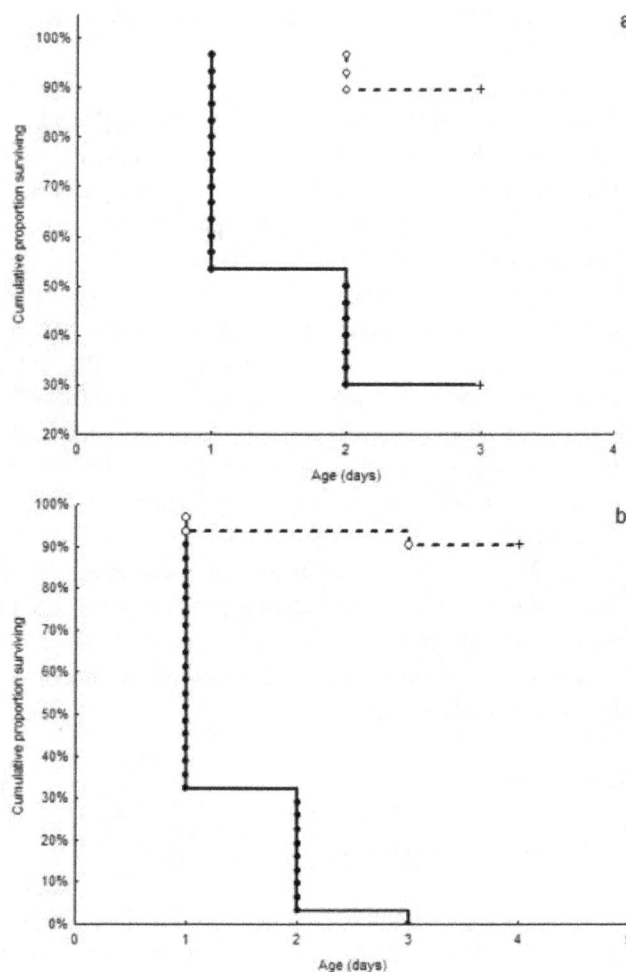

**Fig 1**. Survival curves for workers confined in groups of 10 individuals and submitted to dietary treatments. NSc- bees fed on *S. campanulata* nectar (filled circles); NeC- bees fed on 11% sucrose solution (empty circles). A - *Melipona fasciculata*. B - *Melipona seminigra*.

## Discussion

In this study we were able to prove the effect of *S. campanulata* pollen and nectar in reducing the survival of two meliponine species (*M. fasciculata* and *M. seminigra*) when fed on it. Both pollen and nectar of *S. campanulata* reduced the survival of worker bees undergoing these treatments. Since the food was consumed in the experimental cages, this indicates that the tested bee groups actually died due to *S. campanulata* nectar and pollen ingestion, and due to starvation.

In general, *S. campanulata* nectar and flower bud secretion are referred to as toxic, although little has been studied about its pollen. Calligaris (2001) found *S. postica* and *A. mellifera* survival reduction when *S. campanulata* nectar was added to the bees' diet in bioassays. Portugal-Araújo (1963) attributed the death of insects in *S. campanulata* flowers, including stingless bees, to floral bud secretion toxicity. In a periodic survey, Nogueira-Neto (1997) also reported many Meliponini bees in fallen flowers, highlighting *Plebeia droryana* (Friese), *Tetragonisca angustula* (Latreille), *S. postica*, *Trigona spinipes* (Fabricius) and *Friesella schrottkyi*

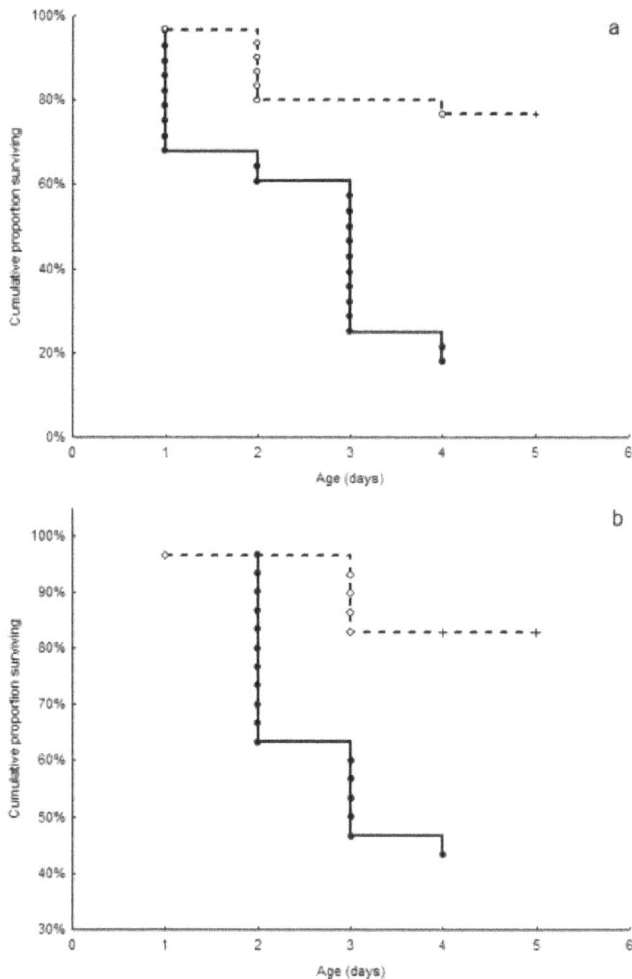

**Fig 2**. Survival curves for workers confined in groups of 10 individuals and submitted to dietary treatments. PSc - bees fed on 11% sucrose solution and *S. campanulata* pollen (filled circles); PoC- bees fed on 11% sucrose solution and its own pollen (empty circles). A - *Melipona fasciculata.* B - *Melipona seminigra.*

(Friese). Stingless bees represented up to 97% of the dead insects in the flowers, especially *S. postica* (Trigo & Santos, 2000).

Trigo and Santos (2000) tested different flower mucilage concentrations on newly emerged *S. postica* workers, in laboratory bioassays. Mucilage at a concentration of 25% reduced bee longevity by 52.9%, while pure mucilage reduced it by 95.2%. However, this last result is considered ambiguous, since there is no evidence that bee feed on pure mucilage.

Both pollen and nectar of *S. campanulata* are considered toxic. Calligaris (2001) did not detect a reduction in the survival of *S. postica* and *A. mellifera* when fed on 5% pollen. However, Oliveira et al. (1991) reported the death of *T. spinipes* bees due to the *S. campanulata* pollen found in its gizzard. In this study we found a marked reduction of *M. fasciculata* and *M. seminigra* survival rates when fed on pure pollen.

Trigo and Santos (2000) suggested the existence of a defense mechanism in *S. campanulata* that protects flower buds from nectar and pollen thieves. Otherwise these resources could be stolen by some meliponine bees, such as *S. postica*, or other efficient pillagers, before flower opening. In this case,

vertebrate pollination would be reduced or even prevented. Indeed, Endress (1994) noted that some plants, including Bignoniaceae, produce a mucilaginous or watery liquid to protect juvenile flower organs before anthesis. Flower bud secretion would thus be a plant defense system, of chemical or physical nature, suffocating the bees (Trigo & Santos, 2000).

Considering these effects and the actual expansion of meliponiculture in Brazil (Contrera et al., 2011; Venturieri et al., 2012), the use of *S. campanulata* trees is not recommended in areas foraged by stingless bees. Such a recommendation has already been made regarding *A. mellifera* (Modro et al., 2011).

## Acknowledgements

We thank the two anonymous referees for the comments that improved our manuscript. We also thank Elisângela Rêgo, Lourival Lucas and Miguel Nascimento for technical assistance. Funding was obtained from the Fundação Amazônia Paraense de Amparo a Pesquisa (FAPESPA) and the Conselho Nacional de Desenvolvimento Cientifico e Tecnológico (CNPq). We also thank the Dean of Research and Post-Graduate studies of UFPA (Pró-Reitoria de Pesquisa e Pós Graduação - PROPESP/UFPA) and Foundation for Research Support and Development (Fundação de Amparo e Desenvolvimento de Pesquisa - FADESP), for the support for through the Program of Qualified Publication Support (Programa de Apoio à Publicação Qualificada - PAPQ).

## References

Adler, L.S. (2000). The ecological significance of toxic nectar. Oikos 91:409-420. doi:10.1034/j.1600-0706.2000.910301.x

Ayensu, E. S. (1974). Plant and bat interactions in West Africa. Ann. Miss. Bot. Gard. 61: 702-727.

Barker, R.J. (1990). Poisoning by plants. In R. A. Morse & R. Nowogrodzki (Eds.), Honey Bee Pests, Predators and Diseases (pp. 309-315). New York: Cornell University Press.

Bueno, O.C., Hebling-Beraldo, M.J., Aulino-Silva, O., Pagnocca, F., Fernandez, J.B. & Vieira, P.C. (1990). Toxic effect of plants on leaf-cutting ants and their symbiotic fungus. In R.K. Jaffé & K.A. Cedeno (Eds.), Applied Myrmecology: a world perspective (pp. 420-426). San Francisco: Westview Press.

Calligaris, I.B. (2001). Toxicidade do néctar e do pólen de *Spathodea campanulata* (Bignoneaceae) sobre operárias de *Apis mellifera* (Hymenoptera: Apidae) e *Scaptotrigona postica* (Hymenoptera: Apidae). Dissertation, Universidade Estadual Paulista Júlio de Mesquita Filho.

Camargo, J.M.F. & Pedro, S.R.M. (2012). Meliponini Lepeletier, 1836. In J.S. Moure, D. Urban & G.A.R. Melo (Orgs) Catalogue of Bees (Hymenoptera, Apoidea) in the Neotropical Region - online version. http://www.moure.cria.org.br/catalogue. (accessed date 27 August, 2013).

Carvalho, A.C.P. & Message, D. (2004). A scientific note on the toxic pollen of *Stryphnodendron polyphyllum* (Fabaceae, Mimosoideae) which causes sacbrood-like symptoms. Apidologie 35:89-90. doi: 10.1051/apido:2003059

Cintra P., Malaspina, O. & Bueno, O.C. (2005). Plantas tóxicas para abelhas. Arq. Inst. Biol. 72:547-551.

Contrera, F.A.L., Menezes, C. & Venturieri, G.C. (2011). New horizons on stingless beekeeping (Apidae, Meliponini). Rev. Bras. Zootec. 40(suppl. esp.): 48-51.

Corlett, R. T. (2005). Interactions between birds, fruit bats and exotic plants in urban Hong Kong, South China. Urban Ecosyst. 8: 275-283.

Costa, L. & Venturieri, G.C. (2009). Diet impacts on *Melipona flavolineata* workers (Apidae, Meliponini). J. Apicult. Res. 48(1): 38-45. doi: 10.3896/IBRA.1.48.1.09.

Crawley, M.J. (2007). The R book. Wiley & Sons Ltd, West Sussex, 1051p.

Dafni, A. & Kevan, P.G. (2003). Field Methods in Pollination Ecology. Cambridge: Ecoquest.

Endress, P.K. (1994). Diversity and evolutionary biology of tropical flowers. Cambridge: Cambridge University Press, 511p.

Erdtman, G. (1960). The acetolysis method. A revised description. Sven. Bot. Tidskr. 54:561-564.

Francis, J.K. (1990). African tulip tree (*Spathodea campanulata* Beauv.) Res. Note SO-ITF-SM-32. http://www.fs.fed.us/global/iitf/Spathodeacampanulata.pdf. (acessed date 1 February, 2013).

Johnson, S.D., Hargreaves, A.L. & Brown, M. (2006). Dark, bitter-tasting nectar functions as a filter of flower visitors in a bird-pollinated plant. Ecology 87: 2709–2716.

Levin, D.A. (1976). The chemical defenses of plants to pathogens and herbivores. Annu. Rev. Ecol. Syst. 7:121-159.

Michener, C.D. (2007). The Bees of the World. 2nd Ed. Baltimore: The Johns Hopkins University Press.

Modro, A.F.H., Message, D., Luz, C.F.P. & Meira-Neto, J.A.A. (2011). Flora de importância polinífera para *Apis mellifera* (L.) na região de Viçosa, MG. Rev. Árvore 35: 1145-1153. doi:10.1590/S0100-67622011000600020.

Nogueira-Neto, P. (1997). Vida e criação de abelhas indígenas sem ferrão. São Paulo: Ed. Nogueirapis, 446p.

Oliveira, R.M., Giannotti, E., Machado, V.L.L. (1991). Visitantes florais de *Spathodea campanulata* Beauv. (Bignoniaceae). Bioikos 5:7-30.

Ott, J. (1988). The delphic bee: bees and toxic honeys as pointers to psychoactive and other medicinal plants. Econ. Bot., 52 (3): 260-266.

Portugal-Araújo, V. (1963). O perigo de dispersão da tulipeira do gabão (*Spathodea campanulata* Beauv.). Chácaras e Quintais 107: 562.

Rangaiah, K., Rao, P.S. & Raju, A.J.S. (2004). Bird-pollination and fruiting phenology in *Spathodea campanulata* Beauv. (Bignoniaceae). Beitr. Biol. Pflanz. 73(3):395-408.

Roubik, D.W. (1989) Ecology and Natural History of Tropical Bees. New York: Cambridge University Press.

Santoro, K.R., Vieira, M.E.Q., Queiroz, M.L., Queiroz, M.C. & Barbosa, S.B.P. (2004). Efeito do tanino de *Stryphnodendron* spp. sobre a longevidade de abelhas *Apis mellifera* L. Arch. Zootec. 53:281-291.

Trigo, J.R. & Santos, W.F. (2000). Insect mortality in *Spathodea campanulata* Beauv. (Bignoniaceae) flowers. Rev. Bras. Biol. 60:537-8.

Venturieri, G.C., Alves, D.A., Villas-Boas, J.K., Carvalho, C.A.L., Menezes, C., Vollet-Neto, A., Contrera, F.A.L., Cortopassi-Laurino, M., Nogueira-Neto, P. & Imperatriz-Fonseca, V.L. (2012). Meliponicultura no Brasil: situação atual e perspectivas futuras para uso na polinização agrícola. In: V.L. Imperatriz-Fonseca, D.A.L. Canhos, D.A. Alves & A.M. Saraiva (Orgs.), Contribuição e perspectivas para a biodiversidade, uso sustentável, conservação e serviços ambientais (pp.213-236). São Paulo: EDUSP.

# 27

# Repellent Effects of *Annona* Crude Seed Extract on the Asian Subterranean Termite *Coptotermes gestroi* Wasmann (Isoptera:Rhinotermitidae)

MN Acda

*University of the Philippines Los Banos, Philippines.*

**Keywords**
Annonaceae, repellent, acetogenins, *Coptotermes gestroi*, feeding deterrence

**Corresponding author**
Menandro N Acda
Department of Forest Products and Paper Science, College of Forestry and Natural Resources
University of the Philippines
Los Banos
E-Mail: mnacda@yahoo.com

**Abstract**

Crude seed extract of three tropical fruits belonging to the family Annonaceae, viz., sweetsop (*Annona squamosa* L.), soursop (*A. muricata* L.) and biriba (*Rollinia mucosa* Baill.) were investigated for their repellent effects on the Asian subterranean termite *Coptotermes gestroi* Wasmann (Isoptera: Rhinotermitidae). Results of laboratory feeding bioassay (choice and no-choice) indicated that crude extract of *A. squamosa, A. muricata* and *R. mucosa* had feeding deterrent effects on *C. gestroi*. Termites showed significant avoidance behavior to filter paper treated with extracts of the three Annona species investigated. Soil barrier test revealed that Annona extracts were able to limit penetration of *C. gestroi* in laboratory tunneling test. The results suggest that *Annona* seed extracts may offer an alternative source of natural insecticide against subterranean termites.

## Introduction

Plant extracts have been studied in the past as potential sources of botanical insecticides to control a variety of arthropod pests. Their use for pest management has been regarded as an alternative to synthetic chemical insecticides (Logan et al., 1990; Isman, 2006). A large pool of plant based materials belonging to the families Meliaceae, Rutaceae, Asceraceace, Labiateae, Piperaceae and Annonaceae, among others, have been investigated for their insecticidal properties (Jacobson, 1975; Schoonhoven, 1982; Grainge et al., 1984; Arnason et al., 1989; Van Beek & Breteler, 1993). Plant parts used for screening and evaluation include leaves, flowers, roots, stem, fruit peeling, seeds and bulbs. Studies have shown that plant extracts containing limonoids and terpenoids are effective against various pests such as the Colorado potato beetle (Alford et al., 1987), the brine shrimp larva (*Culex quinquefascintus*) (Magadula et al., 2009), the larvae of the mosquitoes *Aedes aegypti* (Hoe at al., 1998) and *Anopheles stephensi* (Saxena et al., 2004), the diamondback moth, *Plutella xylostella* (Leatemia and Isman 2004a; 2005), the cabbage looper *Trichoplusia*

*ni* (Leatemia & Isman, 2004b), the lepidopteran *Spodoptera frugiperda* (Alvarez Colom et al., 2007), the German cockroach (*Blattella germanica*) (Alali et al. 2000), the adult and egg masses of the snail *Biomphalaria glabrata* (Dos Santos & Santa Ana, 2010).

One potential application of botanical insecticides is the control of subterranean termites. Termites are serious structural pests of homes and wood structures in tropical and sub-tropical regions of the world (Lee, 1971). Total worldwide damage and repair costs due to termite activity had been estimated to be about USD 40 billion annually (Rust & Su, 2012). Recent studies showed that plant derivatives such as pyrethrins, terpenoids, azadirachtin and flavanoids have excellent termiticidal activity (Sharma et al., 1994; Cornelius et al., 1997; Chen et al., 2001; Zhu et al., 2001a, b). One of the best sources of botanical insecticides is tropical Annona species, i.e., members of the custard apple family (Annonaceae). Annona species are important sources of fruit juices, frozen pulp, jelly and ice cream in Southeast Asia and thousands of tons of seeds from these fruits are generated annually from commercial processing plants (Isman, 2006, SCUC 2006).

Seeds of *Annona* species yield upon extraction a mixture of long fatty acid derivatives known as acetogenins (Chang et al., 1998; Polo et al,. 1998; Pettit et al., 2008). Annonaceous acetogenins are widely reported for its high insecticidal, molluscicidal and nematicidal properties (Rupprecht et al., 1990; McLaughlin et al., 1997; Isman, 2006). However, limited studies have been conducted on the efficacy of annonaceous acetogenins against subterranean termites. The present paper reports on the repellent effects of the crude seed extracts of *Anonna squamosa* L., *A. muricata* L. and *Rollinia mucosa* Baill. on the Asian subterranean termite *Coptotermes gestroi* Wasmann (Isopetra: *Rhinotermitidae*).

## Materials and Methods

### Plant Extracts

Ripe fruits of sweetsop (*A. squamosa*), soursop (*A. muricata*) and biriba (*R. mucosa*) were purchased from several market locations in Laguna province, Philippines. The fruits were harvested from small farms in the area cultivated by local farmers free of applied insecticide. Extraction of seeds to separate bioactive crude components was based on the method described by Leatemia and Isman (2004a). Briefly, seeds from various locations were pooled, washed with water, air dried and ground in a Wiley mill (1.0 mm). One hundred gram of each ground samples were extracted in 200 mL of 95% ethanol (5x) over 5 days by soaking. The suspension was stirred at intervals to facilitate uniform extraction. The supernatants were filtered (Whatman #1) and concentrated in a rotary evaporator under vacuum (35°C). The concentrated extracts were re-suspended in 95% ethanol then transferred to pre-weigh vials. The ethanol was dried in a fume hood and yield determined by re-weighing the vials to determine extract weight.

### Termites

Secondary nests of three active field colonies of the Asian subterranean termite *C. gestroi* were collected from infested buildings in the University of the Philippines Los Banos campus. *C. gestroi* (formerly known as *C. vastator* in the Philippines) is a major structural pest of wood structures in Southeast Asia (Acda, 2004; Lee et al., 2007). The nests were placed in black garbage bags, transported to the laboratory and placed inside 100 liter plastic containers with lids kept in a room at 25° C for three days. Distilled water was sprayed on the sides of the container to keep the relative humidity above 80%. Mature worker (pseudergates beyond the third instar as determined by size) and soldier termites were separated from nest debris by breaking apart and sharply tapping materials into plastic trays containing moist paper towels. Termites were then sorted using a soft bird feather and used for bioassay within one hour of extraction and segregation.

### Feeding Test

A no-choice and choice feeding bioassay was performed to determine toxicity and repellency of crude *Annona* seed extracts on the Asian subterranean termite *C. gestroi*. Dry crude extracts were used to prepare stock solutions in 95% ethanol. The no-choice feeding test was conducted using Petri dishes (5.5 cm in diameter) containing sterilized moist sand (1 mm) and a 2.54 x 2.54 cm filter paper (Whatman #1) impregnated with 80 µL of extract solution (12.4 µL/cm²). Extract concentrations tested were 0, 1, 5, 10, 20% (w/v). Concentrations used were based on preliminary screening. Filter paper wetted with distilled water served as control. Fifty worker termites plus 5 soldiers of *C. gestroi* from three active field colonies were placed in each dish. Experimental units containing the termites were placed in an incubator maintained at 28°C and 85% relative humidity and force-fed on treated paper for 14 days. After the prescribed exposure period, percent mortality was determined by examining the experimental units for dead or moribund termites. Workers were considered moribund when they no longer walk or stand when probed with forceps. Dead or moribund workers were recorded and removed from each experimental unit daily. Termite mortalities were corrected by Abbott's formula (1925). The test was replicated three times for each colony with a total of nine replicates for each concentration. The amount of filter paper consumed by the termites was determined by estimating the loss in surface area (mm2) of treated paper. Paper consumption reported was the average of the evaluation of three individuals. Data from the feeding test were fitted in a completely randomized design and evaluated by analysis of variance (ANOVA) using Statgraphics Centurion 16.1 software (2010). Treatment means were separated by Tukey's Honest Significance Difference (HSD) test ($\alpha = 0.05$).

Choice feeding test was performed similar to the no-choice feeding test described above except that two filter papers, one treated with crude extract and the other untreated, were placed in the center 1.0 cm apart. Papers were impregnated with 80 µL of extract solution (0, 1, 5, 10, 20% (w/v). Fifty workers plus 5 soldiers of *C. gestroi* were introduced to each experimental unit and placed in an unlit incubator as described above for 14 days. Two untreated papers wetted with water were used as control. Percent termite mortality was monitored daily. The test was replicated three times for each colony with a total of nine replicates for each concentration. Consumption of treated paper by termites was determined as described above.

### Soil Barrier Test

Tunneling and penetration of *C. gestroi* through soil treated with crude *Annona* seed extracts were evaluated using method similar to that described by Su and Scheffrahn (1990) with some modifications. The tunneling units consisted of a 12 cm glass tube (1.5 cm diameter) containing a 5 cm segment of treated soil sandwiched between 2 cm layer of 10% non-nutrient agar. Sterilized loamy soil was thoroughly mixed with the crude seed extract corresponding to 0, 1, 5, 10 and 20% [weight (crude extract)/weight (soil)] concentration. Treated soil was left to stand in a fume hood for 24 hours to evaporate the ethanol solvent. Soil wetted with distilled water served as control. Fifty workers plus 5 soldier termites were introduced into the void space and a piece of moistened filter paper was included to serve as food source. Both ends of the glass tube were sealed with aluminum foil and then placed vertically in an unlit incubator maintained at 28°C and 85% relative humidity. Termites were allowed to tunnel freely for 7 days and cumulative tunneling distance was monitored and recorded daily. The assembly was disassembled after 7 days, number of surviving insects was counted and termite mortality calculated. The test was replicated three times for each colony with a total of nine replicates for each concentration.

## Results and Discussion

### Feeding Tests

No choice forced feeding of *C. gestroi* on filter paper treated with crude seed extract of *A. squamosa*, *A. muricata* and *R. mucosa* resulted in significant increase ($p < 0.001$) in termite mortality with increasing extract concentration (Fig. 1). Termites feed on paper treated with 1 and 5% crude seed extract of all three Annona species investigated resulting in 23-68% mortalities after 14 days of exposure. Treated filter papers showed termite nibbling along the edges causing about 5.5–12 mm² reduction of surface area. However, paper treated with 1 and 5% extract was not sufficient to produce high mortality in *C. gestroi*. In addition, final mortalities at 5% concentration were observed to rise only after 6-7 days after exposure. Mortalities on the first 5 days for the three Annona species were about 8-15%. The reason for this observation is not clear but suggests delayed toxicity of Annona seed extracts on *C. gestroi*. Acetogenins, the bioactive component of the plant family Annonaceae, was reported to act as a slow acting stomach poison against chewing insects (Guadano et al., 2000; Leatemia & Isman, 2004a, b). Further study on the delayed toxicity aspect, however, needs be investigated. Termite mortalities at 10-20% crude extract of *A. squamosa* and *A. muricata* were relatively higher at 87-100% after 14 days of exposure (Fig.1). Papers treated with *R. mucosa* showed lower mortalities (78-85%) at the same extract concentration. However, examination of filter papers treated with 10-20% extract for all three samples revealed that the material was undamaged and not consumed by the termites. In addition, termites partially or completely covered the treated filter papers with sand. Apparently, the insects died of starvation and burying papers in sand could be an attempt to avoid contact with treated material. The results suggest that atleast 10% crude Annona seed extracts had repellent or

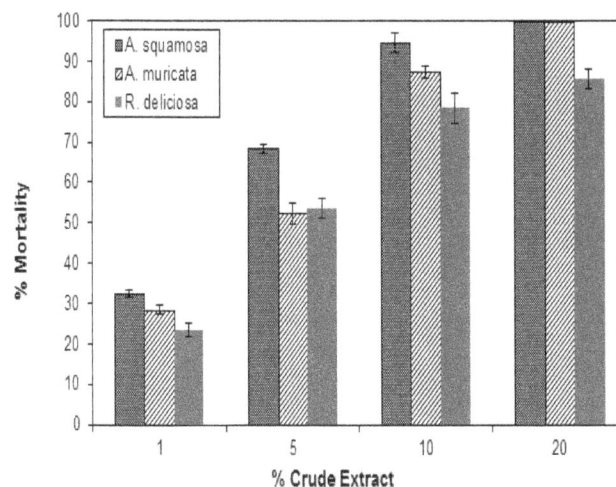

**Fig. 1**. Mortality of *C. gestroi* in no choice feeding with crude *Annona* seed extracts after 14 days of exposure.

feeding deterrent effects on *C. gestroi*. Toxicity by extract ingestion of treated paper could not be judged in these experiments.

Choice feeding test showed that *C. gestroi* consumed both untreated and treated papers up to 5% extract concentration (Table 1). Termite mortality with papers treated with 1 to 5% extract of *A. squamosa*, *A. muricata* and *R. mucosa* was relatively low at about 3-14% after 14 days of exposure. Paper consumption with untreated and those treated up to 5% extract was about 8-12 mm2 and 6-15 mm2, respectively, indicating healthy feeding on both materials. This would agree with the results of the no choice feeding above that low concentration of Annona seed extracts was not toxic by ingestion to cause high mortality in *C. gestroi* nor repellent to prevent feeding of treated materials. However, termites showed significant avoidance behavior on filter papers treated with 10% extract as indicated by non-consumption of treated papers (Table 1). Termite mortality at 10-20% extract was relatively low at about 11-18% for all Annona species investigated with untreated paper consumption of about 15–36 mm2. Termites also buried or covered treated papers with sand similar to the no-choice test above to prevent contact with the extract. Repellent or anti-feeding effect of plant extracts on *C. formosanus* Shiraki, *Reticulitermes santonensis* De Feytaud, *R. virginicus* Banks such as limonoids (Serit et al., 1992), flavanoids (Ohmura et al., 2000), nootkatone and vetiver oil (Zhu et al., 2001a;b), isoborneol and cedar oil (Blaske & Hertel, 2001; Blaske et al., 2003), matrine and oxymatrine (alkaloids) (Mao and Henderson, 2007), essential oils (Sakasegawa et al., 2003; Simaron et al., 2009), extracts from the heartwood of various hardwoods (Chang et al., 1997; Watanabe et al., 2005; Roszaini et al., 2013) had been reported in the literature. Although the repellency levels reported in this study was relatively high in comparison with synthetic insecticides, the availability of significant amount of cheap raw materials (waste seeds) may offset extraction cost to be commercially feasible. In addition, purified components of

**Table 1.** Average mortality and paper consumption of *C. gestroi* in choice feeding test with filter paper treated with various concentration of *Annona* seed extract.

| % Extract Concentration | *A. squamosa* | | *A. muricata* | | *R. mucosa* | |
|---|---|---|---|---|---|---|
| | % Mortality | Consumption (mm²) | % Mortality | Consumption (mm²) | % Mortality | Consumption (mm²) |
| 1 | 8.45 ± 2.13a | 10.23 ± 2.65 | 3.63 ± 1.12a | 5.78 ± 1.22 | 12.03 ± 2.64a | 9.34 ± 2.44 |
| Control | | 12.45 ± 1.83 | | 12.57 ± 3.76 | | 10.25 ± 1.13 |
| 5 | 11.38 ± 1.47a | 12.43 ± 0.75 | 7.48 ± 0.56b | 8.26 ± 1.58 | 14..42 ± 1.47ab | 15.12 ± 2.66 |
| Control | | 8.48 ± 1.27 | | 7.97 ± 2.68 | | 8.573 ± 0.46 |
| 10 | 15.34 ± 1.68b | 0 | 12.34 ± 1.32c | 0 | 11.34 ± 3.17a | 0 |
| Control | | 36.34 ± 3.45 | | 22.53 ± 3.52 | | 30.34 ± 3.55 |
| 20 | 18.56 ± 3.18c | 0 | 15.75 ± 2.75c | 0 | 13.24 ± 3.68a | 0 |
| Control | | 25.45 ± 2.78 | | 15.48 ± 4.35 | | 19.62 ± 2.43 |

aEach value is the mean of 9 replicates each; numbers within a column followed by the same letter are not significantly different (Tukey's HSD test, α = 0.05).

*Annona* extracts (i.e. acetogenins) may be more potent against termites and should be further investigated.

### Soil Barrier

Untreated and soil treated with 1% crude extract of *A. squamosa*, *A. muricata* and *R. mucosa* were fully breached by *C. gestroi* penetrating the full 5 cm barrier after 7 days in tunneling and penetration tubes (Table 2). However, soil treated with 5% crude extract of *A. squamosa* and *A. muricata* were able to limit termite penetration to about 0.62-1.24 cm during the 7 day exposure period. Extract of *R. mucosa* at 5% showed longer penetration of about 3.3 cm but prevented *C. gestroi* from breaching the barrier. Penetration of soil treated with at least 10% extract prevented termite tunneling and penetration of treated soil. No dead termites were found in treated soil. Termite mortality in all concentration tested was relatively low at about 7-16%. The result indicated that Annona crude seed extracts could potentially be used as soil barrier to prevent tunneling and penetration of *C. gestroi*. Apparently, the mechanism involved in preventing penetration of *C. gestroi* appeared to be repellency or avoidance of treated soil. Reduction in penetration and tunneling of subterranean

termites were also reported in laboratory trials in sand or soil treated with plant extracts such as isoborneol (Cornelius et al., 1997; Blaske & Hertel, 2001; Blaske et al., 2003), nootkatone and vetiver oil (Maistrello et al., 2001), essential oils (Peterson & Wilson, 2003) and other plant oils (Yoshida et al., 2003; Acda, 2009).

## Conclusions

In general, the study showed that ethanolic crude seed extracts of sweetsop (*A. squamosa*), soursop (*A. muricata*) and biriba (*R. mucosa*) had repellent or feeding deterrent effects on the Asian subterranean termite *C. gestroi*. No-choice and choice feeding tests indicated that crude extract of *A. squamosa*, *A. muricata* and *R. mucosa* prevented termite feeding on treated filter paper. Termites showed significant avoidance behavior to filter paper treated with at least 10% Annona crude seed extract. Soil treated with 5-10% crude extract of the three *Annona* species investigated prevented tunneling and penetration of *C. gestroi*. Termite behavior in both feeding and soil penetration tests indicated that extracts of *A. squamosa*, *A. muricata* and *R. mucosa* seed extracts had anti-feeding or repellent effect on *C. gestroi* indicating

**Table 2.** Average mortality and penetration distance of *C. gestroi* after 7 days in tunneling tubes containing soil treated with various concentration of *Annona* seed extract.

| % Extract Concentration | *A. squamosa* | | *A. muricata* | | *R. mucosa* | |
|---|---|---|---|---|---|---|
| | % Mortality | Penetration (cm) | % Mortality | Penetration (cm) | % Mortality | Penetration (cm) |
| 0 | 5.23 ± 0.75a | 5.0a | 8.45 ± 1.26a | 5.0a | 8.56 ± 1.13a | 5.0a |
| 1 | 8.34 ± 1.45a | 5.0a | 9.75 ± 2.52a | 5.0a | 10.17 ± 1.55ab | 5.0a |
| 5 | 12.43± 1.78b | 0.62 ± 0.11b | 10.17 ± 1.35ab | 1.24 ± 0.68b | 14.53 ± 1.43b | 3.30 ± 0.74b |
| 10 | 15.17 ± 2.42c | 0b | 12.5 ± 2.11b | 0.53 ± 0.14c | 13.23 ± 2.24b | 0.65 ± 0.12c |
| 20 | 16.26 ± 3.67c | 0b | 14.01 ± 1.34c | 0c | 15.18 ± 1.24b | 0c |

aEach value is the mean of 9 replicates each; numbers within a column followed by the same letter are not significantly different (Tukey's HSD test, α = 0.05).

that these compounds may be potentially useful in the development of an alternative source of natural insecticide or in combination with other control methods against subterranean termites. Purification, identification and testing of bioactive components of seed extracts of Annonaceae species used in this study against termites are now underway.

## Acknowledgments

The author wishes to thank the National Research Council of the Philippines (NRCP), Department of Science and Technology for providing funding support for this project; and Ms. Melania Gibe of the Department of Forest Products and Paper Science, University of the Philippines Los Banos for assistance in sample preparation and seed extraction.

## References

Abbott, W. S. (1925). A method of computing the effectiveness of an insecticide. Journal of Economic Entomology, 18: 265-267.

Acda, M. N. (2004). Economically important termites (Isoptera) of the Philippines and their control. Sociobiology, 43: 159-169.

Acda, M. N. (2009). Toxicity, tunneling and feeding behavior of the termite, Coptotermes vastator, in sand treated with oil of the physic nut, Jatropha curcas. Journal of Insect Science, 9: 1-8.

Alali, F. Q., Kaakeh, W., Bennett, G. W. & McLaughlin, J. L. (2000). Annonaceous acetogenins as natural pesticides: potent toxicity against insecticide-susceptible and resistant German cockroaches (Dictyoptera: Blattellidae). Journal of Natural Products, 63: 773-776.

Alford, A. R., Cullen, J. A., Storch, R. H. & Bentley, M. D. (1987). Antifeedant activity of limonin against the Colorado potato beetle (Coleoptera: Chrysomelidae). Journal of Economic Entomology, 80: 575-578.

Alvarez Colom, O. A., Neske, S., Popich & Bardon, A. (2007). Toxic effects of annonaceous acetogenins from Annona cherimolia (Magnoliales: Annonaceae) on Spodoptera frugiperda (Lepidoptera: Noctuidae). Journal of Pest Science, 80: 63–67.

Arnason, J. T., Philogene, B. J. R. & Morand, P. (Eds.) (1989). Insecticides of Plant Origin, ACS Symposium Series 387.

Blaske, V. U. & Hertel, H. (2001). Repellent and toxic of effects of plant extracts on subterranean termites (Isoptera: Rhinotermitidae). Journal of Economic Entomology, 96: 1267-1274.

Blaske, V. U., Hertel, H. & Forschler, B. (2003). Repellent effects of isoborneol on subterranean termites (Isoptera: Rhinotermitidae) in soils of different composition. Journal of Economic Entomology, 94: 1200-1208.

Chang, F.R., Chen, J. L., Chiu, H.F., Wu, M. J. & Wu, Y. C. (1997). Acetogenins from the seeds of Annona reticulata.

Phytochemistry, 47: 1057-1061.

Chen, F., Zungoli, P. A. & Benson, B. (2001). Screening of natural insecticides from tropical plants against fire ants, termites and cockroaches. Final Report. Clemson University Integrated Pest Management Program, Clemson, SC.

Cornelius, M., Grace, J. K. & Yates, J. R. (1997). Toxicity of monoterpenoids and other natural products to the Formosan subterranean termites. Journal of Economic Entomology, 90: 320-325.

Dos Santos, A.F. & Santa Ana, A. E. G. (2010). Molluscicidal properties of some species of Annona. Malaysian Journal of Science, 29: 153-159.

Grainge, M. S., Ahmed, S., Mitchel, W. C. & Hylin, J. W. (1984). Plant species reportedly possessing pest control properties – a database. Resource System Institute, East-West Center, Honolulu, Hawaii.

Guadano, A., Gutierrez, C., Pena, E., Cortes, D. & Coloma, A. (2000). Insecticidal and mutagenic evaluation of two Annonaceous acetogenins. Journal of Natural Products, 63: 773-776.

Hoe, P. K, Yiu, P. H., Ee, G. C. L., Wong, S. C., Rajan, A. & Bong, C. F. J. (1998). Biological activity of Annona muricata seed extracts. Journal of Economic Entomology, 91: 641-649.

Isman, M. B. (2006). Botanical insecticides, deterrents and repellents in modern agriculture and increasingly regulated world. Annual Review of Entomology, 51: 45-66.

Jacobson, M. (1975). Insecticides from plants: a review of literature. 1954-1971. Agriculture Handbook 461, USDA, Washington, DC. 138 pp.

Leatemia, A. J. & Isman, M.B. (2004a). Insecticidal activity of crude seed extracts of Annona spp., Lansium domesticum and Sandoricum koetjape against lepidopteran larvae. Phytoparasitica, 32: 30-37.

Leatemia, J. A. & Isman, M. B. (2004b). Toxicity and antifeedant activity of crude seed extracts of Annona squamosa (Annonaceae) against lepidopteran pests and natural enemies. International Journal of Tropical Insect Science, 24: 150–158.

Leatemia, J. A. & Isman, M. B. (2005). Efficacy of crude seed extracts of Annona squamosa against diamondback moth, Plutella xylostella L. in the greenhouse. International Journal of Pest Management, 50: 129–133.

Lee, K. E. & Wood, T.G. (1971). Termites and Soils. Academic Press, New York.

Lee, C. Y., Vongkaluang, C., Lenz, M., 2007. Challenges to subterranean termite management of multi-genera faunas in Southeast Asia and Australia. Sociobiology, 50: 213-221.

Logan, J. W. M., Cowie, R. H. & Wood, T.G. (1990). Termite control in agriculture and forestry by non-chemical methods:

a review. Bulletin of Entomological Research, 80: 309-330.

Magadula, J. J., Innocent, E. & Otieno, J. N. (2009). Mosquito larvicidal and cytotoxic activities of 3 *Annona* species and isolation of active principles. Journal of Medicinal Plants Research, 3: 674-680.

Maistrello, L., Henderson, G. & Laine, R.A. (2001). Efficacy of vetiver oil and nootkatone as soil barriers against Formosan subterranean termite (Isoptera: Rhinotermitidae). Journal of Economic Entomology, 94: 1532-1537.

Mao, L. & Henderson, G. (2007). Antifeedant activity and acute and residual toxicity of alkaloids from *Sophora flavescens* (Leguminosae) against Formosan subterranean termites (Isoptera: Rhinotermitidae). Journal of Economic Entomology, 100: 866-870.

McLaughlin, J. L., Zeng, L., Oberlies, N. H., Alfonso, D., Johnson, H. A. & Cummings, B. (1997). Annonaceous acetogenins as new pesticides: recent progress, *In* Phytochemicals for Pest Control, Hedin, P. A., Hollingworth, R. M., Master, E. P., Miyamoto, J., Thompson, D.G., (Eds.), ACS Symposium Series 658. Washington, DC. pp. 117-133.

Ohmura, W., Doi, S., Aoyama, M. & Ohara, S. (2000). Antifeedant activity of flavonoids and related compounds against the subterranean termite *Coptotermes formosanus* Shiraki. Journal of Wood Science, 46: 149-153.

Peterson, C. J. & Wilson J. M. (2003). Catnip essential oil as a barrier to subterranean termites (Isoptera: Rhinotermitidae) in the laboratory. Journal of Economic Entomology, 96: 1275-1282.

Pettit, G. R., Venugopal, J. R. V., Mukku, G. C., Herald, D. L., Knight, J. C. & Herald, C. L. (2008). Antineoplastic agents 558, *Ampelocissus* sp. cancer cell growth inhibitory constituents. Journal of Natural Products, 71: 130–133.

Polo, M. C. Z., Figadere, B., Allardo, T., Tormo, J. R. & Cortes, D. (1998). Natural acetogenins from *Annonaceace*, synthesis and mechanisms of action. Phytochemistry, 48: 1087-1117.

Roszaini, K., Nor Azah, M.A., Mailina, J., Zaini, S. & Mohammad Faridz, Z. (2013). Toxicity and antitermite activity of the essential oils from *Cinnamomum camphora*, *Cymbopogon nardus*, *Melaleuca cajuputi* and *Dipterocarpus* sp. against *Coptotermes curvignathus*. Wood Science and Technology, 47: 1273-128.

Rupprecht, J. K., Hui, Y. H. & McLaughlin, J. L. (1990). Annonaceous acetogenins: review. Journal of Natural Products, 53: 237-276.

Rust, M. K. & Su, N. Y. (2012). Managing social insects of urban importance. Annual Review of Entomology, 57: 355–375.

Saxena, R.C., Harshan, V., Saxena, A., Sukumaran, P., Sharma, M.C. & Kumar, M. L. (2004). Larvicidal and chemosterilant activity of *Annona squamosa* alkaloids against *Anopheles stephensi*. International Journal of Tropical Insect Science, 24: 150-158.

Schoonhoven, L. M. (1982). Biological aspects of antifeedants. Entomologia Experimentalis et Applicata, 31: 57-69.

Sakasegawa, M., Hori, K., Yatagi, M., 2003. Composition and antitermite activities of essential oils from *Melaleuca* species. Journal of Wood Science, 49: 181–187.

Serit, S.I., Ishida, M., Hagiwara, N., Kim, M., Yamamoto, T. & Takahashi, S. (1992). Meliaceae and Rutaceae limonoids as termite antifeedants evaluated using *Reticulitermes speratus* Kolbe (Isoptera: Rhinotermitidae). Journal of Chemical Ecology, 18: 593-603.

Siramon, P., Ohtani, Y. & Ichiura, H. (2009). Biological performance of *Eucalyptus camaldulensis* leaf oils from Thailand against the subterranean termite *Coptotermes formosanus* Shiraki. Wood Science, 55: 41–46.

SCUC-Southampton Centre for Underutilised Crops, 2006. Annona: *Annona cherimola, A. muricata, A. reticulata, A. senegalensis* and *A. squamosa*, Field Manual for Extension Workers and Farmers, University of Southampton, Southampton, UK.

Sharma, R. N., V. B. Tungikar, Pawar, P. V. & Vartak, P.H. (1994). Vapor toxicity and repellency of some oils and terpenoids to the termite *Odontotermes brunneus*. Insect Science Application, 15: 495-498.

Statgraphics Centurion XVI: User's Manual (2010). Manugistics Inc., Rockville, MD.

Su, N. Y. & Scheffrahn, R. H. (1990). Comparison of eleven soil termiticides against the Formosan subterranean termite and Eastern subterranean termite (Isoptera: Rhinotermitidae). Journal of Economic Entomology, 83: 1918-1924.

Van Beek, T. A. & Breteler, H., (Eds) (1993). Phytochemistry and Agriculture. Clarendon Press, Oxford, UK.

Watanabe, Y., Mihara, R., Mitsunaga, T. & Yoshimura, T. (2005). Termite repellent sesquiterpenoids from *Callitris glaucophylla* heartwood. Journal of Wood Science, 51: 514–519.

Yoshida, S., Nakagaki, T., Igarashi, A. & Enoki, A. (2003). An anti-termite formulation for soil treatment with natural products and its efficacy against *Coptotermes formosanus* Shiraki. IRG/WP 03-30319. The International Research Group on Wood Preservation, Stockholm, Sweden.

Zhu, B. C. R., Henderson G., Chen F., Maistrello L. & Laine, R.A. (2001a). Nootkatone is repellent to Formosan subterranean termite (*Coptotermes formosanus*). Journal of Chemical Ecology, 27: 523-531.

Zhu, B. C. R., Henderson G., Chen F., Fei, H. X. & Laine, R.A. (2001b). Evaluation of vetiver oil and seven insect active essential oil against the Formosan subterranean termite. Journal of Chemical Ecology, 27: 1617-1625.

# 28

# Termite assemblages in dry tropical forests of Northeastern Brazil: Are termites bioindicators of environmental disturbances?

AB Viana Junior[1], VB Souza[2], YT Reis[1], AP Marques-Costa[1]

1 - Universidade Federal de Sergipe, São Cristóvão, SE, Brazil.

2 - Universidade Tiradentes, Aracaju, SE, Brazil.

**Keywords**

bioindicator; Caatinga; environmental variables; feeding groups; Isoptera.

**Corresponding author**

Arleu Barbosa Viana-Junior
Universidade Federal de Sergipe
Programa de Pós-Grad. em Ecologia e Conservação
Av. Marechal Rondon, s/n
Jardim Rosa Elze
São Cristóvão, SE, Brazil
49100-000
E-mail: arleubarbosa@yahoo.com.br

**Abstract**

Termites exhibit several characteristics that emphasize their potential as bioindicators of habitat quality for use in environmental monitoring studies, but little is known about this group in vegetations of semi-arid regions of Brazil. The present study was conducted in three areas of Caatinga under different levels of anthropogenic disturbance, in the High Backwoods of Sergipe State, aiming to verify whether termite communities create different groups associated with the conservation of the area, by analyzing richness, abundance, and composition. Twelve transects of 65 x 2 m were set up in each area, where each one consisted of five plots of 5 x 2 m, making it possible to collect termites in all potential nesting and foraging sites. Five feeding groups of termites were sampled: (WF) wood-feeders, (SF) soil-feeders, (SWF) soil/wood interface-feeders, (LF) litter-foragers, and (SPF) specialized-feeders. Soil samples were collected from each plot in order to measure the environmental variables particle size, moisture percentage, and soil pH. Overall, richness and abundance were significantly different in the three studied areas. Wood-feeders were the most dominant in number of species and number of encounters collected at all sites, whereas the composition of termites in each area, given the environmental disturbances, was distinct. The environmental variables reinforced that the areas are different in terms of degree of conservation. The agreement between environmental variables and ecological data for species composition fortifies the potential of termites as biological indicators of habitat quality in areas of Caatinga of Northeastern Brazil.

## Introduction

Termites are among the most abundant insects in tropical ecosystems (Bignell & Eggleton, 2000). They are considered 'ecosystem engineers' because they have the ability to greatly modify their own habitats and to alter the structure of the ecosystems in which they live (Jones et al., 1994; Ferreira et al., 2011).

The ecological importance of termites is based on several aspects, which have been observed in arid and semi-arid ecosystems. They participate in the decomposition and in the flow of carbon and nutrients (Bignell & Eggleton, 2000; Bandeira & Vasconcellos, 2002), moving particles at different depths (Jouquet et al., 2011), increasing soil porosity (Holt & Lepage, 2000), and consequently increasing water retention, affecting directly the vegetation structure and the local primary pro-

ductivity (Nash & Whitford, 1995). For these reasons, they are considered key organisms to maintaining the structure and functional integrity of ecosystems (Holt & Coventry, 1990; Whitford, 1991) and have been considered for ecological monitoring analysis (Brown Jr, 1997).

Termites had a score of 20, on a scale from 0 to 24 established by Brown Jr (1991), in analysis of potential bioindicators using some animal groups (being after butterflies and ants). This score was given following the attributes such as taxonomic and ecological diversity, easy identification, widespread geographic distribution, functional importance, sedentarism, and good response to disturbances (Brown Jr, 1991; 1997).

However, still little is known about the effects of environmental disturbances in areas of Caatinga affecting termites. It is common knowledge that Caatinga, despite being a Brazilian endemic biome, has been changed over decades. It used to occupy an area of about 840,000 km² (Santos et al., 2011), corresponding to 54% of the Northeastern Region and 11% of the national territory (Alves et al., 2009). Currently, about 45.3% of its area is degraded, being ranked as the third Brazilian biome most changed by humans, after the Atlantic Forest and the Cerrado (Leal et al., 2005), and it may have its status changed to the second most disturbed, according to Castelletti et al. (2003). Caatinga biome presents only eleven strictly protected areas, which correspond to less than 1% of the biome under legal protection, being the biome with fewer and lesser extent of protected areas in the country (Leal et al., 2005).

This study was conducted in three areas under different levels of anthropogenic disturbance, in two municipalities of Sergipe State, aiming to verify if termite communities are associated with areas of different levels of conservation. Some abiotic variables were also evaluated, in order to check, for example, if there was some sort of association between these variables and the established environmental structures. Also, it was investigated if termites could reflect, on their community composition, the variation of the habitat, as observed in other environments, like in humid (Eggleton et al., 2002; Jones et al., 2003) and dry tropical forest (Vasconcellos et al., 2010; Alves et al., 2011).

## Material and Methods

### Study site

This study was conducted from April to May 2012 and from November 2012 to January 2013, in two municipalities of the State of Sergipe, Northeastern Brazil: at the protected area named Monumento Natural Grota do Angico (9°39'S/37°41'W), in Poço Redondo, and at the permanent protection area named Fazenda São Pedro (10°02'S/37°24'W), in Porto da Folha. Both areas are characterized as dry tropical forests, with predominance of xeric vegetation, presence of cacti [*Pilosocereus gounellei* (A. Weber ex. K. Schum.) Bly. ex. Rowl., *Melocactus zehntneri* (Britton & Rose) Luetzelb, *Selenicereus grandiflorus* (L.) Britton & Rose], shrubs and trees [*Poincianella pyramidalis* (Tul.) Queiroz, *Aspidosperma pyrifolium* Mart., *Sideroxylon obtusifolium* (Humb. ex. Roem. & Schult.) TD Penn], and bromeliads [*Bromelia laciniosa* Mart. ex. Schult., *Bromelia pinguin* L.]. The average annual temperature goes up to 26°C, with annual average rainfall of 550 mm, being the period of rain concentrated from March to June. The choice of study areas was based according the land use and historic conservation; thus, the areas were classified as: A1, abandoned pasture area, characterized by sparse vegetation with predominance of herbaceous and shrubs plants, with presence of cattle; A2, area in regeneration process for five years, located in the Monumento Natural Grota do Angico, characterized by being more heterogeneous than the area A1, with presence of trees 4-6 m high, and presence of bromeliads; and A3, located at the Fazenda São Pedro, without disturbance for over 30 years, characterized by dense vegetation, with trees 20 m high, and presence of bromeliads.

### Termite sampling

A protocol similar to that described by Jones and Eggleton (2000) was applied to each sampling site. Twelve transects of 65 x 2 m were installed in each area, subdivided into five plots of 5 x 2 m, spaced from each other by 10 m and alternating to the right and left, for a total sample of 60 spots and area of 600 m²/area. Distances of 100 m between each transect and 50 m from the edge of the fragment was maintained. The number of plots where a given species was present was used as estimation for relative abundance (Jones, 2000; Bignell & Eggleton, 2000; Oliveira et al., 2013). As standard time scale, each plot was explored for 1h/person, making it possible to collect termites in all potential nesting and foraging sites. The specimens were identified to generic level with the aid of identification keys (Constantino, 1999), and to species level by comparisons with samples previously identified and housed at the Collection of the Laboratório de Agricultura e Pragas Florestais of the Universidade Federal de Sergipe, and at the Museu de Zoologia da Universidade de São Paulo (MZUSP).

### Feeding groups

All genera and species identified were classified into five feeding groups, according to the classification suggested by Swift and Bignell (2001), and to information in the literature (Gontijo & Domingos, 1991; DeSouza & Brown, 1994; Mélo & Bandeira, 2004; Reis & Cancello, 2007; Vasconcellos et al., 2010; Alves et al., 2011), namely: (WF) wood-feeders (termites feeding on wood and wood litter, including dead branches still attached to trees); (SF) soil-feeders (termites feeding deliberately on mineral soil, with higher proportions of soil organic matter and silica, and lower proportions of recognizable plant tissue than in other groups); (SWF) soil/wood interface-feeders (termites feeding in highly decayed wood which has become friable and soul-like, or predominantly within soil under logs or soil plastered on the surface or inside of rotting logs or mixed with leaf litter in stilt-root complexes); (LF) litter-foragers (termites foraging on leaves and small woody items, often taken back and stored temporarily in the nest); and (SPF) specialized-feeders (species of termites that feed on fungi, algae, lichens on the bark of trees, manure and vertebrate carcasses).

### Abiotic variables

The following abiotic variables related with soil were analyzed: particle size, moisture, and pH. For particle size analysis, 300 grams of soil were collected from each plot using an auger, totaling 30 samples/area. All samples were taken to the Instituto de Tecnologia e Pesquisa de Sergipe (ITPS), where the percentages of sand, silt and clay were measured. In each plot, moisture and soil pH measurements were taken using a pH meter (pH Instrutherm 2500).

*Statistical analyses*

An One-way ANOVA with Tukey's test a posteriori was conducted to verify significant differences in relative abundance and mean richness per transect between the three areas sampled.

Species richness was estimated for each site using the non-parametric richness estimator Jacknife1 (Colwell & Coddington, 1994), considered one of the best tools to estimate this parameter (Palmer, 1990; Walther & Moore, 2005). Based on samples from each plot, accumulation curves were constructed, with 1000 randomizations to compare richness between sampling sites.

To analyze the species composition, a non-metric multidimensional scaling (NMDS), with data ordered by transects, and a matrix of presence/absence of the species, were performed. An analysis of similarity (ANOSIM) was carried out with a significance level of 95%, to verify the existence of significant differences in species composition between sampling sites; Jaccard similarity index was used to measure distance (Muellerdombois & Ellenberg, 1974; Hammer et al., 2001).

Furthermore, a principal component analysis (PCA) with the abiotic data was used to verify if there was spatial segregation between the respective areas (Clarke & Warwick, 2001), for which the data were logarithmized. In the PCA, the variable sand was excluded due to high correlation with silt and clay (0.86 and 0.79, respectively). This procedure is advisable when there is a strong correlation between the variables analyzed, causing no loss of information (Clarke & Warkick, 2001). To verify that the variables differ statistically between the areas, an One-way ANOVA was performed with all five environmental variables (Tukey post hoc test, p < 0.05), and with the values of the first PCA components.

The statistical software R (R Development Core Team, 2008) was used to make ANOVA and Tukey's test. The richness accumulation curve was performed using the program EstimateS 9.1.0 (Colwell, 2009). The NMDS, ANOSIM and the PCA were performed with the aid of statistical software PAST (Hammer et al., 2001).

## Results

One hundred and eighty samples of termites, classified in three families, 12 genera and 16 species, were found at the experimental sites. Termitidae was the most abundant and richest family, with 14 species collected, followed by Kalo-

termitidae and Rhinotermitidae, with one species each. Only three species were common for all areas, namely: *Nasutitermes macrocephalus* (Silvestri), *Heterotermes sulcatus* (Mathews), and *Amitermes amifer* Silvestri. Two of the sixteen species collected were exclusive of the area A1 (*Anoplotermes* sp. and *Amitermes* sp.), and five of the area A3 (*Rugitermes* sp., *Ruptitermes* sp., *Cylindrotermes* sp., *Inquilinitermes fur* (Silvestri), and *Microcerotermes* cf. *indistinctus* Mathews (Table 1).

Significant differences in relative abundance ($F_2$, 33=12.70; p < 0.01), and in the richness of termites ($F_2$, 33 = 10.04; p < 0.01), were found among the studied areas (Fig. 1).

From the values obtained by the non-parametric estimator Jacknife1, the estimated species richness was close to the observed (Table 1). From the observation of non-overlapping confidence intervals, it was found a difference in the richness between the areas A1 and A3 (Fig. 2). In relation to feeding groups, the wood-feeders were the most dominant in number of species and number of encounters collected at all areas, representing about 50% of the total fauna found. The most conserved area (A3) was the only one where all groups were found (Fig. 3).

The NMDS analysis showed differences in the species composition of termites when the sites A1 and A3 were compared (Fig. 4), while ANOSIM showed significant difference between all the three areas (Table 2).

The richness and relative abundance of termites varied positively with the degree of conservation of the area, i.e. the more conserved the area, more abundant and diverse was the termite community. The level of disturbance significantly altered the composition of species and the feeding groups present.

The abiotic factors associated with soil analyses (particle size, moisture, and pH) showed significant difference in at least one area (Table 3). Separation between sampling areas was evident through the principal component analysis (PCA) (Fig. 5), indicating that such areas can be considered different; ANOVA of the first components showed significant differences between the studied areas ($F_2$, 15 = 20.25; p < 0.001).

## Discussion

Results found by Araujo (1970) and Constantino (1998) showed Termitidae as the richest family of Isoptera with regards to number of species, abundance, and diversity in ecological terms, which is consistent with the results found here. In another hand, Kalotermitidae, a family that includes all drywoods and some dampwoods termites that do not require soil contact to survive, presented low abundance in the areas sampled. In this family, different species have different temperature and moisture requirements, but generally speaking, they inhabit and eat various types of dead wood (Cancello, 1996) and may nest in the treetops, hindering their sampling (Roisin et al., 2006), which could explain the lack of specimens in our samples.

**Table 1.** Termites collected in three areas of Caatinga, Sergipe State, Brazil: A1 (pasture area), A2 (scrub forest), and A3 (arboreal forest); Feeding groups: WF (wood-feeders), SF (soil-feeders), LF (litter foragers), SWF (soil/wood interface-feeders), SPF (specialized- feeders).

| Species | Area A1 | Area A2 | Area A3 | # of species found | Feeding group |
|---|---|---|---|---|---|
| Kalotermitidae | | | | | |
| *Rugitermes sp.* | 0 | 0 | 2 | 2 | WF |
| Rhinotermitidae | | | | | |
| *Heterotermes sulcatus* (Mathews) | 10 | 18 | 8 | 36 | WF |
| Termitidae | | | | | |
| Apicotermitinae | | | | | |
| *Anoplotermes* sp. | 4 | 0 | 0 | 4 | SF |
| *Ruptitermes* sp. | 0 | 0 | 4 | 4 | SPF |
| Nasutitermitinae | | | | | |
| *Constrictotermes cyphergaster* (Silvestri) | 0 | 5 | 7 | 12 | SPF |
| *Diversitermes* sp. | 0 | 5 | 14 | 19 | LF |
| *Nasutitermes corniger* (Motschulsky) | 0 | 3 | 4 | 7 | WF |
| *Nasutitermes macrocephalus* (Silvestri) | 1 | 8 | 2 | 11 | WF |
| Termitinae | | | | | |
| *Amitermes amifer* Silvestri | 7 | 11 | 25 | 43 | SWF |
| *Amitermes nordestinus* Mélo & Fontes | 4 | 5 | 0 | 9 | SWF |
| *Amitermes* sp. | 1 | 0 | 0 | 1 | SWF |
| *Cylindrotermes* sp. | 0 | 0 | 2 | 2 | WF |
| *Inquilinitermes fur* (Silvestri) | 0 | 0 | 1 | 1 | SF |
| *Microcerotermes* cf. *exiguus* (Hagen) | 0 | 1 | 8 | 9 | WF |
| *Microcerotermes* cf. *indistinctus* Mathews | 0 | 0 | 11 | 11 | WF |
| *Termes* sp. | 0 | 2 | 7 | 9 | SWF |
| Richness | 6 | 9 | 13 | 16 | |
| Number of encounters (relative abundance) | 27 | 58 | 95 | 180 | |
| Estimated richness (Jackknife 1) | 7.97±2.76 | 9.98±1.96 | 13.98±1.96 | | |

Kalotermitidae can also be absent in a disturbed environment, as it would have the number of trees and dead wood reduced, decreasing the likely nesting sites of the group (Vasconcellos et al., 2010). Even considering richness and relative abundance between the different areas analyzed, some species were common and had relatively high frequency.

The assemblages of termites include a great relative abundance of *H. sulcatus*, which appears to be one of the most important species to the wood cycle in dry areas and it is very resistant to disturbance (Alves et al., 2011). Melo and Bandeira (2007) measured the influence of this species in wood consumption in the Caatinga, and found that it was a generalist species that could consume wood in different stages of decomposition, even when already attacked by other species of termites.

The species *A. amifer* has been recorded in the Caatinga (Alves et al., 2011; Vasconcellos et al., 2010), Atlantic Forest, and Cerrado (Mélo & Bandeira, 2004), and may be considered of wide distribution.

*N. macrocephalus* is also considered to have a wide dis-

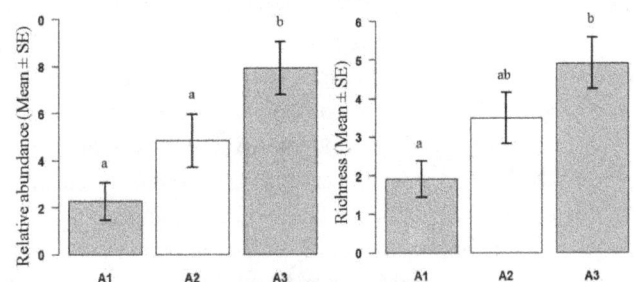

**Fig 1.** (A) Richness average; and (B) abundance per transect in three areas of Caatinga, Sergipe State, Brazil. Different letters indicate significant differences given by the Tukey test (p<0.05).

**Table 2.** Analysis of similarity (ANOSIM) for the species composition of termites communities sampled in three areas (A1, A2, and A3) of Caatinga, Sergipe State, Brazil.

| Area | R-value | P-value |
|---|---|---|
| A1 x A2 | 0.14 | 0.01 |
| A1 x A3 | 0.39 | <0.01 |
| A2 x A3 | 0.36 | <0.01 |

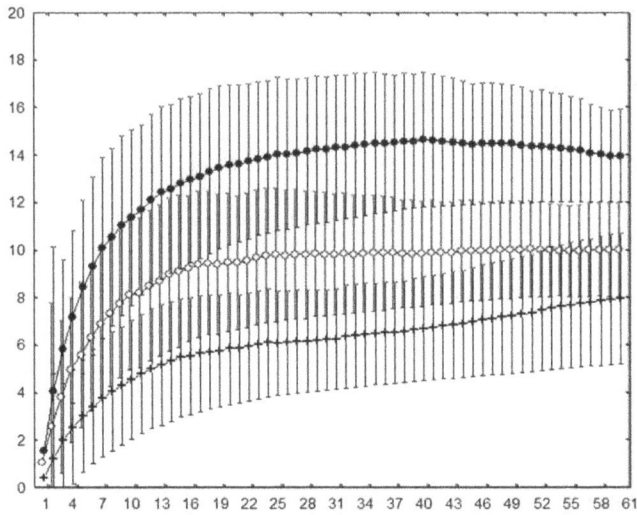

**Fig 2**. Accumulation species curves estimated for termites with confidence interval of 95%, in three areas of Caatinga, Sergipe State, Brazil, with different levels of disturbance. Area A1 (crosses); area A2 (open circles), and area A3 (closed circles).

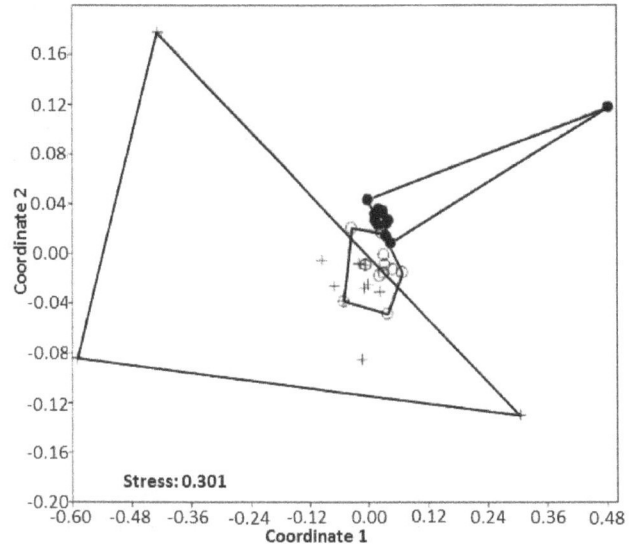

**Fig 4**. Analysis of non-metric multidimensional scaling (NMDS) for the species composition of termites in three areas of Caatinga, Sergipe State, Brazil: area A1 (crosses); area A2 (open circles), and area A3 (closed circles).

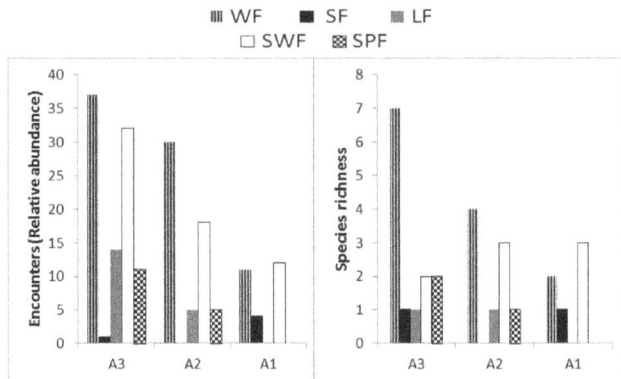

**Fig 3**. Termite species richness and relative abundance (encounters) by feeding group, in three areas of Caatinga under different levels of disturbance, Sergipe State, Brazil. A1, abandoned pasture area; A2, regeneration area; A3, conservation area. Feeding groups: WF, wood-feeders; SF, soil-feeders; LF, litter-feeders; SWF, soil/wood interface-feeders; SPF, specialized-feeders.

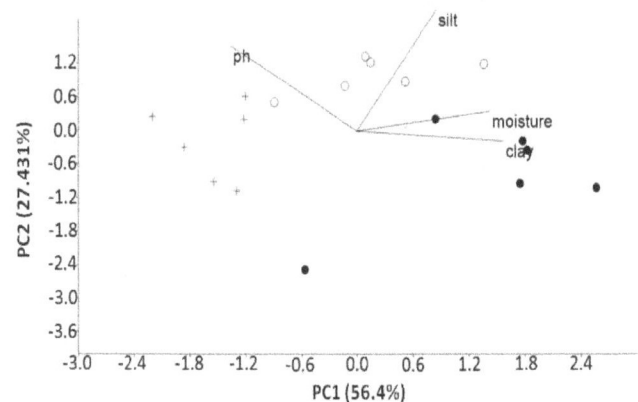

**Fig 5**. Principal component analysis (PCA) for environmental variables in three areas of Caatinga, Sergipe State, Brazil: area A1 (crosses); area A2 (open circles), and area A3 (closed circles).

tribution, since it was recorded in all three fragments sampled in this study, being already previously recorded in areas of Caatinga (Mélo & Bandeira, 2004), Amazon, Atlantic Forest, and Cerrado (Constantino, 2005; Alves et al., 2011; Souza et al., 2012). To date, there is no work with regards to its biology or ecology.

From the results of relative abundance, average richness per transect, and the accumulation curve, it is possible to suggest that the richness and abundance of termites can be changed by altering habitat, being directly related to the conservation of the area, in agreement with results observed in other researches on termites, in Savannah (Dosso et al., 2012; Cunha & Orlando, 2011; Carrijo et al., 2009; Brandão & Souza, 1998), Humid Tropical Forests (Ackerman et al., 2009; Eggleton et al., 1996; DeSouza & Brown, 1994; Bandeira & Torres, 1985) or Caatinga (Alves et al., 2011; Vasconcellos et al., 2010; Bandeira et al., 2003).

The abundance of wood-feeders, when compared to other groups, is already a result well discussed in the literature (Souza et al., 2012; Alves et al., 2011; Vasconcellos et al., 2005). Soil-feeders and soil/wood interface-feeders are more sensitive to environmental disturbances and natural weather fluctuations than wood-feeders in humid forests (DeSouza & Brown, 1994; Bandeira et al., 2003), but also, the small amount of organic matter in the soil (consequence of low leaf productivity) could be a contributing factor to the low abundance of those groups in the Caatinga environment (Mélo & Bandeira, 2004). However, the wood-feeders were the group most affected by disturbance in the Cerrado and Caatinga environments (Carrijo et al., 2009; Vasconcellos et al., 2010), as shown in the present study.

The variation in species composition in the three areas studied corroborates the hypothesis that anthropogenic changes

**Table 3.** Mean values (± standard error) of environmental variables collected in three areas of Caatinga, Sergipe State, Brazil. Different letters indicate significant differences between the means of the variables obtained by the Tukey test (p<0.05).

| Area | pH (Mean±SE) | Moisture (Mean±SE) | Sand (Mean±SE) | Clay (Mean±SE) | Silt (Mean±SE) |
|------|-------------|--------------------|-----------------|-----------------|-----------------|
| A1 | 6.99 ± 0.03 a | 0.66 ± 2.34 a | 54.41 ± 1.41 a | 11.59 ± 0.70 a | 33.98 ± 0.95 a |
| A2 | 6.95 ± 0.05 a | 11.26 ± 3.31 b | 46.59 ± 2.00 b | 13.65 ± 0.99 b | 39.69 ± 1.35 b |
| A3 | 6.58 ± 0.05 b | 47.53 ± 3.31 c | 46.63 ± 2.00 b | 17.58 ± 0.99 c | 35.77 ± 1.35 a |

(including deforestation and/or land use) may be responsible for changes in the availability of plant material and/or ecological niches for species (Junqueira et al., 2008). As a result, a direct and positive relationship between the local termite diversity and the conservation status of the area could be assumed, i.e., termites responded to disturbances occurred in Caatinga sites of the High Backwoods of Sergipe, mapped through the differences in richness, abundance and composition of species in the areas sampled.

Alves et al. (2011) analyzing three areas of Caatinga quite similar with regards to disturbance levels, observed, through a PCA (using eleven environmental variables), no significant difference on termites assemblage composition, whereas Vasconcellos et al. (2010) found that areas with different levels of disturbance presented different communities of termites. Thus, the results presented here, combined with data from literature, reinforce the potential of termites as biological indicators of environmental quality in the areas of Caatinga, which could be applied to other ecosystems.

## Acknowledgments

We would like to thank Drs. Alexandre Vasconcellos (UFPB) and Genésio Ribeiro (UFS) for the important contributions on the master's dissertation of the first author, which generated this paper; to Dr. Maurício Rocha (MZUSP) for the identification of termites specimens included in this study. We also thank to Dr. Ana Paula Albano Araújo for help in the statistical analysis; to Dr. Mirian Watts and John Watts for their suggestions and revision of the paper; and to Rony Peterson, Sidieris da Costa, and Brisa Marina, for their essential help during fieldwork. We also thank the funding agency CAPES for the master's scholarship given to the first author.

## References

Ackerman, I. L., Constantino, R., Gauch Jr., H. G., Lehmann, J.; Riha, S. J. & Fernandes, E.C.M. (2009). Termite (Insecta: Isoptera) species composition in a Primary Rain Forest and Agroforests in Central Amazonia. Biotropica, 41: 226-233. doi: 10.1111/j.1744-7429.2008.00479.x

Alves, J. J. A., Araújo, M. A. de & Nascimento, S. S. do. (2009). Degradação da Caatinga: uma investigação ecogeo-gráfica. Revista Caatinga, 22: 126-135.

Alves, W. de F., Mota, A. S., Lima, R. A. A. de, Bellezoni, R. & Vasconcellos, A. (2011). Termites as bioindicators of habitat quality in the Caatinga, Brazil: is there agreement between structural habitat variables and the sampled assemblages? Neotropical Entomology, 40: 39-46. doi: 10.1590/S1519-566X2011000100006

Araujo, R. L. (1970). Termites of the Neotropical region. In K. Krishna & F.Weesner (Eds.), Biology of Termites, (pp. 527-571). New York, Academic Press.

Bandeira, A. G. & Torres, M. F. P. (1985). Abundância e distribuição de invertebrados do solo em ecossistemas amazônicos. O papel ecológico dos cupins. Boletim do Museu Paraense Emílio Goeldi, Ser. Zool., 2: 13-38.

Bandeira, A. G. & Vasconcellos, A. (2002). A quantitative survey of termites in a gradient of disturbed high and forest in northeastern Brazil (Isoptera). Sociobiology, 39: 429-439.

Bandeira, A. G., Vasconcellos, A., Silva, M. P. & Constantino, R. (2003). Effects of habitat disturbance on the termite fauna in a highland humid forest in the Caatinga domain, Brazil. Sociobiology, 42: 117-127.

Bignell, D. E. & Eggleton, P. (2000).Termites in ecosystems. In T. Abe; D. E. Bignell & M. Higashi (Eds.), Termites: evolution, sociality, symbioses, ecology (pp. 363-387). Netherlands: Kluwer Academic Publishers.

Brandão, D. & Souza, F. (1998). Effects of deforestation and implantation of pastures on the termite fauna in the Brazilian "Cerrado" region. Tropical Ecology, 39: 175-178.

Brown Jr., K. S. (1991). Conservation of Neotropical environments: insects as indicators. In N. M. Collins, J. A. Thomas (Eds.), The conservation of insects and their habitats (pp. 349-404). Academic Press, London.

Brown Jr., K. S. (1997). Diversity, disturbance, and sustainable use of Neotropical forests: insects as indicators for conservation monitoring. Journal of Insect Conservation, 1: 25-42. doi: 10.1023/A:1018422807610

Cancello, E. M. (1996).Termite diversity and richness in Brazil - an overview. In C. E. de M., Bicudo & N. A., Menezes, (Eds.), Biodiversity in Brazil - a first approach (pp.173-182).

São Paulo, CNPq.

Carrijo, T. F., Brandão, D., Oliveira, D. E. de, Costa, D. A. & Santos, T. (2009). Effects of pasture implantation on the termite (Isoptera) fauna in the central Brazilian savanna (Cerrado). Journal of Insect Conservation, 13: 575-581. doi: 10.1007/s10841-008-9205-y

Castelletti, C. H. M., Santos, A. M. M., Tabarelli, M. & Silva, J. M. C. (2003). O quanto ainda resta da Caatinga? Uma estimativa preliminar. In I. R. Leal, M. Tabarelli & J. M. C. Silva (Eds.), Ecologia e conservação da caatinga (pp. 777-796). Univ. Federal de Pernambuco, Recife.

Clarke, K. R. & Warwick, R.M. (2001). Change in marine communities: an approach to statistical analyses and interpretation. PRIMER-E: Plymouth, 91 p.

Colwell, R. K. & Coddington, J. A. (1994). Estimating terrestrial biodiversity through extrapolation. Philosophical Transactions of the Royal Society, London, B Biol. Sci., 345: 101-118. doi: 10.1098/rstb.1994.0091.

Colwell, R.K. (2009). EstimateS: statistical estimation of species richness and shared species from samples. Versão 8.2.0.University of Connecticut, USA. Retrived from: http://viceroy.eeb.uconn.edu/estimates

Constantino, R. (1998). Catalog of the living termites of the New World (Insecta: Isoptera). Arquivos de Zoologia, 35: 135-231.

Constantino, R. (1999). Chave ilustrada para a identificação dos gêneros de cupins (Insecta: Isoptera) que ocorrem no Brasil. Papéis Avulsos de Zoologia, 40: 387-448.

Constantino, R. (2005). Padrões de diversidade e endemismo de térmitas no bioma Cerrado. In A. Scariot; J. C. S. Silva; J. M. Felfili (Eds.), Cerrado: ecologia, biodiversidade e conservação. (pp. 319-333). Brasília: Ministério do Meio Ambiente.

Cunha, H. F. & Orlando T. Y. S. (2011). Functional composition of termite species in areas of abandoned pasture and in secondary succession of the Parque Estadual Altamiro de Moura Pacheco, Goiás, Brazil. Bioscience Journal, 27: 986-992.

DeSouza, O. F. F. & Brown, V. K. (1994). Effects of habitat fragmentation on Amazonian termite communities. Journal of Tropical Ecology, 10: 197-206. doi: 10.1017/S0266467400007847.

Dosso, K., Yéo, K., Konaté, S. & Linsenmair, K. E. (2012). Importance of protected areas for biodiversity conservation in central Côte d'Ivoire: Comparison of termite assemblages between two neighboring areas under differing levels of disturbance. Journal of Insect Science, 12: 1-18. doi: 10.1673/031.012.13101

Eggleton, P., Bignell, D. E., Sands, W. A., Mawdsley, N. A., Lawton, J. H., Wood, T. G. & Bignell, N.C. (1996). The diversity, abundance and biomass of termites under differing levels of disturbance in the Mbalmayo Forest Reserve, Southern Cameroon. Philosophical Transactions of the Royal Society. of London, B Biol. Sci., 351: 51-68.

Eggleton, P., Bignell. D. E., Hauser. S., Dibog, L., Norgorve, L. & Madong, B. (2002). Termite diversity across an anthropogenic disturbance gradient in the humid forest zone of West Africa. Agriculture, Ecosystems and Environment. 90: 189-202. doi: 10.1016/S0167-8809(01)00206-7

Ferreira, E. V. O., Martins V., Inda-Junior, A. V., Giasson, E. & Nascimento, P. C. (2011). Ação dos termitas no solo. Ciência Rural, 41: 804-911. doi: 10.1590/S010384782011005000044

Gontijo, T. A. & Domingos, D. J. (1991). Guild distribution of some termites from cerrado vegetation in southeast Brazil. Journal of Tropical Ecology, 7: 523-529. doi: 10.1017/S0266467400005897

Hammer, O., Harper, D. A. T., & Ryan. P. D. (2001). PAST: Paleontological Statistics Software Package for Education and Data Analysis. Palaeontologia Electronica. http://palaeo-electronica.org/2001_1/past/issue1_01.htm. 9 p.

Holt, J. A. & Coventry, R. J. (1990). Nutrient cycling in Australian savannas. Journal of Biogeography, 17: 427-432. doi: 10.2307/2845373

Holt, J. A. & Lepage, M. (2000).Termites and soil properties. In T. Abe; D. E. Bignell & M. Higashi (Eds.), Termites, evolution, sociality, symbiosis, ecology (pp. 389-407). Dordrecht, Kluwer Academic.

Jones, C. G., Lawton, J. H. & Shachak, M. (1994).Organisms as ecosystem engineers. Oikos, 69: 373-386.

Jones, D. T. & Eggleton, P. (2000). Sampling termite assemblages in tropical forests: testing a rapid biodiversity assessment protocol. Journal of Applied Ecology, 37: 191-203. doi: 10.1046/j.1365-2664.2000.00464.x.

Jones, D. T. (2000). Termite assemblages in two distinct montane forest types at 1000 m elevation in Maliau Basin, Sabah. Journal of Tropical Ecology, 16: 271-286. doi: 10.1017/S0266467400001401

Jones, D. T., Susilo, F. X., Bignell, D. E., Hardiwinoto, S., Gillison, A.N. & Eggleton, P. (2003).Termite assemblage collapse along a land-use intensification gradient in lowland central Sumatra, Indonesia. Journal of Applied Ecology, 40: 380-391. doi: 10.1046/j.1365-2664.2003.00794.x

Jouquet, P., Traoré, S., Choosai, C., Hartmann, C. & Bignell, D. (2011). Influence of termites on ecosystem functioning. Ecosystem services provided by termites. European Journal of Soil Biology, 47, 215-222. doi: 10.1016/j.ejsobi.2011.05.005

Junqueira, L. K.; Diehl, E. & Berti-Filho, E. (2008). Termites in eucalyptus forest plantations and forest remnants: an ecological approach. Bioikos, 22: 3-14.

Leal, I. R., Silva, J. M. C., Tabarelli, M. & Júnior, T. E. L. (2005). Mudando o curso da conservação da biodiversidade na Caatinga do Nordeste do Brasil. Megadiversidade, 1: 139-146.

Mélo, A. C. S. & Bandeira, A. G. (2004). A qualitative and quantitative survey of termites (Isoptera) in an open shrubby Caatinga in Northeast Brazil. Sociobiology, 44: 707-716.

Mélo, A. C. S. & Bandeira, A. G. (2007) Consumo de madeira por *Heterotermes sulcatus* (Isoptera: Rhinotermitidae) em ecossistema de Caatinga no Nordeste do Brasil. Oecologia Brasiliensis, 11: 350-355.

Mélo, A. C. S. & Fontes, L. R. (2003). A new species of *Amitermes* (Isoptera, Termitidae, Termitinae) from Northeastern Brazil. Sociobiology, 41: 411-418.

Muellerdombois, D. & Ellenberg, H. (1974). Aims and methods of vegetation ecology. New York: Jonh Wiley, 547 p.

Nash, M. H. & Whitford , W. G. (1995). Subterranean termites: regulators of soil organic matter in the Chihuahuan. Desert. Biology and Fertility of Soils, 19: 15-18. doi: 10.1007/BF00336340

Oliveira, D. E., Carrijo, T. F. & Brandao, D. (2013). Species composition of termites (Isoptera) in different cerrado vegetation physiognomies. Sociobiology, 60: 190-197. doi: 10.13102/sociobiology.v60i2.190-197.

Palmer, M. (1990). The estimation of species richness by extrapolation. Ecology, 71: 1195-1198.

R Development Core Team. (2010). R: A Language and Environment for Statistical Computing. Vienna, Austria. Retrieved from http://www.r-project.org/

Reis, Y. T. & Cancello, E. M. (2007). Riqueza de cupins (Insecta, Isoptera) em áreas de Mata Atlântica primária e secundária do sudeste da Bahia. Iheringia, Série Zoologia, 97: 229-234. doi: 10.1590/S0073-47212007000300001

Roisin, Y., Dejean, A., Corbara, B., Orivel, J., Samaniego, M. & Leponce, M. (2006). Vertical stratification of the termite assemblage in a Neotropical rain forest. Oecologia, 149: 301-311. doi: 10.1007/s00442-006-0449-5

Santos, J. C., Leal, I. R., Almeida-Cortez, J. S., Fernandes, G. W. & Tabarelli, M. (2011). Caatinga: the scientific negligence experienced by a dry tropical Forest. Tropical Conservation Science, 4: 276-286.

Souza, H. B. A., Alves, W. F. & Vasconcellos, A. (2012). Termite assemblages in five semideciduous Atlantic Forest fragments in the northern coastland limit of the biome. Revista Brasileira de Entomologia, 56: 67-72. doi: 10.1590/S0085-56262012005000013

Swift, M. J. & Bignell, D. (2001). Standard methods for assessment of soil biodiversity and land use practice. ASB Lecture Note 6B, Bogor, Indonesia.

Vasconcellos, A., Bandeira, A. G., Moura F. M. S., Araújo V. F. P. & Constantino, R. (2010). Termite assemblages in three habitats under different disturbance regimes in the semi-arid Caatinga of NE Brazil. Journal of Arid Environments, 74: 298-302. doi: 10.1016/j.jaridenv.2009.07.007

Vasconcellos, A.; Mélo, A. C. S.; Segundo, E. M. V. & Bandeira, A. G. (2005). Cupins de duas florestas de restinga do nordeste brasileiro. Iheringia, Série Zoologia, 95: 127-131. doi: 10.1590/S0073-47212005000200003

Walther, B. A. & Moore, J. L. (2005). The concepts of bias, precision and accuracy, and their use in testing the performance of species richness estimators, with a literature review of estimator performance. Ecography, 28: 815-829. doi: 10.1111/j.2005.0906-7590.04112.x

Whitford, W.G. (1991). Subterranean termites and long-term productivity of desert range lands. Sociobiology, 19: 235-242.

# Permissions

All chapters in this book were first published in Sociobiology, by Universidade Estadual de Feira de Santana; hereby published with permission under the Creative Commons Attribution License or equivalent. Every chapter published in this book has been scrutinized by our experts. Their significance has been extensively debated. The topics covered herein carry significant findings which will fuel the growth of the discipline. They may even be implemented as practical applications or may be referred to as a beginning point for another development.

The contributors of this book come from diverse backgrounds, making this book a truly international effort. This book will bring forth new frontiers with its revolutionizing research information and detailed analysis of the nascent developments around the world.

We would like to thank all the contributing authors for lending their expertise to make the book truly unique. They have played a crucial role in the development of this book. Without their invaluable contributions this book wouldn't have been possible. They have made vital efforts to compile up to date information on the varied aspects of this subject to make this book a valuable addition to the collection of many professionals and students.

This book was conceptualized with the vision of imparting up-to-date information and advanced data in this field. To ensure the same, a matchless editorial board was set up. Every individual on the board went through rigorous rounds of assessment to prove their worth. After which they invested a large part of their time researching and compiling the most relevant data for our readers.

The editorial board has been involved in producing this book since its inception. They have spent rigorous hours researching and exploring the diverse topics which have resulted in the successful publishing of this book. They have passed on their knowledge of decades through this book. To expedite this challenging task, the publisher supported the team at every step. A small team of assistant editors was also appointed to further simplify the editing procedure and attain best results for the readers.

Apart from the editorial board, the designing team has also invested a significant amount of their time in understanding the subject and creating the most relevant covers. They scrutinized every image to scout for the most suitable representation of the subject and create an appropriate cover for the book.

The publishing team has been an ardent support to the editorial, designing and production team. Their endless efforts to recruit the best for this project, has resulted in the accomplishment of this book. They are a veteran in the field of academics and their pool of knowledge is as vast as their experience in printing. Their expertise and guidance has proved useful at every step. Their uncompromising quality standards have made this book an exceptional effort. Their encouragement from time to time has been an inspiration for everyone.

The publisher and the editorial board hope that this book will prove to be a valuable piece of knowledge for researchers, students, practitioners and scholars across the globe.

# List of Contributors

**GK Souza**
Universidade Federal de Viçosa, Viçosa - MG, Brazil

**TG Pikart**
Universidade Federal de Viçosa, Viçosa - MG, Brazil

**GC Jacques**
Universidade Federal de Viçosa, Viçosa - MG, Brazil

**AA Castro**
Universidade Federal de Viçosa, Viçosa - MG, Brazil

**MM De Souza**
Instituto Federal de Educação, Ciência e Tecnologia Sul de Minas Gerais, Pouso Redondo-MG, Brazil

**JE Serrão**
Universidade Federal de Viçosa, Viçosa - MG, Brazil

**JC Zanuncio**
Universidade Federal de Viçosa, Viçosa - MG, Brazil

**TMR Santos**
Universidade Federal de Mato Grosso do Sul, Campo Grande – MS, Brazil

**JT Shapiro**
Universidade Federal de Mato Grosso do Sul, Campo Grande – MS, Brazil

**PS Shibuya**
Universidade Federal de Mato Grosso do Sul, Campo Grande – MS, Brazil

**C Aoki**
Universidade Federal de Mato Grosso do Sul, Aquidauana – MS, Brazil

**WS Li**
Key Laboratory of Natural Pesticide and Chemical Biology, Ministry of Education, South China Agricultural University, Guangzhou, China, 510642

**Y Zhou**
Key Laboratory of Natural Pesticide and Chemical Biology, Ministry of Education, South China Agricultural University, Guangzhou, China, 510642

**H Li**
Key Laboratory of Natural Pesticide and Chemical Biology, Ministry of Education, South China Agricultural University, Guangzhou, China, 510642

**K Wang**
Key Laboratory of Natural Pesticide and Chemical Biology, Ministry of Education, South China Agricultural University, Guangzhou, China, 510642

**DM Cheng**
Key Laboratory of Natural Pesticide and Chemical Biology, Ministry of Education, South China Agricultural University, Guangzhou, China, 510642
Department of Plant Protection, Zhongkai University of Agriculture and Engineering, Guangzhou, China, 510225

**ZX Zhang**
Key Laboratory of Natural Pesticide and Chemical Biology, Ministry of Education, South China Agricultural University, Guangzhou, China, 510642
State Key Laboratory for Conservation and Utilization of Subtropical Agro-Bioresources, Guangzhou, China, 510642

**ECF Gomes**
Universidade Federal de Sergipe (UFS), São Cristóvão, Sergipe, Brazil

**GT Ribeiro**
Universidade Federal de Sergipe (UFS), São Cristóvão, Sergipe, Brazil

**TMS Souza**
Universidade Federal de Sergipe (UFS), São Cristóvão, Sergipe, Brazil

**L Sousa-Souto**
Universidade Federal de Sergipe (UFS), São Cristóvão, Sergipe, Brazil

**LC Santos-Junior**
Universidade Federal da Grande Dourados, Dourados, Mato Grosso do Sul, Brazil
Universidade Estadual do Mato Grosso do Sul, Mato Grosso do Sul, Brazil

**JM Saraiva**
Universidade Estadual do Mato Grosso do Sul, Mato Grosso do Sul, Brazil

**R Silvestre**
Universidade Federal da Grande Dourados, Dourados, Mato Grosso do Sul, Brazil

**WF Antonialli-Junior**
Universidade Federal da Grande Dourados, Dourados, Mato Grosso do Sul, Brazil
Universidade Estadual do Mato Grosso do Sul, Mato Grosso do Sul, Brazil

**P de J Conceição**
Federal University of Recôncavo da Bahia (UFRB), Cruz das Almas-BA, Brazil

**CM de L Neves**
Federal University of Recôncavo da Bahia (UFRB), Cruz das Almas-BA, Brazil

**G da S Sodré**
Federal University of Recôncavo da Bahia (UFRB), Cruz das Almas-BA, Brazil

**CAL de Carvalho**
Federal University of Recôncavo da Bahia (UFRB), Cruz das Almas-BA, Brazil

**AV Souza**
Federal University of Recôncavo da Bahia (UFRB), Cruz das Almas-BA, Brazil

**GS Ribeiro**
Federal University of Recôncavo da Bahia (UFRB), Cruz das Almas-BA, Brazil

**R de C Pereira**
Federal University of Recôncavo da Bahia (UFRB), Cruz das Almas-BA, Brazil

**BF Bartelli**
Universidade Federal de Uberlândia, Instituto de Biologia, Uberlândia, Minas Gerais, Brazil

**AOR Santos**
Universidade Federal de Uberlândia, Instituto de Biologia, Uberlândia, Minas Gerais, Brazil

**FH Nogueira-Ferreira**
Universidade Federal de Uberlândia, Instituto de Biologia, Uberlândia, Minas Gerais, Brazil

**P Rojas1,**

**C Fragoso**
Instituto de Ecología A.C. (INECOL). Xalapa, México

**WP Mackay**
University of Texas at El Paso, El Paso, USA

**Biqiu Wu**
South China Agricultural University, Guangzhou, China
Guangxi Province Academy of Agricultural Sciences, Guangxi, China

**Lei Wang**
South China Agricultural University, Guangzhou, China

**Guangwen Liang**
South China Agricultural University, Guangzhou, China

**Yongyue Lu**
South China Agricultural University, Guangzhou, China

**Ling Zeng**
South China Agricultural University, Guangzhou, China

**JJ Resende**
Universidade Estadual de Feira de Santana, Feira de Santana, Bahia, Brazil

**PEC Peixoto**
Universidade Estadual de Feira de Santana, Feira de Santana, Bahia, Brazil

**EN Silva**
Universidade Estadual de Feira de Santana, Feira de Santana, Bahia, Brazil

**JHC Delabie**
Universidade Estadual de Santa Cruz / CEPLAC/ CEPEC, Ilhéus-Itabuna, Bahia, Brazil
CEPLAC/Centro de Pesquisas do Cacau, Itabuna, Bahia Brazil

**GMM Santos**
Universidade Estadual de Feira de Santana, Feira de Santana, Bahia, Brazil

**E Diehl**
Universidade Federal do Rio Grande do Norte, Natal, RN, Brazil

**E Diehl-Fleig**
Universidade Federal do Rio Grande do Norte, Natal, RN, Brazil

**EZ de Albuquerque**
Museu de Zoologia, Universidade de São Paulo, São Paulo, SP, Brazil

**LK Junqueira**
Pontifícia Universidade Católica de Campinas, Campinas, SP, Brazil

**AC Neves**
Universidade Federal de Minas Gerais, Belo Horizonte, MG, Brazil

**CT Bernardo**
Universidade de Brasília, Brasília, DF, Brazil

**FM Santos**
Universidade Estadual de Campinas, Campinas, SP, Brazil

**L. Ríos-Casanova**
UBIPRO, FES-Iztacala, Universidad Nacional Autónoma de México, Tlalnepantla, Estado de México, México

**G. Castaño**
Facultad de Ciencias, Campus Juriquilla, Universidad Nacional Autónoma de México, Querétaro, México

**V. Farías-González**
UBIPRO, FES-Iztacala, Universidad Nacional Autónoma de México, Tlalnepantla, Estado de México, México

**P. Dávila**
UBIPRO, FES-Iztacala, Universidad Nacional Autónoma de México, Tlalnepantla, Estado de México, México

**H. Godínez-Alvarez**
UBIPRO, FES-Iztacala, Universidad Nacional Autónoma de México, Tlalnepantla, Estado de México, México

**I Burikam**
Kasetsart University, Nakhon Pathom, Thailand

**D Kantha**
Kasetsart University, Nakhon Pathom, Thailand

**G Li**
Huazhong Agricultural University, Wuhan, China

**X Zou**
Huazhong Agricultural University, Wuhan, China

**C Lei**
Huazhong Agricultural University, Wuhan, China

**Q Huang**
Huazhong Agricultural University, Wuhan, China

**M Ramalho**
Universidade Federal da Bahia (UFBA), Salvador, BA, Brazil

**M Silva**
Universidade Federal da Bahia (UFBA), Salvador, BA, Brazil
Faculdade de Tecnologia e Ciências, Salvador, BA, Brazil

**G Carvalho**
Faculdade de Tecnologia e Ciências, Salvador, BA, Brazil

**PSM Montine**
Centro Universitário de Volta Redonda (UniFOA), Volta Redonda, Rio de Janeiro, Brazil

**NF Viana**
Centro Universitário de Volta Redonda (UniFOA), Volta Redonda, Rio de Janeiro, Brazil

**FS Almeida**
Universidade Federal Rural do Rio de Janeiro, Três Rios, Rio de Janeiro, Brazil

**W Dáttilo**
Universidad Veracruzana, Xalapa, Veracruz, Mexico

**AS Santanna**
Centro Universitário de Volta Redonda (UniFOA), Volta Redonda, Rio de Janeiro, Brazil

**L Martins**
In memorian

**AB Vargas**
Centro Universitário de Volta Redonda (UniFOA), Volta Redonda, Rio de Janeiro, Brazil

**MC Gallego-Ropero**
University of Cauca, Department of Biology, Popayán, Colombia

**RM Feitosa**
Departamento de Zoologia, Universidade Federal do Paraná, Curitiba, Brazil

**TA Sales**
PPGCB: Comportamento e Biologia Animal, Universidade Federal de Juiz de Fora, Juiz de Fora, Minas Gerais, Brazil

**IN Hastenreiter**
PPGCB: Comportamento e Biologia Animal, Universidade Federal de Juiz de Fora, Juiz de Fora, Minas Gerais, Brazil

**LF Ribeiro**
PPGCB: Comportamento e Biologia Animal, Universidade Federal de Juiz de Fora, Juiz de Fora, Minas Gerais, Brazil

**JFS Lopes**
PPGCB: Comportamento e Biologia Animal, Universidade Federal de Juiz de Fora, Juiz de Fora, Minas Gerais, Brazil

**MD Silva**
Universidade Federal da Bahia, Salvador, BA, Brazil
Instituto Federal de Educação, Ciência e Tecnologia Baiano, Governador Mangabeira, BA, Brazil

**M Ramalho**
Universidade Federal da Bahia, Salvador, BA, Brazil

**EM Silva**
Universidade Estadual de Feira de Santana, Feira de Santana, Bahia, Brazil

**AM Medina**
Universidade Estadual de Feira de Santana, Feira de Santana, Bahia, Brazil

**IC Nascimento**
Universidade Estadual do Sudoeste da Bahia, Campus Jequié, Bahia, Brazil

**PP Lopes**
Universidade Estadual de Feira de Santana, Feira de Santana, Bahia, Brazil

**KS Carvalho**
Universidade Estadual do Sudoeste da Bahia, Campus Jequié, Bahia, Brazil

**GMM Santos**
Universidade Estadual de Feira de Santana, Feira de Santana, Bahia, Brazil

**WP Silva**
Universidade Estadual de Feira de Santana (UEFS), Feira de Santana, BA, Brazil

**M Gimenes**
Universidade Estadual de Feira de Santana (UEFS), Feira de Santana, BA, Brazil

**G. Moya-Raygoza**
Universidad de Guadalajara, CUCBA, Jalisco, Mexico

**K. J. Larsen**
Department of Biology, Luther College, Decorah, USA

**FH Oda**
Universidade Estadual de Maringá (UEM), Maringá, PR, Brazil

**AF Oliveira**
Universidade Federal de Mato Grosso do Sul (UFMS), Campo Grande, MS, Brazil

**C Aoki**
Universidade Federal de Mato Grosso do Sul (UFMS), Aquidauana, MS, Brazil

**ACM Queiroz**
Embrapa Amazônia Oriental, Belém, PA, Brazil

**FAL Contrera**
Universidade Federal do Pará (UFPA), Belém, PA, Brazil

**GC Venturieri**
Embrapa Amazônia Oriental, Belém, PA, Brazil

**MN Acda**
University of the Philippines Los Banos, Philippines

**AB Viana Junior**
Universidade Federal de Sergipe, São Cristóvão, SE, Brazil

**VB Souza**
Universidade Tiradentes, Aracaju, SE, Brazil

**YT Reis**
Universidade Federal de Sergipe, São Cristóvão, SE, Brazil

**AP Marques-Costa**
Universidade Federal de Sergipe, São Cristóvão, SE, Brazil